THERMODYNAMICS – SYSTEMS IN EQUILIBRIUM AND NON-EQUILIBRIUM

Edited by **Juan Carlos Moreno-Piraján**

Thermodynamics – Systems in Equilibrium and Non-Equilibrium
Edited by Juan Carlos Moreno-Piraján

Published by InTech
Janeza Trdine 9, 51000 Rijeka, Croatia

Publishing Process Manager Viktorija Zgela
Technical Editor Teodora Smiljanic
Cover Designer Jan Hyrat

First published September, 2011
Printed in Croatia

Additional hard copies can be obtained from orders@intechweb.org

Thermodynamics – Systems in Equilibrium and Non-Equilibrium,
Edited by Juan Carlos Moreno-Piraján
p. cm.
ISBN 978-953-307-283-8

For EU product safety concerns:
IN TECH d.o.o., Prolaz Marije Krucifikse Kozulić 3, 51000 Rijeka, Croatia,
info@intechopen.com or visit our website at intechopen.com.

Meet the editor

Dr. Juan Carlos Moreno-Piraján is a Full Professor, Department of Chemistry at the Universidad de los Andes (Bogotá, Colombia) and a Director of the Research Group of Porous Solids and Calorimetry. He is a chemist by profession who obtained his Ph.D. from the Universidad Nacional de Colombia, where his work was concentrated on the construction of the Tian-Calvet microcalorimeter, which was used for the characterization of activated carbons. Professor Moreno has spent his research career developing new materials for water purification. These interesting materials have been synthesized and he has also designed several microcalorimeters that have contributed to these studies. As a result of his research, Professor Moreno-Piraján has 160 publications. His main contributions can be found in the thermodynamic interpretation of solids and liquids, and in the information obtained from calorimetric measurements.

Contents

Preface

Thermodynamics is one of the most exciting branches of physical chemistry which has greatly contributed to the modern science. Since its inception, great minds have built their theories of thermodynamics. One should name those of Sadi Carnot, Clapeyron Claussius, Maxwell, Boltzman, Bernoulli, Leibniz etc. Josiah Willard Gibbs had perhaps the greatest scientific influence on the development of thermodynamics. His attention was for some time focused on the study of the Watt steam engine. Analysing the balance of the machine, Gibbs began to develop a method for calculating the variables involved in the processes of chemical equilibrium. He deduced the phase rule which determines the degrees of freedom of a physicochemical system based on the number of system components and the number of phases. He also identified a new state function of thermodynamic system, the so-called free energy or Gibbs energy (G), which allows spontaneity and ensures a specific physicochemical process (such as a chemical reaction or a change of state) experienced by a system without interfering with the environment around it. The essential feature of thermodynamics and the difference between it and other branches of science is that it incorporates the concept of heat or thermal energy as an important part in the energy systems. The nature of heat was not always clear. Today we know that the random motion of molecules is the essence of heat. Some aspects of thermodynamics are so general and deep that they even deal with philosophical issues. These issues also deserve a deeper consideration, before tackling the technical details. The reason is a simple one - before one does anything, one must understand what they want.

In the past, historians considered thermodynamics as a science that is isolated, but in recent years scientists have incorporated more friendly approach to it and have demonstrated a wide range of applications of thermodynamics.

These four volumes of applied thermodynamics, gathered in an orderly manner, present a series of contributions by the finest scientists in the world and a wide range of applications of thermodynamics in various fields. These fields include the environmental science, mathematics, biology, fluid and the materials science. These four volumes of thermodynamics can be used in post-graduate courses for students and as reference books, since they are written in a language pleasing to the reader.

They can also serve as a reference material for researchers to whom the thermodynamics is one of the area of interest.

Juan Carlos Moreno-Piraján
Department of Chemistry
University of the Andes
Colombia

1

Thermodynamics of Seed and Plant Growth

Vesna Dragicevic and Slobodanka Sredojevic
Maize Research Institute "Zemun Polje"
Serbia

1. Introduction

Living systems are open, irreversible systems, determined by inheritance and dependent on temperature and time. They exchange substances with the environment and they need free energy for life. Living systems transform energy and matter during metabolism, which could be described as a controlled capacity to transform energy, by the First Law of Thermodynamics. Nevertheless, energy transformation includes the loss of some free energy as heat, by the Second Law of Thermodynamics, which as a consequence increases disorder - entropy. Plant cells are simultaneously characterized by two opposing types of reactions: endergonic, such as photosynthesis (occurring in green plastids) and exergonic, such as respiration (present in mitochondria). Since the metabolic reactions are controlled, they need activation energy that is provided by biological catalysts – enzymes, which lower the activation energy without its consumption. Nevertheless, the limits in the application of thermodynamics in the biochemistry of living systems are the non-existence of time and total reversibility, as a category, which could be surpassed in plant systems by introducing of temperature sums as important factor of plant development.
A living system assimilates high-enthalpy, low-entropy compounds from its surroundings, transforms them into a more useful form of chemical energy and returns low-enthalpy, high-entropy compounds to environment. From this point of view, a living organism must be ordered and cannot be at equilibrium. Steady state in an open system is the analogue of equilibrium in a closed system. From the standpoint of thermodynamics, the normal functions of living systems are enabled by the concomitant presence of two opposing tendencies: the preservation of a steady state and the aspiration to spontaneously transcend a non-equilibrium state. A steady state, *i.e.*, near-equilibrium state is maintained based on minimal energy expenditure (Taiz & Zeiger, 2010). A steady inward flow of energy is the most stable state that an open system can achieve. Furthermore, the last ten years of 20th century were marked by the application of thermodynamics to research of functional (such as erythrocytes) and reproducible (such as *Methylobacterium extorquens*) cell growth (Holzhütter, 2004). In higher plants, the functional and reproducible parts in seed are connected by the irreversible transfer of hydrolysed monomers from an endosperm (functional) to an embryo (reproducible). The product of seed germination - a plant, consists of two reciprocally reversible segments: a root and a shoot, which grow by the simultaneous presence of two processes: cell elongation and cell division. Water plays an important role in growing processes. At the end of the 1960s, Boyer (1969) introduced the energy concept, to quantify water transport into plants. In addition, the input of water was determined as energy input in an essay with seedlings of different crops (Manz et al., 2005; Kikuchi et al., 2006).

Plants are open systems which can directly use (transform) light energy to convert CO_2 and H_2O into glucose, which cellular respiration converts into ATP. They reproduce and surviving the unfavourable conditions in the form of seeds. A seed is a living system with a low water content and metabolism reduced to the minimum. It contains genetic information which enables the life of a new plant. The most critical period for a seed is imbibition and the beginning of germination, which represents a shifting of the system from a latent state (steady state). Seed storage (ageing factor) induces qualitative and quantitative changes, which could have as a consequence loss of viability. If deterioration is not significant, the system results in a new plant by rehydration and substance allocation present in processes of hydrolysis and biosynthesis. During its lifetime cycle, a plant dissipates energy gradients from the point of growth and development. Environmental stresses increase the internal entropy of a plant, moving it closer to equilibrium. In response, plants employ repair systems, requiring additional energy for the recovery processes, having as a consequence a lowering of the energy available for work and an increase in entropy. Decrease of entropy of any living system towards equilibrium, having as a consequence death (Shimokawa & Ozawa, 2005).

2. The thermodynamics of seed and the maintenance of seed viability

A seed is a biological system in the state of anhydrobiosis with living processes reduced to the minimum to maintain the germination ability (viability) - the crucial biological aspect. The term is derived from Greek and indicates "life without water". Anhydrobiosis is a highly stable state of suspended energy due to desiccation pending recovery by rehydration. This state seems always to be characterized by the cessation of measurable metabolism.

From this point, seed viability could be maintained during long periods owing to their glass structure, which was defined by Buitink & Leprince (2004) as a thermodynamically unstable state, with high viscosity (enabled by low tissue moisture and low temperature), so as the viscosity is so high that diffusional movement is effectively prevented for time periods or practical utility. Sun (2000); Walters (2007); Buitink & Leprince (2008) ascertained that glass stability is not upheld *per se*, it is based on groups of different biomolecules, linked by hydrogen bonds with water molecules. Bryant et al. (2001) and Benson (2008) suggested that the formation of a glassy matrix (i.e., vitrification) in could represent a strategy for desiccation tolerance and storage stability, in general. Vitrification achieves a high viscosity without a great deal of molecular reorganization (Hatanaka & Sugawara, 2010) and, therefore, limits major changes in the cellular structures (Buitink & Leprince, 2008; Walters et al., 2010). Glasses exhibit temperature-dependent transitions during which they pass from a glassy mechanical solid to a state with a markedly decreased viscosity. This can be detected by a change in the heat capacity or by direct measurement of mechanical relaxation of viscosity (Walters et al., 2010).

2.1 Seed structure and water as bearer of equilibrium

Water is of great importance for living systems; either it is a reaction medium or a reactant. The main characteristic of seed is low water content, which could vary depending on plant species, environment and seed condition (Beardmore et al., 2008; Siddiqui et al., 2008). Seed water consists of two components, bound and bulk water (Krishnan et al., 2004c).

Ageing is a characteristic of all living systems, seeds included. Irrespective to fact that vitrification presents a conservation state for seed systems (close to a steady state), silent metabolic processes are present, with a lower ability to counteract developed injuries. This means that vitrification has a double nature: conservation (low entropy and enthalpy) with a low ability of recovery, as opposed to the hydrated state, where the entropy and enthalpy are large with high mobility and high recovery ability.

Based on ability to maintain viability during long periods and to endure dessication, Pammentner & Berjak (1999) divided the seeds of all plant species into two groups:

1. Orthodox – which could be stored for a long time and endure extreme desiccation;
2. Recalcitrant - which could not be stored for a long time and could be injured by desiccation.

From the point of glass stability, water and temperature play significant roles in determining the storage longevity of orthodox seeds. Some models have demonstrated that the effects of water and temperature on seed aging are interdependent (Beardmore et al., 2008), indicating desiccation as the main contributor to loss of seed viability. On the other hand, Walters et al. (2001; 2010) considered that seed ageing, as the main result of free radical production, is the most important factor for viability loss. Furthermore, Vertucci (1989) established that an increase of seed moisture over 0.25 g g^{-1} dry weight increases seed respiration. One of the basic mechanisms in energy transmission during desiccation (induced by ageing) is the redox state of system (Kranner & Birtić, 2005). This gives a more complex view to the maintenance of seed viability.

Desiccation of plant tissues presents a shift of the water from the liquid to the vapour phase (Sun, 2002). Temperature influences evaporation, as well as the partial water vapour pressure in the air and the energy status of water in plant tissue, both in dry and hydrated plant tissue. An increase in temperature results in a decrease in the equilibrium water content at a given relative humidity (water activity) or an increase in the equilibrium water activity for a given tissue water content (Fig. 1). Water activity can be described as the 'effective' water content, which is thermodynamically available for various physiological processes in cells. The temperature dependence of the isotherm shift is described by the Clausius-Clapeyron Equation:

$$\ln = \frac{a_{w2}}{a_{w1}} = \frac{q+\lambda_w}{R} = \left[\frac{1}{T_2} - \frac{1}{T_1}\right] \quad (1)$$

where q is the excess heat of sorption; λ_w is the latent heat of vaporization for water (44.0 kJ kg^{-1} at 25°C); R is the gas constant; a_{w1} and a_{w2} are the water activities for a given equilibrium water content at temperature T_1 and T_2, respectively.

The structural changes observed during the process of seed ageing consider disturbances of the glass structure (Hoekstra et al., 2001; Walters et al., 2010) and the increase of the oxidative activity (Walters et al., 2001; Dussert et al., 2006) as a consequence of increased respiration. Based on thermodynamics, the change of internal energy of a system represents the maximal work which could be achieved.

The alterations of external (temperature and/or humidity) and internal factors (glass stability) influence the respiration of dry seeds (Walters et al., 2001) and their energy status, which could be calculated by use of the sorption isotherm, as suggested by Vertucci & Leopold (1984):

$$\Delta H = \frac{R\,T_1\,T_2}{T_2 - T_1} \ln\left(\frac{a_{w1}}{a_{w2}}\right) \quad (2)$$

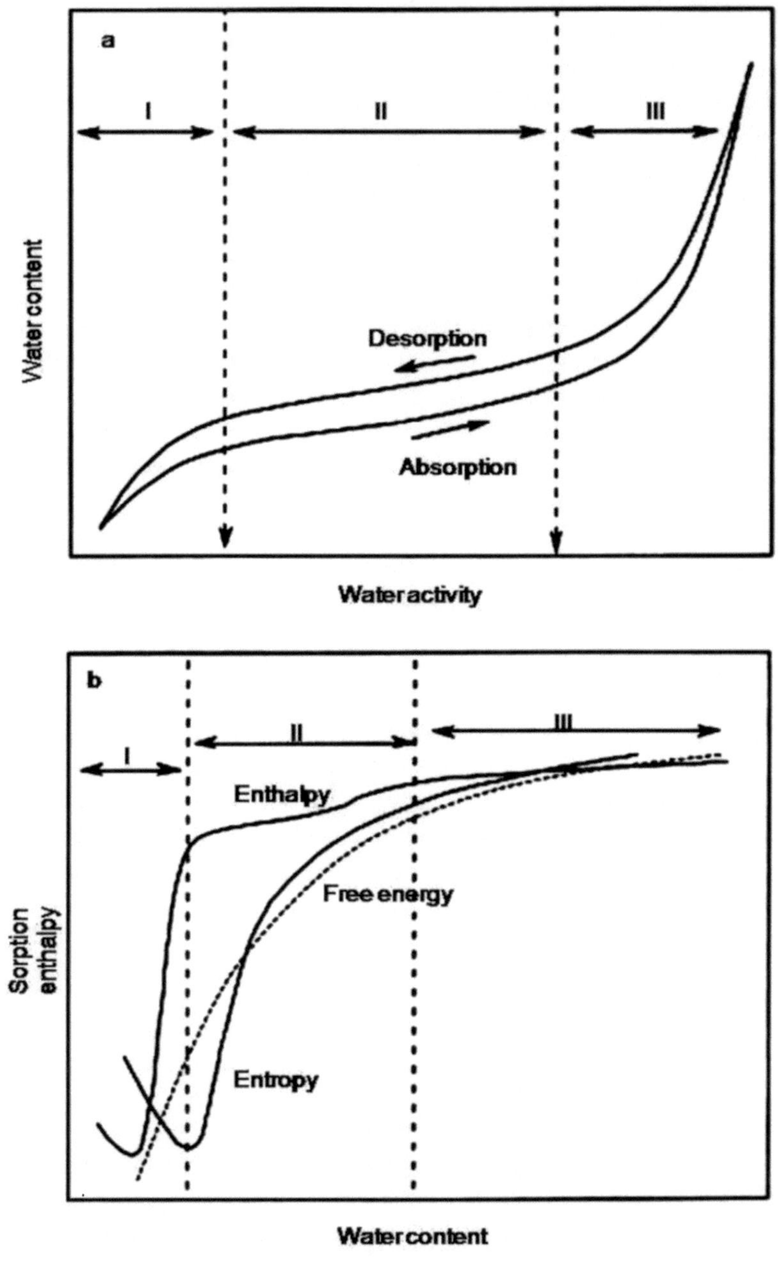

Fig. 1. Analysis of the water sorption isotherms. (a) Typical shapes of the desorption and adsorption curves of plant tissues. The difference between these two curves shows hysteresis, indicating the irreversibility of water sorption in the tissues during dehydration and rehydration. The sigmoid shape of sorption curves is presumably due to the existence of three types of water-binding sites in tissues (strong (I), weak (II) and multilayer molecular sorption sites (III)). (b) Differential enthalpy (ΔH), free energy (ΔG) and entropy (ΔS) of hydration. Data from Sun (2002).

$$\Delta G = RT\ln(a_w) \quad (3)$$

$$\Delta S = \frac{\Delta H - \Delta G}{T} \quad (4)$$

where, at a given tissue water content, a_{w1} and a_{w2} are the relative humidity at the lower and higher temperatures: T_1 and T_2, respectively, ΔH is the differential enthalpy of hydration, ΔG is the differential free energy and ΔS is the differential entropy, while R is the gas constant (8.3145 J mol^{-1} K^{-1}).

Irrespective of the fact that the observed experiments were performed on tissue during hydration/dehydration, it is well known that even minimal water content in seeds could have an important function in altering the glassy matrix. Molecular mobility was found to be inversely correlated with storage stability. With decreasing water content, the molecular mobility reached a minimum, but increased again at very low water contents. This correlation suggests that storage stability might be at least partially controlled by molecular mobility (Buitink et al., 2000).

Krishnan et al. (2004b) ascertained that the thermodynamic properties of seed water determine the reaction kinetics during seed deterioration. The thermodynamic properties showed a critical upper limit, with tolerant species having higher values than susceptible species. In general, the values of the critical limits of the thermodynamic parameters decreased with increasing temperature. The differential enthalpy and entropy increased in seeds with period of storage and became asymptotic as the seeds lost their viability. The importance of temperature, as a seed deterioration factor was also emphasised by Dragicevic (2007), with the increased values of the differential free energy found during accelerated ageing of susceptible (sugary genotypes) and tolerant (dent genotypes) maize seeds (Fig. 2). The radical increase in the ΔG values indicates intensification of endergonic reactions and consumption of relative high amounts of energy (Davies, 1961; Sun, 2002). It should be mentioned that the observed research data were calculated using temperature sums, which have a significant function in plant development. From this point of view, a time category was introduced in the plant thermodynamics and Gladyshev's (2010) postulate on hierarchical thermodynamics that the thermodynamics of a system considers only the initial and final states (the importance of whether the process under study occurs under equilibrium or non-equilibrium conditions) could be enhanced. Parallel to the results of Krishnan et al. (2004b), the differential ΔS and ΔH increased (Fig. 3) with the period of accelerated ageing. Whereas the entropy presents capacity, which means that the system is holding under conditions of limited energetic capacity and relative stability ($\Delta S \leq 0$), characteristic for seed glasses (Sun, 2002; Walters, 2007; Buitink & Leprince, 2008), in addition, the higher entropy values, present in dent seeds, could indicate a higher capacity of the system to undergo change. Concomitantly, the ΔH values, as a measure of total energy, have sigmoid shapes, with values $\Delta H > 0$ for dent seeds, as tolerant genotypes. Meanwhile, the radical decrease in germination corresponds with the trend of enthalpy decrease, with values present on the negative part of the scale, indicating a shift of the system from a relatively ordered state to a random state (Davies 1961; Sun, 2002). The observed results are in accordance with the data of Krishnan et al. (2004a, 2004b) concerning seeds with a low germination potential, which is characterized by low relaxation, as well as values below the boundary of enthalpy equilibrium.

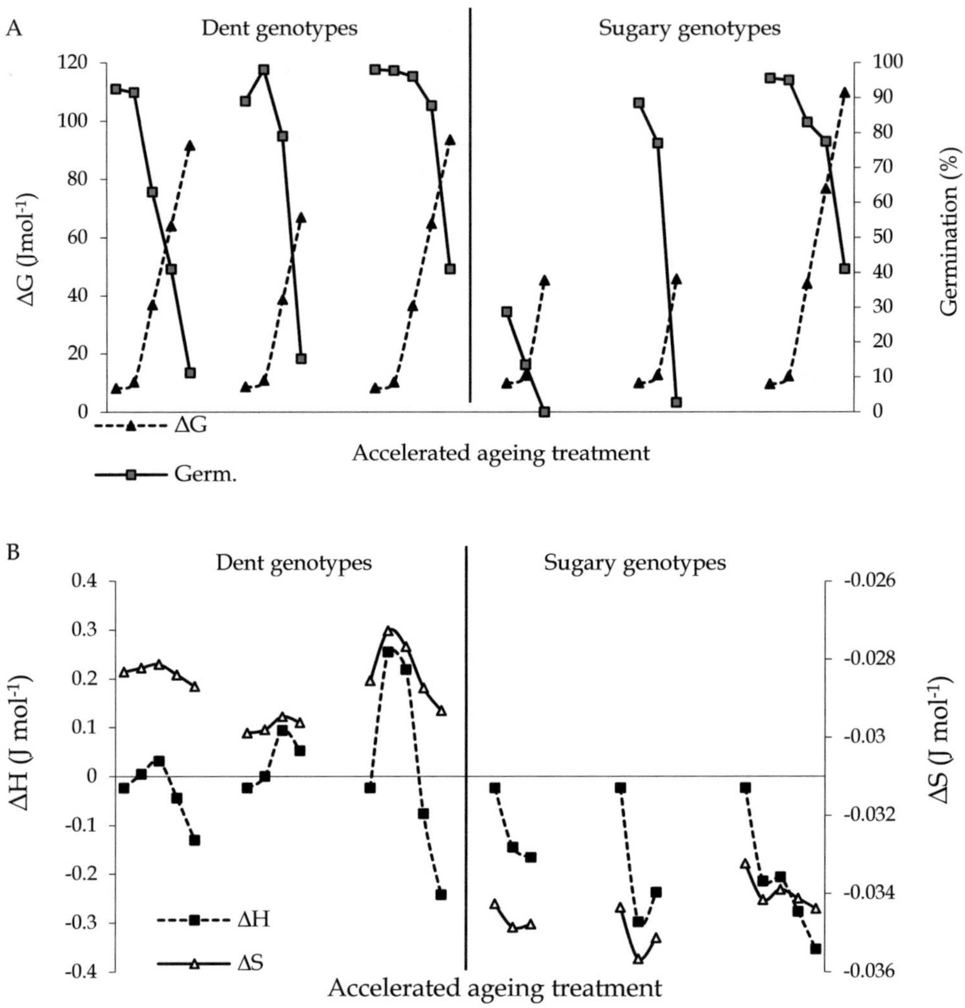

Fig. 2. Differential free energy (ΔG) entropy (ΔS) and enthalpy (ΔH) and decrease in germination of dent maize genotypes and sugary genotypes as functions of time of accelerated ageing

2.2 Importance of redox equilibrium

Life depends on a balance between entropy and enthalpy. For plants, the required energy for maintaining an ordered state is achieved by oxidation (respiration) of photosynthesized substances. Mitochondrial respiration provides energy for biosynthesis and its balance with photosynthesis determines the rate of plant biomass accumulation (Millar et al., 2011). The result of oxidation is an overall reducing environment in cells. During oxidation, photo-oxidative stress and photorespiration, the production of reactive oxygen species (ROS) is an unavoidable consequence (Brosche et al., 2010). ROS production requires or releases some quantities of energy (the voltage of an electrochemical cell is directly related to the change of the Gibb's energy, Fig. 3), which was briefly described by Vitvitskii (1969) and Buettner (1993):

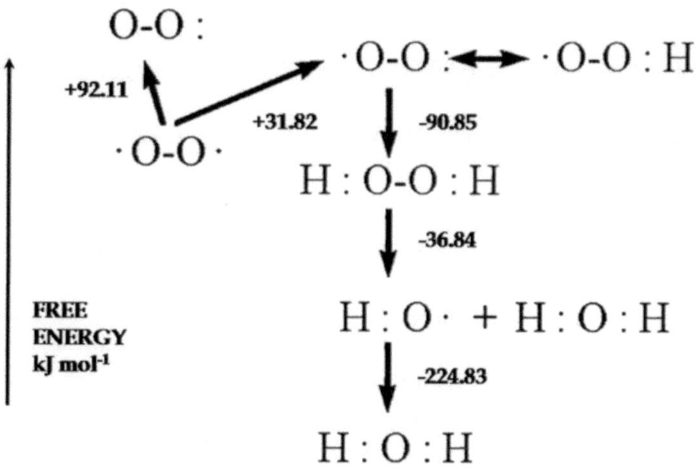

Fig. 3. The free energy of different reactive oxygen species

Redox reactions require redox couples which are responsive to electron flow, which contribute to the distinct reducing/oxidizing environment. Sets of redox couples can be independent from other sets if the activation energies for the reactions are high and there are no enzymatic systems to link them kinetically, which is commonly the case in seeds. Schafer & Buettner (2001) defined the redox environment of a linked set of redox couples, as found in a biological fluid, organelle, cell, or tissue, as the summation of the products of the reduction potential and reducing capacity of the linked redox couples present.
The reduction potential can be described as the voltage (reducing capacity) present in the number of available electrons. The reducing capacity could be estimated by determining the concentration of the reduced species in a redox couple using the Nernst equation:

$$\text{Redox environment} = \sum_{i=1}^{n\,(\text{couple})} E_i\ [\text{reduced species}]_i \qquad (5)$$

where, E_i is the half-cell reduction potential for a given redox pair and (reduced species) is the concentration of the reduced species in that redox pair.
The Nernst equation has a wide range of applications in biology because many biochemical reactions in living organisms involve electron transfer reactions. These reactions are responsible for energy production. The voltage of an electrochemical cell is directly related to the change in the Gibbs energy:

$$\Delta G^0 = -nF\Delta E^0 \qquad (6)$$

where n is the number of electrons exchanged in the chemical process, F is the Faraday constant and ΔE° is the electromotive force under standard conditions, i.e., the difference in the standard reduction potentials of the two half-cells involved in the process.
Under non-standard conditions, the relationship can be derived from a process such as:

$$aRed_1 + bOx_2 \rightleftarrows cOx_1 + dRed_2 \qquad (7)$$

$$\Delta G = \Delta G^0 + RT\ln k \qquad (8)$$

$$k = \frac{[Ox_1]^c\,[Red_2]^d}{[Ox_2]^b\,[Red_1]^a} \tag{9}$$

where k is the mass action expression, RT ln k is a "correction" factor, necessary because of the non-standard conditions.

Using Eq. 5, the voltage of an electrochemical cell can be expressed by the Nernst equation:

$$\Delta E = \Delta E^0 - \frac{RT}{nF}\ln k \tag{10}$$

Understanding the dynamics of the redox elements in biological systems remains a major challenge for redox signalling and oxidative stress research (Schafer & Buettner, 2001). The reduction potential of various redox couples in the cell could be viewed as triggers to activate a cellular switchboard that move the cell through different physiological phases: from proliferation (Fig. 4) through various stages of differentiation and, when stressed or damaged in such a way that the redox environment cannot be maintained, into apoptosis. Necrosis is the complete loss of the ability to activate and/or respond to changes in these nano-switches.

With the exception of stress situations, this phenomenon is connected with planned dismissing of individual parts of an organism, which lose functionality (programmed cell death), such as necrosis of aleurone and seed rest, after shifting to autonomic nutrition (Mrva et al., 2006). This approach represents the first step into a new area of quantitative biology.

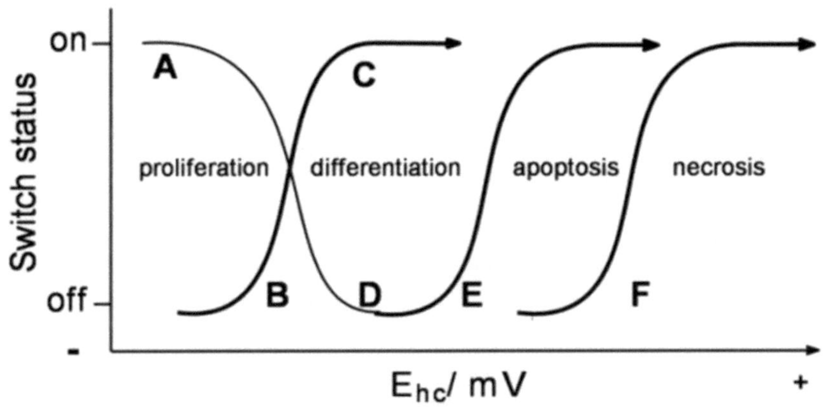

Fig. 4. Reduction potential-driven nano-switches move cells through different biological stages. The redox environment of a cell changes throughout its life cycle. During proliferation the E_{hc} for the GSSG/2GSH couple has its most negative value. (A) The switches for proliferation are fully on. (B) When E_{hc} becomes more positive, the differentiation switches can be turned on while proliferation decreases. (C) The more positive the E_{hc} becomes, the more differentiation switches are turned on until they reach a maximum, when nearly all cells are differentiating. (D) Cells that are not terminally differentiated could undergo proliferation with an appropriate signal and associated redox environment. (E) If the E_{hc} values become too positive, then death signals are activated and apoptosis is initiated. (F) Very high values of E_{hc}, resulting from severe oxidative stress, leave only necrosis as the path to cell death. Data from Schafer & Buettner (2001).

The redox couples and quantification of physiological states are not only important for seeds, but also for all phases of any living system. They contribute to stress tolerance, increase the enthalpy of a system, manage development and, as such, the life cycle on the cellular level. From that point of view plants develop a series of pathways at different levels that combat with environmental stress, which produces more ROS (Shao et al., 2008). These pathways include the phtorespiratory pathway, enzymatic and non-enzymatic pathways, corresponding responsive-gene regulation and anatomical ways, which includes drought, salinity, low temperature, UV-B and others.

3. Thermodynamics of germination and plant growth

The introduction of universal thermodynamic parameters could enable a better understanding of the processes of growth and reactions, such as hydrolysis and biosynthesis, incorporated in seed germination. Boyer (1969) quantified water transport into plant based on the energy concept. In addition, the input of water was determined as energy input (Manz et al., 2005; Kikuchi et al., 2006). Recently, Sun (2002) recognized free energy input by water as the presumable factor of plant growth. Moreover, plant growth is also the result of biomass (substance) assimilation and from that point of view Hansen et al. (1998) and Smith et al. (2006) proposed thermodynamic model to describe relation between plant growth and respiration rates (metabolism efficiency). When considered together, simultaneously measured values of CO_2 production rate, O_2 use rate and metabolic heat rate provide a link between cellular and whole-plant processes. 25 KJ mol^{-1} is taken as total enthalpy change per mole of carbon incorporated into biomass.

3.1 Germination as the double phase shifting of water

During the imbibition, seeds absorb high water quantities during a relative short period, which depend on the species, *i.e.*, the chemical composition of seeds and their condition (Copeland & McDonald, 2001; Boyd & Acker, 2004). The time curve of water absorption has a sigmoid shape (Beardmore et al., 2008; Siddiqui et al., 2008). The rapid entrance of water by the laws of diffusion and osmosis, present during phase I of imbibition (Fig. 1), is followed by enthalpy domination and an increase of the free energy status, present during phase II of imbibition (Sun, 2002). Moreover, imbibition and germination are thermally dependent processes (Sun, 2002; Nascimento, 2003;Taylor, 2004). The energy required for their activation is provided by the temperature of the environment and, in the next step, the energy of the double phase shifting of the water front (Osborne et al., 2002; Volk et al., 2006), which enters through channels under defined temperature conditions (Heimburg, 2010). After water access, the energy necessary for biochemical reactions preceding germination is produced by intensive respiration, activated during the first hours of imbibition (Sanchez-Nieto et al., 2011). The entered and on this way produced energy activates a whole range of reactions, including hydrolysis and biosynthesis (Copeland & McDonald, 2001), as concomitant processes, thereby further increasing the energy, which is required for growth. From this point of view, the free energy can be defined by both: the water volume in the seed/seedling system and the constants of hydrolysis of condensed seed substances and biosynthesis of the plant *de novo*.

Sun (2002) delineated that water sorption is an exothermic event (Fig. 1). A high negative enthalpy value at low water content suggests the strong affinity of water molecules to polar sites. As the water content increases, the enthalpy becomes less negative. As process based on volume increasing, sorption is temperature dependent category. Moreover, Vertucci & Leopold (1984); Manz et al. (2005); Kikuchi et al. (2006) clarified that the strongest negative enthalpy value occurs at about 7 % water content, and a small negative enthalpy is also observed at moisture contents between 8 % and 25 % (Fig.5). Within this region, the entropy approximates zero. The lack of measurable respiration at moisture contents below 8 % is consistent with the lack of activity for most enzymes at such dry conditions. The region between 8 and 25 % moisture has been termed the region of "restricted metabolism". This is the range in which liquid water first appears and where the differential entropy values indicate the first solution effects. Within this region of hydration, there are great changes in the ability of the seed to endure excessive and rapid imbibition. The enthalpy is low, but still negative. In the final wetting range, at moisture contents between 24 and 32 %, respiration begins to expand rapidly in response to moisture, when resistance to leakage and chilling injury is established. Damage due to imbibing water is the greatest when the initial seed moisture contents are in the region of strongest water binding. Damage is reduced and finally absent when the seed moisture contents are increased to the second and then to the third level of water affinity. The primary hydration process is considered to be completed when the differential enthalpy of hydration approaches zero. The entropy change reflects the relative order and its peak (Fig. 1 and 5) is presumably associated with the saturation of all primary hydration sites. No consistent differences in the water sorption characteristics has been found between recalcitrant and orthodox seed tissues (Sun, 2000).

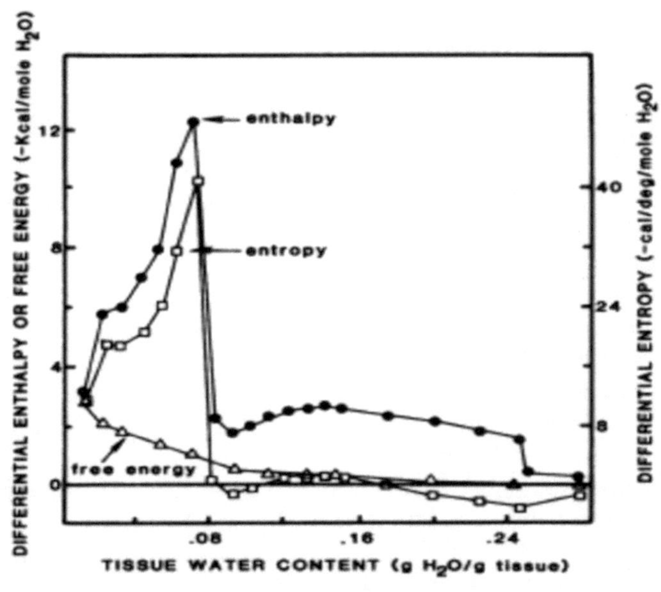

Fig. 5. Differential energies and entropies of water sorption at different moisture contents of ground soybean embryos. Data from Vertucci & Leopold (1984).

Contrary to experimental data realized on viable seeds, Krishnan et al. (2004a, 2004c) established that germinating and non-germinating seeds contained three types of water (bound, bulk and free water) in phase I of hydration. During phase II of hydration, the bulk water of non-germinating seeds disappeared completely, resulting in two types of water. However, three types of water were observed in germinating seeds in phase II. The rapid hydration in phase III was observed only in germinating seeds. The observed data indicate the importance of free water during the imbibition process, as an activator and bearer of the germination process. The water front brings energy into seed, contributing to an increase in the thermal energy necessary for the commencement of endergonic reactions (activation energy).

3.2 Growth as conversion of energy and substance

When biochemical reactions in imbibed seeds attain a critical point, inducing cell division, germination commences, giving as the product a new plant, resulting from the genome which was stored during shorter or longer periods in the seed (embryo). It is important to underline that one of the most important factors in this moment is triggering of different developmental phases by reduction potential-driven nano-switches. The next phase is characterized by water and substance distribution (allocation) from hydrolysed seed substances, denoted as the plant growth - biosynthesis. Although living systems are non-linear thermodynamic systems, which are far removed from equilibrium (Trepagnier et al., 2004), it is necessary to hypothetically define an energy balance for partial phases and processes. This could be enabled through the introduction of the basic processes of germination: hydrolysis and biosynthesis, as well as the status of their free energy, derived from the reaction constants.

$$\Sigma Hy = DWseed - DWseed\ rest \quad (11)$$

$$\Sigma Bs = DWroot + DWshoot \quad (12)$$

$$Cc = \Sigma Hy / \Sigma Bs \quad (13)$$

$$Dev = DWseed - (\Sigma Bs - DWseed\ rest) \quad (14)$$

where ΣHy is the sum of hydrolysis, ΣBs is the sum of biosynthesis, DW is the dry weight of seed, as well as root, shoot and seed rest, as parts of seedling in monocotyledonous plants (in dicotyledonous plants, cotyledons are seed rests and they are photosynthetically active), Cc is coefficient of seed substance conversion, Dev is devastated substance.

$$G_{Wc} = - RT \ln (W_c) \quad (15)$$

$$kHy = \Sigma Hy \times DWsed\ rest / DWseed \quad (16)$$

$$kBs = \Sigma Bs \times Dev / \Sigma Hy \quad (17)$$

$$G_{Hy, Bs} = -RT \ln (k) \quad (18)$$

where G_{Wc} is the free energy, based on water volume according to the Clausius-Clapeyron equation for the heat of vaporization (Eq. 1), G_{Hy} is the free energy of hydrolysis and G_{Bs} is the free energy of biosynthesis, based on the reaction constants, R is the gas constant and T is the germination temperature, rendered as the sum of the average daily temperatures.

After the imbibition, the initial growth (germination process) is followed by substance conversion: from the hydrolysed seed substance into biosynthesised substance of the root and shoot, since the young plant is not capable to produce its own substance by photosynthesis (chlorophyll has not been synthesised or it quantity is under a critical amount). Regarding the fact that the seedling's parts grow unequally (Vysotskaya, 2005; Rauf et al., 2007), which is a particular characteristic of the starting phases of germination, so as their free energy is unequally distributed. Experiments on seed ageing (Dragicevic et al., 2007, Dragicevic, 2007) demonstrated that the highest energy potential of 7-day old seedlings was present on the shoot level, then on the seed rest level and finally on the root level. In comparison to energy introduced by the hydration process, the free energy of hydrolysis and particularly of biosynthesis had significantly lower absolute values. Moreover, the consequences of deterioration processes on the seed/seedling level are present in the absorption of high water quantities and it was defined as "water induced growth" by Boyer (1969), which is a negative state for the system. Contrary to the high energy status introduced into the seed/seedling system by water, the free energy of hydrolysis tends to have minimal variations, while the free energy of biosynthesis had values closest to equilibrium, tending upon higher order of biosynthetic reactions. It is important to emphasise that the domination of exergonic reactions (Davies, 1961; Sun, 2002) is important for the release of the necessary energy and the more intensive they are, the greater is the growth potential of the system. According to the Hess Law, free energy is cumulative, irrespective of its origin; hence, all the potential energy present in a plant system is given by the sum of the individual energy states, resulting from the double phase shifting of water and that released from all the hydrolysis and biosynthesis reactions.

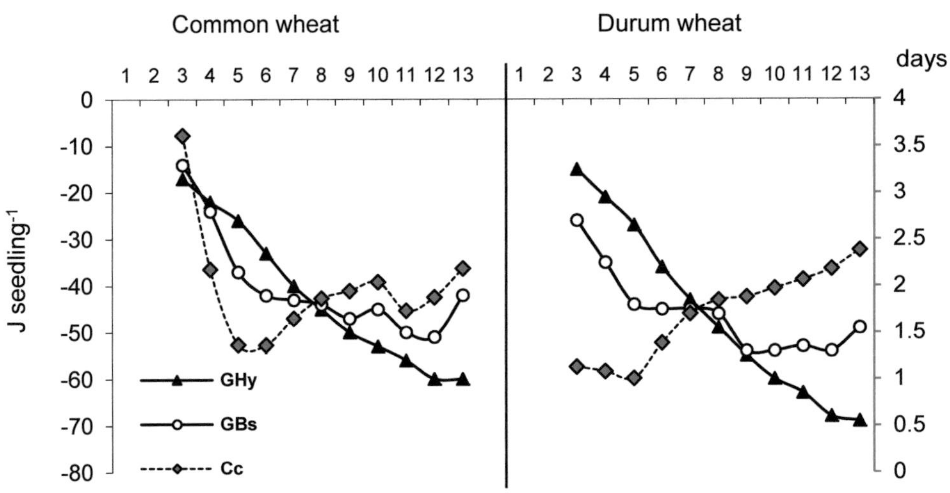

Fig. 6. Dynamics of the coefficient of conversion (*Cc*) of the free energy of hydrolysis (G_{Hy}) and biosynthesis (G_{Bs}) during 13 days of germination

After germination, the growth process is followed by less or more intensive water absorption, as well as substance synthesis. Experiments with wheat (Dragicevic et al., 2008) showed that the free energy of hydrolysis increased linearly (Fig. 6), while the free energy of biosynthesis fluctuated, with the coefficient of conversion. The observed non-linearity could indicate a change in the balance between exergonic and endergonic reactions. Moreover, the inputted and released energies are the result of process and reactions which minimize the energies of a given system. In living, as highly hydrated systems, energy is inputted by water and the total energy of the reactions has to be in a stable equilibrium.

The important points of germination and growth processes are substance losses, which Tukey (1970) defined as devastated substance. This substance includes losses through respiration (Kuzyakov & Larionova, 2005; Hill et al., 2007), as well as exudates of roots (Tukey, 1970; Jones et al., 2009) and it is calculated as illustrated in Eq. 8. The leaching (devastation) of hydrolysed substance from a seed-seedling system also means the permanent lost of energy and it starts with imbibition. Sredojevic et al. (2008) established that for soybean about 36 % and for sunflower about 46 % of free energy was lost between the 1st and 8th, i.e., 10th day, of the germination process by the leaching process alone.

The rate of plant growth is proportional to the product of the metabolic rate and the metabolic efficiency for the production of anabolic products. Over much of the growth temperature range, the metabolic rate is proportional to the mean temperature and the efficiency is proportional to the reciprocal temperature variability (Criddle et al., 2005), what could be considered as improvement in understanding of energy and biomass conversions, according to previous model purposed by Hansen et al. (1998) and Smith et al. (2006).

3.3 Growth as a flow process

In living systems, the largest flow of water is from soil, through plants to the atmosphere, the so-called by Yeo & Flowers (2007) soil-plant-atmosphere continuum. The driving force for this water continuum is the difference in free energy between liquid water in the soil and water vapour in the atmosphere. The driving forces diverge upon the different parts of the system. Water movement in soil depends on depth of the soil profile and on the forces that bind water in the capillaries (between soil particles). Water can move in plants through a matrix composed of capillaries (the cell walls) or by tubes (the xylem and phloem) where bulk flow of water occurs under pressure gradients. Water movement between cells depends on the properties of the membranes, which are differentially permeable.

In plants, water moves passively by the potential gradient of water (through semi-permeable membranes) or the potential gradient of pressure (without semi-permeable membranes). Solutes moving across semi-permeable membranes due to the potential gradients of water, combining both: the solute potential and the hydrostatic pressure, with decreasing of free energy of water.

Plants can change intracellular solute potentials, thereby influencing water flow. Water can move in a plant against its water potential only when coupled to the movement of solutes, decreasing the free energy of the solute and when the general net change of free energy (solute and water) is negative. Flow across membranes is passive in response to differences in the water potential and occurs primarily through aquaporins (integral membrane proteins) rather than directly through an impermeable membrane. Aquaporins can be gated reversibly, so that plants may be able to control their plasma membrane water permeability.

Transportation of an uncharged solute uphill against its concentration gradient (Taiz & Zeiger, 2010) from lower to higher concentrations decreases entropy and requires the input of free energy:

$$\Delta G = 2.3RT \log ((C2)/(C1)) \qquad (19)$$

If $C2 > C1$, then $\Delta G > 0$ and work is required to make this transfer. However, movement such as by diffusion can proceed spontaneously from C2 to C1 when $C2 > C1$, since this increases entropy and $\Delta G < 0$.

The change in free energy when moving one mole of a substance or ion against the membrane potential when considering the work required or performed arises from both the voltage ("electro") and concentration ("chemical") gradients.
Ions tend to flow from areas of higher to lower electrochemical potential so that ΔG is negative, which defines the maximal work output a reaction can perform. Cellular functions are ultimately linked to metabolic fluxes brought about by thousands of chemical reactions and transport processes.
The direction of flux is dictated by the change of Gibbs free energy which can be expressed through the thermodynamic equilibrium constant K_j^{equ} as follows (Holzhutter, 2004):

$$\Delta G_j^{(0)} = -RT\ln(K_j^{equ}) \tag{20}$$

where $\Delta G_j^{(0)} \leq 0$ is the change of free energy under the condition that all reactants are present at unit concentrations, $K_j^{equ} \geq 1$.
With accumulating concentrations of the reaction products (appearing in the nominator) and/or vanishing concentrations of the reaction substrates (appearing in the denominator) (Eq. 13), the concentration-dependent term (Eq. 14) may assume negative values, i.e., in principle, the direction of a chemical reaction can always be reverse provided that other reactions in the system are capable of accomplishing the required change in the concentration of the reactants. From this point of view, the introduction of irreversible thermodynamics (Yeo & Flowers, 2007) imposes upon research of biological fluxes, as more general view, which includes various forces. It applies the parameters used in classical thermodynamics to non-equilibrium conditions, i.e. in situations where there is the net flux of a substance, although the system must be close to equilibrium.

4. Conclusion

Plants are non-linear systems determined by inheritance and dependent on temperature and time, they exist owing to the opposing metabolic processes:

1. Those which consume energy and produce substances (photosynthesis);
2. Those which utilize substances and release stored energy to increase enthalpy, as well as to maintain their own status close to a steady state, which increases entropy.

Every plant attempts to maintain structure with the minimal expenditure of energy: they are able to conserve energy in substances; hence they form structures with stored energy to surpass negative conditions and to increase their reproduction ability. The consequence is the formation of seeds in higher plants.
A seed is a biological system in a vitrified state. Regardless of the fact that vitrification presents conservation state for seed systems (close to a steady state), silent metabolic processes, with lower abilities to counter developed injuries, are present. This means that vitrification has a double nature: conservation (low entropy and enthalpy) with low ability of recovery. Long storage could induce deterioration, which is related to changes in the internal energy of the system. In general, the values of the critical limits of the thermodynamic parameters decrease with increasing temperature. The differential enthalpy and entropy increase in seeds with storage time and became asymptotic as the seeds lose their viability. A radical drop in germination follows the trend of ΔG increase and ΔH

decrease (with values < 0 J mol^{-1}), indicating intensification of endergonic reactions, as well as a parallel shift of the system from a relatively ordered to a random state.

One of the important factors present during any deterioration or stress is electron transfer, inducing the production of free radicals, which are responsible for energy production. A reduction of the potential of various redox couples in a cell could be viewed as a trigger to activate cellular switchboards that move the cell from proliferation through various physiological stages into apoptosis and finally, necrosis (equilibrium). The redox couples are important for all living systems, vitrified or hydrated: they contribute to stress tolerance, and manage development and by this, the life cycle on a cellular level.

The germination process requires energy, which is provided by the environmental temperature and, in the next step, by the energy of the double phase shifting of the water front, under defined temperature conditions. The entered and in this way produced energy activates a whole range of reactions, including hydrolysis and biosynthesis, as concomitant processes, increasing, as a consequence, also the energy. When the biochemical reactions in imbibed seeds reach a critical point, inducing cell division, germination commences, giving as a product a new plant. The next phase is characterized by the distribution of water and substances (from hydrolysed seed substance), denoted as the plant growth - biosynthesis. In comparison to the energy introduced by hydration processes, the free energy of hydrolysis and, particularly, of biosynthesis have significantly lower absolute values. The domination of exergonic reactions is the release of the necessary energy and the more intensive they are, the higher is the growth potential of the system.

From that point of view, water with its characteristics, redox signals and substance conversion (including environment) are crucial points of processes providing in plants, they are dependent on energy flow and its transformations in plant systems.

Furthermore, plant growth is provided for by water flow and substance accumulation. The driving force for soil-plant-atmosphere water continuum is the difference in free energy between liquid water in the soil and water vapour in the atmosphere. Cellular functions are ultimately linked to metabolic fluxes which direction is dictated by the change of the free energy, too.

The introduction of universal thermodynamic parameters, as well as irreversible thermodynamics could lead to a better understanding of the growth process and consecutive reactions such as hydrolysis and biosynthesis, as parts of seed germination. The free energy input by water is a presumable factor of plant growth. From this point of view, free energy can be defined by the water volume in seedling/plant, as well as the constants of substance conversion. Namely, energy generation in seedling arises from the double phase shifting of the absorbed water and by its liberation *via* hydrolyses and biosyntheses. Consequently, a combined approach of thermodynamics and biochemistry could be established as a method for quantification of physiological processes, with an ecological background in the selection and breeding of new genotypes in crop production.

5. Acknowledgment

This work was supported by Ministry of Science and Technological Development of Republic of Serbia, Project No. TR31068, "Improvement of maize and soybean traits by molecular and conventional breeding".

6. References

Beardmore, T., Wang, B.S.P., Penner, M. & Scheer, G. (2008). Effects of Seed Water content and Storage Temperature on the Germination Parameters of White Spruce, Black Spruce and Lodgepole Pine Seed. *New Forests*, Vol. 36, No. 2, (May 2008), pp. 171-185, ISSN 1573-5095

Benson E.E. (2008). Cryopreservation of Phytodiversity: A Critical Appraisal of Theory & Practice. *Critical Reviews in Plant Sciences*, Vol. 27 No. 3, (May 2008), pp. 141–219, ISSN 1549-7836

Boyd, N.S. & Van Acker, R.C. (2004). Imbibition Response of Green Foxtail, Canola, Wild Mustard, and Wild Oat Seeds to Different Osmotic Potentials. *Canadian Journal of Botany*, Vol. 82, No. 6, (June 2004), pp. 801–806, ISSN 1916-2804

Boyer, J.S. (1969). Measurement of Water Status of Plants. *Annual Review of Plant Physiology*, Vol. 20, (June 1969) pp.351-364, *ISSN:* 0066-4294

Bryant, G., Koster, K.L. & Wolfe, J. (2001). Membrane Behavior in Seeds and Other Systems at Low Water Content: The Various Effects of solutes. *Seed Science and Research*, Vol. 11, No. 1 (March 2001), pp. 17–25, ISSN 0960-2585

Brosche, M.; Overmyer, K., Wrzaczek, M., Kangasjarvi, J. & Kangasjarvi, S. (2010). Stress Signalling III: Reactive Oxygen Species (ROS). In: *Abiotic Stress Adaptation in Plants, Physiological, Molecular and Genomic Foundat*ion, A. Pareek, S.K. Sopory, H.J. Bohnert. (Ed.) pp. 91-102, Springer, ISBN 978-90-481-3112-9, Dordrecht, Netherlands

Buettner, G.R. (1993). The Packing Order of Free Radicals and Antioxidants: Lipid Peroxidation, α-tocopherol, and Ascorbate. *Archives of Biochemistry and Biophysics*, Vol. 300, No. 2, (February 1993), pp. 535-543, ISSN *0003-9861*

Buitink, J., Hoekstra, F.A. & Hemminga, M.A. (2000). Molecular Mobility in the Cytoplasm of Lettuce Radicles Correlates with Longevity. *Seed Science and Research*, Vol. 10, No. 3, (September 2000), pp. 285 -292, ISSN 0960-2585

Buitink, J. & Leprince, O. (2004). Glass Formation in Plant Anhydrobiotes: Survival in the Dry State. *Cryobiology*, Vol. 48, No. 3, (June 2004), pp. 215-228, ISSN 0011-2240

Buitink, J. & Leprince, O. (2008). Intracellular Glasses and Seed Survival in the Dry State. *Comptes Rendus Biologies*, Vol. 331, No. 10, (October 2008), pp. 788-795, ISSN 1631-0691

Copeland, L.O. & McDonald, M.B. (2001). Seed germination In: *Principles of seed science and technology*, 4th Ed., pp. 72-123, Kluwer Academic Publishers Group, Dordrecht, Netherlands, ISBN 0-7923-7322-7

Criddle, R.S., Hansen, L.D., Smith, B.N., Macfarlane, C., Church, J.N., Thygerson T., Jovanovic, T. & Booth, T. (2005). Thermodynamic Law for Adaptation of Plants to Environmental Temperatures *Pure and Applied Chemistry*, Vol. 77, No. 8, (August 2005), pp. 1425–1444, ISSN 1365-3075

Davies, D.D. (1961) Bioenergetics. In: *Intrmediary Metabolism in Plants*, Cambridge Monographs In Experimental Biology, No. 11., T.A. Bennet-Clark, P.B.M.G. Salt, C.H. Waddington, V.B. Wigglesworth (Ed), 35-52 Cambridge University Press., London, Great Britain

Dragicevic, V. (2007). The Influence of Accelerated Ageing and Stimulative Concentrations of 2,4-D on Maize (*Zea Mays* L.) Seeds. PhD, Agricultural Faculty, Univerity of Novi Sad, Serbia, (December 2007), UDC: 633.15:581.48:57.017.6(043.3)

Dragicevic, V., Sredojevic, S., Djukanovic, L., Srebric, M., Pavlov, M. & Vrvic M. (2007). The Stimulatory Effects of 2,4-D as Hormetic on Maize Seedling's Growth. *Maydica* Vol. 52, No. 3 (2007): 307-310, ISSN 0025-6153

Dragicevic, V., Sredojevic, S. & Milivojevic, M. (2008). Dependence of Starting Wheat Growth on Dynamics of Hydrolysis, Biosynthesis and Free Energy. *9th International Conference on Fundamental and Applied Aspects of Physical Chemistry*, Proceedings, pp. 409-411, ISBN 978-86-82475-16-3, Belgrade, Serbia, 24-26 September, 2008

Dussert, S., Davey, M.W., Laffargue, A., Doulbeau, S., Swennen, R. & Etienne, H. (2006). Oxidative Stress, Phospholipid Loss and Lipid Hydrolysis During Drying and Storage of Intermediate Seeds. *Physiologia Plantarum*, Vol. 127, No. 2, (June 2006), pp. 192–204, ISSN 1399-3054

Gladyshev, G.P. (2010). Thermodynamics and Life. *Herald of the International Academy of Science (Russian Section)*, No. 1, pp.6-10, ISSN 1819-5733

Hansen, L.D., Smith, B.N. & Criddle R.S. (1998). Calorimetry of Plant Metabolism: A Means to Rapidly Increase Agricultural Biomass Production. *Pure and Applied Chemistry*, Vol. 70, No. 3, pp. 687-694, ISSN *1365-3075*

Hatanaka, R. & Sugawara, Y. (2010). Development of Desiccation Tolerance and Vitrification by Preculture Treatment in Suspension-Cultured Cells of the Liverwort *Marchantia polymorpha. Planta*, Vol. 231, No. 4, (March 2010), pp. 965-976, ISSN 0032-0935Heimburg, T. (2010). Lipid Ion Channels. *Biophysical Chemistry*, Vol. 150, No. 1-3, (August 2010), pp. 2–22, *ISSN:* 21530378

Hill, P., Kuzyakov, Y., Jones, D. & Farrar, J. (2007). Response of Root Respiration and Root Exudation to Alterations in Root C Supply and Demand in Wheat. *Plant and* Soil, Vol. 291, No. 1-2, (February 2007), pp. 131-141, ISSN 1573-5036

Hoekstra, F.A., Golovina, E.A. & Buitink, J. (2001) Mechanisms of Plant Desiccation Tolerance. *Trends in Plant Science*, Vol. 6, No. 9, (September 2001), pp. 431-438, ISSN 1360-1385

Holzhütter, H.G. (2004). The Principle of Flux Minimization and its Application to Estimate Stationary Fluxes in Metabolic Networks. *Eurocan Journal Of Biochemistry*, Vol. 271 No. 14, (July 2004) pp. 2905–2922, ISSN: 1742-4658

Jones, D.L., Nguyen, C., Finlay, R.D. (2009). Carbon Flow in the Rhizosphere: Carbon Trading at the Soil–Root Interface. *Plant and Soil*, Vol. 321, No. 1-2, (August 2009), pp. 5-33, ISSN: 1573-5036

Kikuchi, K., Koizumi, M., Ishida, N. & Kano, H. (2006). Water Uptake by Dry Beans Observed by Micro-magnetic Resonance Imaging. *Annals of Botany*, Vol. 98, No. 3, (September 2006), pp. 545-553, ISSN 1095-8290

Kranner I. & Birtić S. (2005). A Modulating Role for Antioxidants in Desiccation Tolerance. *Integrative and Comparative Biology*, Vol. 45, No. 5, (November 2005), pp. 734-740, ISSN *1557-7023*

Krishnan, P., Joshi, D.K., Nagarajan, S. & Moharir, A.V. (2004a). Characterization of Germinating and Non-Viable Soybean Seeds by Nuclear Magnetic Resonance (NMR) Spectroscopy. *Seed Science Research*, Vol. 14, No. 4 (December 2004), pp. 355-362, ISSN 0960-2585

Krishnan P., Nagarajan, S. & Moharir, A.V. (2004b). Thermodynamic Characterisation of Seed Deterioration during Storage under Accelerated Ageing Conditions. *Biosystems Engineering* Vol. 89, No. 4, (December 2004), pp. 425-433, ISSn 1537-5129

Krishnan, P., Joshi, D.K., Nagarajan, S. & Moharir, A.V. (2004c) Characterisation of Germinating and Non-Germinating Wheat Seeds by Nuclear Magnetic Resonance (NMR) Spectroscopy. *European Biophysics Journal,* Vol. 33, No. 1 (February 2004) pp. 76–82, *ISSN* 1432-1017

Kuzyakov, Y. & Larionova, A.A. (2005). Root and Rhizomicrobial Respiration: A Review of Approaches to Estimate Respiration by Autotrophic and Heterotrophic Organisms in Soil. *Journal of Plant Nutrition and Soil Science,* Vol. 168, No. 4, (August 2005), 503–520, ISSN 1522-2624

Manz, B., Müller, K., Kucera, B., Volke, F. & Leubner-Metzger, G. (2005). Water Uptake and Distribution in Germinating Tobacco Seeds Investigated in Vivo by Nuclear Magnetic Resonance Imaging. *Plant Physiology,* Vol. 138, No. 3 (July 2005), pp. 1538–1551, ISSN: 1532-2548

Millar, H., Whelan, J., Soole, K.L. & Day D.A. (2011). Organization and Regulation of Mitochondrial Respiration in Plants. *Annual Review of Plant Biology,* Vol. 62, (June 2011), in press *ISSN:* 1543-5008

Mrva, K., Wallwork, M. & Mares, D.J. (2006). α-Amylase and Programmed Cell Death in Aleurone of Ripening Wheat Grains. *Journal of Experimental Botany,* Vol. 57, No. 4, (March 2006), pp. 877–885, ISSN 1460-2431

Nascimento, W.M. (2003). Preventing Thermoinhibition in a Thermosensitive Lettuce Genotype by Seed Imbibition at Low Temperature. *Scientia Agricola,* Vol. 60, No. 3, (July/September 2003) pp. 477-480, ISSN 0103-9016

Osborne D.J., Boubriak I., Leprince O. (2002). Rehydration of Dried Systems: Membranes and the Nuclear genome, In: *Desiccation and Survival in Plants: Drying Without Dying,* Black M. & Pritchard H. W. (Ed.), pp. 343-366, CABI Publishing, New York, USA, *ISBN* 0 85199 534 9

Pammentner, N.W. & Berjak, P. (1999). A Review of Recalcitrant Seed Physiology in Relation to Desiccation Tolerance Mechanisms. *Seed Science and Research,* Vol. 9, No 1, (January 1999), pp. 13-37, ISSN 0960-2585

Rauf, M., Munir, M., ul Hassan, M., Ahmad, M. & Afzal, M. (2007). Performance of Wheat Genotypes under Osmotic Stress at Germination and Early Seedling Growth Stage. *African Journal of Biotechnology,* Vol. 6, No. 8, (April 2007), pp. 971-975, ISSN 1684–5315

Sánchez-Nieto, S., Enríquez-Arredondo, C., Guzmán-Chávez, F., Hernández-Muñoz, R., Ramírez, J. & Gavilanes-Ruíz M. (2011). Kinetics of the H^+-atpase from Dry and 5-Hours-Imbibed Maize Embryos in its Native, Solubilized, and Reconstituted Forms. *Mol. Plant* Vol. 4, No. 2 (March 2011) pp. In Press, ISSN 1752-9867

Schafer, F.Q. & Buettner, G.R. (2001). Redox Environment of the Cell as Viewed Through the Redox State of the Glutathione Disulfide/Glutathione Couple. *Free Radicals in Biology and Medicine,* Vol. 30, No. 11, (June 2001), pp. 1191-1212, ISSN 0891-5849

Shao, H.B., Chu, L.Y., Lu, Z.H. & Kang, C.M. (2008) Primary antioxidant free radical scavenging and redox signaling pathways in higher plant cells. *International Journal of Biological Sciences,* Vol. 4, No 1 pp. 8-14, ISSN: 1449-2288

Shimokawa, S. & Ozawa, H. (2005). Thermodynamics of the Ocean Circulation: A Global Perspactive on the Ocean System and Living Systems. In: *Non-equilibrium thermodynamics and the production of entropy: life, earth and beyond,* A. Kleidon, R.

Lorenz, R. D. Lorenz (Ed.), pp. 121-134, Springer-Verlag, Germany, ISBN 3-540-22495-5

Siddiqui, S.U., Ali, A. & Chaudhary, M.F. (2008). Germination Behavior of Wheat (*Triticum Aestivum*) Varieties to Artificial Ageing under Varying Temperature and Humidity. *Pakistan Journal of Botany*, Vol. 40, No. 3, (Jume 2008), pp. 1121-1127, ISSN 2070-3368

Smith, B.N., Harris L.C., Keller, E.A., Gul, B., Ajmal Khan, M. & Hansen, L.D. (2006). Calorespirometric Metabolism and Growth in Response to Seasonal Changes of Temperature and Salt. In: *Ecophysiology of High Salinity Tolerant Plants*, M.A. Khan and D.J. Weber (Ed.), pp. 115-125, Springer, Netherlands, ISBN 978-1-4020-4018-4

Sredojevic, S., Dragicevic, V., Srebric, M., Peric, V., Nisavic, A. & Djukanovic, L. (2008). The Quantitatrive Determination of Seed Mass Defect During Germination. 1. The Daily Dynamics of Net Supplemental Free Energy. Journal of Scientific Agricultural Research, Vol. 69, No. 4, pp. 63-77, ISSN 0354-5695

Sun, W.Q. (2000). Dielectric Relaxation of Water and Water-Plasticized Biomolecules in Relation to Cellular Water Organization, Cytoplasmic Viscosity and Desiccation Tolerance in Recalcitrant Seed Tissues. *Plant Physiology*, Vol. 124, No. 3 (November 2000), pp. 1203–1215, ISSN: 1532-2548

Sun, W.Q. (2002). Methods for the Study of Water Relations under Desiccation Stress, In: *Desiccation and Survival in Plants: Drying Without Dying*, Black M. & Pritchard H. W. (Ed.), pp. 47-91, CABI Publishing, New York, USA, *ISBN* 0 85199 534 9

Taiz, L. & Zeiger, E. (2010). Solute Transport. Chapter 6, In: *Plant Physiology*, Fifth Edition pp. 87-108, Sinauer Associates, Sunderland, USA, ISBN-13 978-0-87893-507-9

Taylor, G.B. (2004). Effect of Temperature and State of Hydration on rate of Imbibition in Soft Seeds of Yellow Serradella. *Australian Journal of Agricultural Research,* Vol. 55, No. 1, pp. 39 – 45, ISSN 0004-9409

Trepagnier, E.H., Jarzynski, C., Ritort, F., Crooks, G.E., Bustamante, C.J. & Liphardt, J. (2004). Experimental Test of Hatano and Sasa's Noequilibrium Steady-State Equality. Proceedings of the National Academy of Sciences, Vol. 101, No. 42, (19 October 2004), pp. 15038-15041, *ISSN 0027-8424*

Tukey, H.B. (1970). The Leaching of Substances From Plants. Annual Review of Plant Physiology Vol. 21, pp. 305-324 ISSN: 0066-4294

Vertuci, C.W. & Leopold, A.C. (1984). Bound Water in Soybean Seed And its Relation to Respiration and Imbibitional Damage. *Plant Physiology* Vol. 75, No 1, (January 1984), pp. 114-117, ISSN: 1532-2548

Vitvitskii, A.I. (1969). Activation Energy of Some Free-Radical Exchange Reactions. *Theoretical and Experimental Chemistry*, Vol. 5, No. 3, (May 1969), pp. 276-278, *ISSN* 1573-935X

Volk, G.M., Crane, J., Caspersen, A.M., Hill, L.M., Gardner, C. & Walters, C. (2006). Massive Cellular Disruption Occurs during Early Imbibition of *Cuphea* Seeds Containing Crystallized Triacylglycerols. *Planta*, Vol. 224, No. 6, (November 2006) pp. 1415-1426, ISSN 0032-0935

Vysotskaya, L.B. (2005). Mechanisms Coordinating Wheat Seedling Growth Response as Affected by Shoot/Root Ratio. *Russian Journal of Plant Physiology*, Vol. 52, No. 5, pp. 679-684, ISSN 1021-4437

Walters, C. (2007). Glass Formation, Glass Fragility, Molecular Mobility and Longevity of Germplasm Stored at cryogenic Temperatures. *Cryobiology*, Vol. 55, No. 3, (December 2007), pp. 357-358, ISSN 0011-2240

Walters, C., Pammenter, N.W., Berjak P. & Crane, J. (2001). Desiccation Damage, Accelerated Ageing and Respiration in Desiccation tolerant and Sensitive Seeds. *Seed Science and Research*, Vol. 11, No. 2, (Jun 2001) pp. 135 -148, ISSN 0960-2585

Walters, C., Ballesteros, D. & Vertucci, V.A. (2010). Structural Mechanics of Seed Deterioration: Standing the Test of Time. *Plant Science*, Vol. 179, No. 6, (December 2010), pp. 565–573, ISSN 0168-9452

Yeo, A.R. & Flowers, T.J. (2007). The driving forces for water ans solute movement. In: Plant Solute Transport, Yeo, A.R. & Flowers, T.J. (Ed.), pp. 29-46, Blackwell Publishing, Oxford, UK, ISBN-13: 978-1-4051-3995-3

2

The Concept of Temperature in the Modern Physics

Dmitrii Tayurskii[1] and Alain Le Méhauté[2]
[1]*Kazan Federal University*
[2]*Institute International Franco-Québécois*
[1]*Russia*
[2]*France*

1. Introduction

Thermodynamics, or thermodynamic method, has been invented as a way to describe the exchange processes of energy and matter occurring at the molecular and atomic levels without considering the details of molecular motion. The very beginning of it has been inspired by steam engines efficiency and the initial context of thermodynamics concerned the macroscopic systems in equilibrium and/or quasi-equilibrium states. Statistical physics has been appeared as an attempt to describes the same processes with taking into account atomic hypothesis and atomic dynamic parameters, i.e. the velocities and the positions of atoms. The probabilistic language appeared to be very suitable for these purposes. But again the studying has been restricted by the macroscopic systems in quasi-equilibrium states mostly because of using variational principles like the maximum entropy principle.
Today the relevance of thermodynamic formalism and applicability of statistical physics are questionable when the nanosystems, the systems far from equilibrium and the systems with strong interactions begin to be studied. First of all the problems appear for the definition of the temperature that is the key concept in the formalism of thermodynamics and statistical physics. The assumptions used to define the temperature have to be treated very carefully in the cases of nanosystems, systems with strong interactions and other complex systems[1].
The goal of the present paper is to remind the main conditions which has to be satisfied to introduce the physical quantity "temperature" and to discuss the possibilities for introducing the temperature in complex systems. As an example the model system - the system of interacting spins at external magnetic field - will be used to demonstrate the advantages and restrictions of using "spin temperature" concept.

2. What is the temperature?

The thermodynamic or statistical definition of the temperature can be found in any standard physics textbook. Here we reproduce briefly these procedures with emphasizing some points which are usually considered as the given. As far as the exchange processes of energy and

[1] During the preparation of the present paper the very detailed book has been published (Biró, 2011) where the basic concepts at the very heart of statistical physics are presented and their challenges in high energy physics are discussed

matter are the subject of investigations two macroscopic systems in equilibrium are usually considered. If one allows only energy exchange between these two systems the equilibrium means the equality of some physical parameter in this case. We can call this parameter as a temperature. But how we can measure and/or calculate this parameter? It is necessary to mention here that the key-point in the definition of the temperature is the existence of the so-called thermal equilibrium between two systems.
In the framework the phenomenological approach - thermodynamics - the temperature is measured by the monitoring other physical parameters (expansion coefficient, resistivity, voltage, capacity etc.). That is why there are so many different kinds of thermometers. The procedure of temperature measurements consists of the thermal contact (energy exchange) between the system under consideration and a thermometric body the physical state of which is monitored. This thermometric body should be as small as possible in order to do not disturb the state of the system during measurement. In the realm of very small systems such a procedure is rather questionable. What the size should be for the thermometer to measure, for example, the temperature of a nanosystem? Should the thermometric body be an atom or elementary particle in this case? But the states of atoms and elementary particles are essential quantum ones and can not be changed continuously. The excellent treatment of the more sophisticated measurements of temperature (spectral temperature and radiation temperature) the reader can find in the very recent book (Biró, 2011).
The simplest way to determine the temperature in statistical physics is based on the consideration of possible microscopic states for the given macrostate which is determined by the energy of system or other constraints. The number of possible microscopic states $\Omega(E)$ is extremely fast rising function of the energy of system E. For two systems being in the thermal contact the total number of possible microscopic states is given usually by the following product:

$$\Omega_{12} = \Omega(E_1) \cdot \Omega(E_2). \tag{1}$$

An implicit assumption in the notion of thermal contact is that the system-system interaction is vanishingly small, so that the total energy E is simply given by

$$E = E_1 + E_2 = const. \tag{2}$$

The product $\Omega(E_1) \cdot \Omega(E - E_1)$ shows the very sharp maximum (see Figure 1) and it is more convenient to study the extremal conditions for the logarithm $ln(\Omega(E_1) \cdot \Omega(E - E_1))$ from which on immediately get the definition of the statistical temperature:

$$\frac{\partial ln(\Omega(E_1))}{\partial E_1} = \frac{\partial ln(\Omega(E_2))}{\partial E_2} = \beta = \frac{1}{k_B T}, \tag{3}$$

where k_B is the Boltzmann's constant and T denotes the absolute thermodynamics temperature. In fact the taking of logarithm leads us to some additive quantity, and it is the property which is carried by the Boltzmann definition of entropy:

$$S = k_B ln\Omega(E). \tag{4}$$

Considering the system in the contact with the thermal bath (thermal reservoir) the same assumption about neglecting system-bath interaction leads to the existence of canonical (Gibbs) distribution for the probabilities to find the system in the state with energy E_α:

$$P(E_\alpha) \propto exp(-\beta E_\alpha). \tag{5}$$

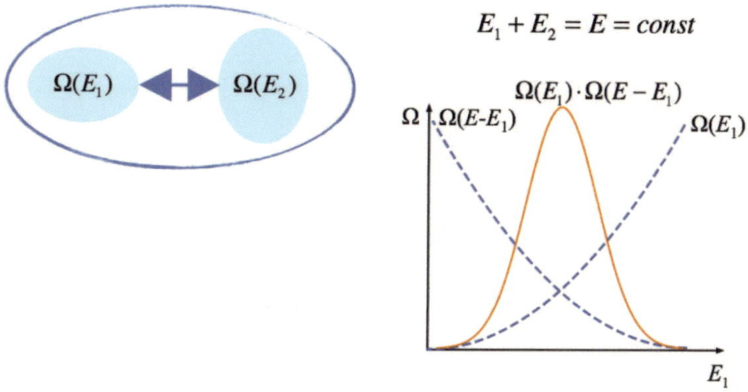

Fig. 1. Two systems in the thermal contact and the number of possible microscopic states for them and for the joint system in dependence on energy. Note that in the definition of thermal contact the energy of interaction between systems is vanishingly small, so the total energy is just $E = E_1 + E_2$.

We have seen that all obtained results are valid only if one neglects by the system-system or system-bath interactions. In fact such neglecting proved to be the consequence of the so-called thermodynamic limit which is reached as the number of particles (atoms or molecules) in a system, N, approaches infinity. For the most systems in the thermodynamic limit the macroscopic extensive variables (energy, entropy, volume) possess the property of additivity like Equation 2. It is necessary to point out here that the possibility to describe the thermodynamic behavior of the system under consideration by the statistical physics methods as N tends to infinity is not automatically granted but depends crucially of the nature of the system. It has been shown many times that the Gibbs canonical ensemble is valid only for sufficiently short range interactions and there are examples - self-gravitating systems, unscreened Coulomb systems - for which the assumed additivity postulate is violated.
Also it should be mentioned that it is possible to give an alternative definition of the thermodynamic limit (Biró, 2011). If a system is so large that it itself can serve as a perfect thermal reservoir (bath) for its smaller parts, then one can consider this system as being in the thermodynamic limit. This definition is not restricted to large volume and large particle number. But again here the energy of spin-bath interactions is neglected and, in practice, such definition also implies the existence of additive quantities.
The more rigorous definition of temperature in statistical physics is based on the maximization of entropy for the system composed from two subsystems (like we consider above) while the energy, volume, particle number etc. are composed from the corresponding subsystems value (see, for example, (Biró, 2011)). Mathematically in denotes that one should look for the maximum of the entropy of the system

$$S(E, V, N, \ldots) = max \tag{6}$$

when the following equations are satisfied

$$\begin{aligned} E &= E_1 \oplus E_2 \\ V &= V_1 \oplus V_2 \\ N &= N_1 \oplus N_2. \end{aligned} \tag{7}$$

Strictly speaking the composition law $\oplus$ in Equations 7 is not restricted only by addition. But finding the maximum of entropy 8

$$dS(E,V,N,...) = \left(\frac{\partial S}{\partial E_1}dE_1 + \frac{\partial S}{\partial E_2}dE_2\right) + \left(\frac{\partial S}{\partial V_1}dV_1 + \frac{\partial S}{\partial V_2}dV_2\right) + \left(\frac{\partial S}{\partial N_1}dN_1 + \frac{\partial S}{\partial N_2}dN_2\right) + ... = 0 \quad (8)$$

when the total energy, volume and particle number are fixed, i.e.

$$dE = \left(\frac{\partial E}{\partial E_1}dE_1 + \frac{\partial E}{\partial E_2}dE_2\right) = 0 \quad (9)$$

$$dV = \left(\frac{\partial V}{\partial V_1}dV_1 + \frac{\partial V}{\partial V_2}dV_2\right) = 0 \quad (10)$$

$$dN = \left(\frac{\partial N}{\partial N_1}dN_1 + \frac{\partial N}{\partial N_2}dN_2\right) = 0 \quad (11)$$

(please note, that here energy, volume and particle number are considered as independent variables!!!) one gets the following expressions:

$$\left(\frac{\partial S}{\partial E_1}\frac{1}{\frac{\partial E}{\partial E_1}}\right) = \left(\frac{\partial S}{\partial E_2}\frac{1}{\frac{\partial E}{\partial E_2}}\right) \quad (12)$$

$$\left(\frac{\partial S}{\partial V_1}\frac{1}{\frac{\partial V}{\partial V_1}}\right) = \left(\frac{\partial S}{\partial V_2}\frac{1}{\frac{\partial V}{\partial V_2}}\right) \quad (13)$$

$$\left(\frac{\partial S}{\partial N_1}\frac{1}{\frac{\partial N}{\partial N_1}}\right) = \left(\frac{\partial S}{\partial N_2}\frac{1}{\frac{\partial N}{\partial N_2}}\right) \quad (14)$$

The left (right) part of Equation 12 will depends on the quantities describing subsystem 1 (2) only in the case, when the simplest addition is taken in 7 as the composition law. So again we conclude that the rigorous definition of the temperature can be done only for the extensive systems in thermodynamic limit.

Now we will demonstrate that the thermodynamic limit does not exist for the systems with strong enough interactions and how this limit can be restored by some averaging procedure. We consider a N-particle system describing by the following Hamiltonian:

$$\mathcal{H} = \sum_{i=1}^{N} \mathcal{H}_{0i} + \sum_{i<j}^{N} \mathcal{U}_{ij}, \quad (15)$$

where $\mathcal{H}_{0i}$ is the Hamiltonian of free particle and $\mathcal{U}_{ij}$ describes the interaction between i-th and j-th particles. It is obviously that because of second term in Equation 15 the energy of system is not additive and the system can not be trivially divided into two- or more independent subsystems. By increasing the number of particles in the system in n times the energy of systems is not increased in n times too. The conditions of thermodynamic limit are violated here. But if the interaction between particles $\mathcal{U}_{ij}$ is a short-range one and a rather small the concept of the mean field can be introduced, so the Hamiltonian 15 can be re-written as:

$$\mathcal{H} = \sum_{i=1}^{N} \mathcal{H}_{0i} + \sum_{i=1}^{N} \langle \mathcal{U}_{ij} \rangle_j. \tag{16}$$

So the energy becomes additive and the conditions of thermodynamic limit are satisfied.
We note here that for nanoscale systems in which the contribution to their energy (or other quantity) from the surface atoms is comparable with that from the bulk volume atoms the non-additivity exists forever. The surface energy is increased in $n^{2/3}$ times when one increases the number of particles in n times. That is why the question about the definition of temperature for nanoscale system is very intriguing not only from the point of view of its physical measurements but also from the estimations for the minimal length scale on which this intensive quantity exists (Hartmann et al., 2004).
Last decades the non-extensive thermodynamics has being developed to describe the properties of systems where the thermodynamic limit conditions are violated. It is not a purpose of this paper to give one more review of non-extensive thermodynamics, its methods and formalism. The reader can found it in numerous papers, reviews and books (see, for example, (Abe & Okamoto, 2001; Abe et al., 2007; Gell-Mann & Tsallis, 2004; Tsallis, 2009)). Here we would like to underline only that the definition of temperature is very close related to the existence of thermal equilibrium and is very sensitive to the thermodynamic limit conditions, so a researcher should be very careful in the prescribing the meaning of temperature to a Lagrange multiplier when entropy maximum is looked for.
In the next part of this work we will show how the thermodynamic formalism can be applied for the system of spins in an external magnetic field and will discuss the existence of two spin temperatures for one spin system. It is a real good example when one has to remember all conditions being used to

3. Spin temperature

An important progress in the description of the behavior of spin degrees of freedom in solids was reached with the use of thermodynamic approach. The relation between electron and nuclear magnetism on the one hand and thermodynamics on the other hand was first established by Casimir and Du Pre (Casimir & Du Pré, 1938). They introduced the concept of spin temperature for a system of non-interacting spins in an external magnetic field. But for a long time spin temperature was being considered an elegant theoretical representation only. Further investigations of spin thermodynamics showed that spin temperature can be experimentally measured and its change is connected with the transfer of heat and change of entropy. For example in the framework of spin temperature concept Bloembergen et al. (Bloembergen et al., 1948; 1959) built the classical theory of saturation and cross-relaxation in spin systems. The excellent account of spin temperature concept the reader can find in (Abragam, 2004; Atsarkin & Rodak, 1972; Goldman, 1970).
In the classical theory of Bloembergen et al. (Bloembergen et al., 1948) the Zeeman levels are considered as being infinitely sharp thus neglecting the broadening due to spin-spin interactions. This is only justified in cases of liquids and gases, where the rapid motion of the atoms or molecules averages the spin-spin interactions to zero and the spins can be considered as being independent of each other. However in solids the spin-spin interactions are normally so strong that the whole ensemble of spins acts as a collective system with many degrees of freedom. The next main step in the understanding of spin thermodynamics in solids was made by Shaposhnikov (Shaposhnikov, 1947; 1948; 1949), whose works were far in advance of the experimental possibilities of their verifying. In these little known works Shaposhnikov

pointed out the significance of taking into account the interactions inside spin systems. As long as the time taken for internal thermal equilibrium to be established is finite then to describe any state of spin system one needs to determine two thermodynamic coordinates, which are magnetization and spin temperature. As distinct from the theory of Casimir and Du Pré these coordinates are not connected between each other by Curie law. Moreover, the spin temperature and total energy of spin system were proven to be just thermodynamically conjugated variables, and the magnetization characterizes a state of spin system in an external magnetic field.

Independently in 1955 to investigate the saturation in system of interacting spins Redfield (Redfield, 1955) introduced the hypothesis that under strong saturation the whole spin system stayed in internal equilibrium, thus permitting its description by one single temperature. The problem for an arbitrary degree of saturation was solved in 1961 by Provotorov who developed Shaposhnikov's ideas about taking into account spin- spin interactions. In well-known works (Provotorov, 1962a;b; 1963) Provotorov showed that under some conditions an energy of spin-spin interactions that are small compared with the interaction with strong external magnetic field can be extracted into a separate thermodynamic subsystem called by the reservoir of spin-spin interactions. This thermodynamic subsystem has its own temperature different from the temperature of spin system in an external magnetic field determined in Gasimir and Du Pré theory. Therefore according to the Provotorov's theory any state of spin system can be described by two temperatures. The concept of two temperatures turned out to be fruitful and it was confirmed experimentally in electron as well as in nuclear magnetism. This concept led to the revision of some representations in the theory of magnetic resonance and relaxation and to the prediction of a number of unexpected physical effects.

Today the two-temperature formalism being unusual from the point of view of statistical thermodynamics forms the basic framework for the theory of magnetic resonance in solids. But because of conceptual and mathematical difficulties the Provotorov's theory was well developed only in the so-called high-temperature approximation when the heat energy of spin greatly exceeds the energy of spin in external magnetic field.

The attempts to extend the theory towards low temperatures (the energy of spin in external field is more than heat energy) demonstrated the principal difficulties in the choosing of thermodynamic variables and in the understanding of energy redistribution inside spin-system (see, for review (Tayurskii, 1989; 1990) and references therein). But there are many experiments at low temperatures to interpret of which it is necessary to have a theory describing the spin thermodynamics and kinetics in this case. Among them we point out dynamic polarization experiments (Abragam & Goldman, 1982), nuclear ordering in solids Bonamour et al. (1991); Lounasmaa (1989). The object of this paper is to give a description of some theoretical approaches to the studying of spin-system in solids at low temperatures. In the next section we will summarize the concept of spin temperature, as far as it is very important for the understanding of low temperature thermodynamics of spin-system in solids.

3.1 High and low temperatures

If an electron (or nuclear) spin S is subjected to a magnetic field H_0 in the direction z then the Zeeman interaction

$$\mathcal{H}_z = \omega_0 S^z \tag{17}$$

establishes a set of $2S + 1$ sublevels with energy $E_m = m\omega_0$ (here and further the units $\hbar = 1$ are used); $\omega_0 = \gamma H_0$ is the Larmor frequency, γ is the gyromagnetic ratio, m is the magnetic quantum number ($m = -S, -S + 1, \dots S - 1, S$). In Equation 17 one must take the negative

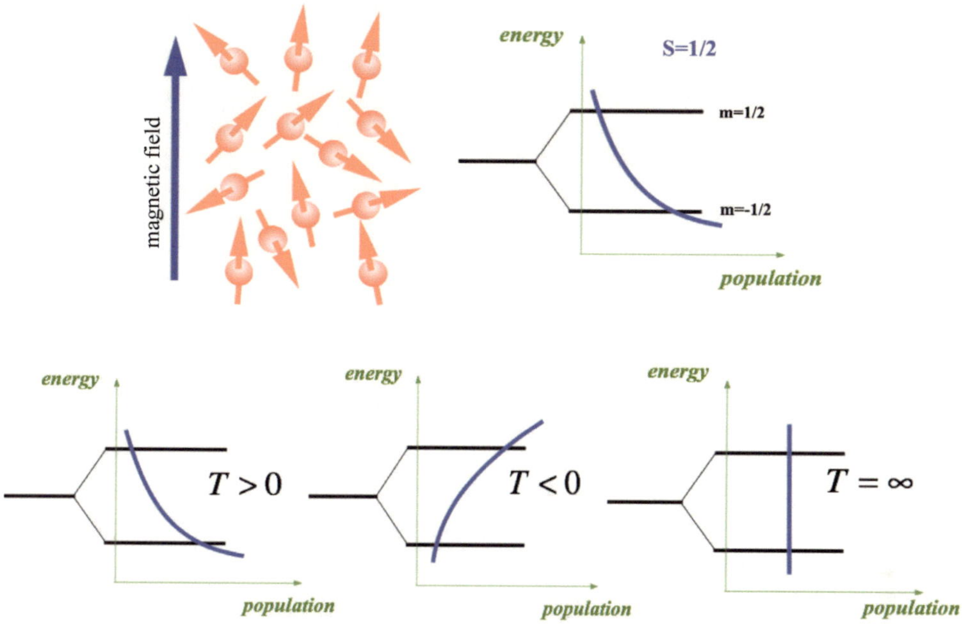

Fig. 2. Illustration of spin temperature concept for the case $S = 1/2$. In the lower row the populations of level are shown for the case of positive, negative and infinite spin temperatures.

sign if S is a nuclear spin. For an ensemble of N identical spins $S = 1/2$ we can introduce the Zeeman level populations n_+ and n_- ($n_+ + n_- = N$), where n_+ is the number of spins in the state $m = +1/2$ and n_- is the number of spins with $m = -1/2$. If the spins are in equilibrium with lattice the distribution of the different spins over the magnetic levels is determined by the Boltzmann law

$$\frac{n_-}{n_+} = \exp\{\frac{\omega_0}{k_B T_L}\} \tag{18}$$

where T_L is a lattice temperature. Such an equilibrium can be violated, for example, after saturation by radio-frequency field. In this case one can describe spin system introducing spin temperature T_s that is distinguished from lattice temperature and is determined by the expression

$$\frac{n}{n_+} = \exp\{\frac{\omega_0}{k_B T_s}\} \tag{19}$$

It should be noted that in the case of thermal equilibrium with the lattice the levels of highest energy are always less populated than the lower ones. At the saturation it is possible to create situations in which the highest levels are more populated than the lower ones. If one assumes Equation 19 to be still valid this situation corresponds to negative spin temperatures. The spin system is not then of course in equilibrium with lattice because negative temperature is only defined for systems with an upper bound in the energy spectrum. The Figure 2 illustrates the concept of spin temperature for the case $S = 1/2$ and positive, negative and infinite values of temperature.

In this part we shall discuss the thermodynamics of spin systems at low temperatures. We shall define "low temperature" as a temperature at which the energy of spin in strong external magnetic field is of the same order as the average heat energy or exceeds it, i.e. the following condition is true:

$$\omega_0 \gtrsim k_B T_L \tag{20}$$

In the case of localized electron spins in insulators this condition denotes a magnetic field of about 50 kG and a temperature about 1 K. Concerning the energy of spin-spin interactions the assumption is made that it is small compared to the energy of spins in the external magnetic field. As we are not interested in the effects connected with phase transition into magnetically ordered state then we can assume that

$$E_{ss} \ll k_B T_L \tag{21}$$

where E_{ss} is an average energy of spin-spin interactions. The temperatures at which an average heat energy of spins becomes more than the energy of a spin in an external field are called "high temperatures". The transition from the range of high temperatures into the range of low temperatures is accompanied first of all by the essential changes of the thermodynamic properties of spin system.

3.2 High-temperature thermodynamics of spin system

Let a consider the regular lattice of spins in an external magnetic field. The corresponding Hamiltonian of spin system has the following form

$$\mathcal{H}_s = \mathcal{H}_z + \mathcal{H}_{ss} \tag{22}$$

where

$$\mathcal{H}_z = \omega_0 \sum_j S_j^z \tag{23}$$

is the Hamiltonian describing the Zeeman interaction of spins with the constant magnetic field directed along z-axis, S_j^z is a longitudinal component of j-th spin. In Equation 22 $\mathcal{H}_{ss}$ describes the interactions of spins between themselves. For the sake of simplicity we will not take into account further interactions such as hyperfine interactions or interaction with crystal field. Temporarily we omit the interactions such as spin-phonon interactions and interactions with other external fields. The energy levels of the Hamiltonian $\mathcal{H}_z$ are strongly degenerated because for any eigenvalue of the Hamiltonian one can find many combinations in which the eigenvalues of operators S_j^z can be taken. This degeneracy is removed by the spin-spin interactions. Further we will suppose the external field to be large compared to the internal fields induced by spin-spin interactions. Therefore the Hamiltonian of spin-spin interactions $\mathcal{H}_{ss}$ can be considered as a perturbation. In the first order of the perturbation theory only those terms of $\mathcal{H}_{ss}$ which don't cause the change of any eigenvalue of the Hamiltonian $\mathcal{H}_z$ will give a contribution into the splitting of the energy level corresponding to this eigenvalue. So in the first order in the perturbation only part of $\mathcal{H}_{ss}$ that commutates with $\mathcal{H}_z$ will give a contribution in the broadening of the Zeeman levels. Usually this part is called the secular part and is presented as (4)

$$\mathcal{H}_{ss} = \frac{1}{2} \sum_{i,j} (A_{ij} S_i^z S_j^z + B_{ij} S_i^+ S_j^-) \tag{24}$$

where S_i^+ and S_i^- are the transverse components of j-spin. Further we shall use the notation $\mathcal{H}_{ss}$ exactly for the secular part of the Hamiltonian of spin-spin interactions. In this section we don't need the explicit form of spin-spin interaction constants A_{ij} and B_{ij} . Thus the energy spectrum of the Hamiltonian $\mathcal{H}_s$ represents quasi-continued equidistant bands of energy levels and these bands are separated in the energy by equal intervals ω_0 .

The distribution of spins over these levels is the result of spin-spin interactions during some characteristic time T_2 that is of order of inverse width of the magnetic resonance line shape. As it follows from the experience usually in solids this time T_2 is significantly shorter than spin-lattice relaxation time T_1. The distribution of spins over the energy levels entirely determines the thermodynamic state of a spin system. It was appeared to be very convenient to introduce two spin temperatures for describing this distribution.
Formally the idea of two spin temperatures can appear from the following considerations. As long as spin system placed into strong external magnetic field has two first integrals - they are the Zeeman energy and the energy of the secular part of spin-spin interactions - then in order to describe a sufficiently slow evolution of any thermodynamic state it is natural to choose these first integrals as the thermodynamic coordinates. Such a choice becomes very attractive at high temperatures because the Zeeman energy and the energy of the secular part of spin-spin interaction are statistically independent in this case. The statistical independence denotes that one can change a value of one energy without changing a value of the other, i.e. spin system is decoupled to two subsystems the energies of which are additive.
To explain the latter now we consider a quasi-equilibrium distribution, described by the following density matrix:

$$\begin{aligned} \rho_s &= Q_s^{-1} \exp\{-\alpha \mathcal{H}_z - \beta \mathcal{H}_{ss}\} \\ Q_s &= Sp \exp\{-\alpha \mathcal{H}_z - \beta \mathcal{H}_{ss}\} \end{aligned} \tag{25}$$

Such a form of quasi-equilibrium distribution takes place due to the fact of the availability of two invariants of motion. In Equation 25 parameters α and β linked to the operators $\mathcal{H}_z$ and $\mathcal{H}_{ss}$ are thermodynamically conjugative parameters for the Zeeman energy and the energy of spin-spin interactions respectively. We can expand the exponent in Equation 25 in powers of $\alpha\mathcal{H}_z$ and $\beta\mathcal{H}_{ss}$ and keep only the linear terms. As we shall see later such a linearization corresponds to the high temperature approximation. In the linear approximation in $\alpha\mathcal{H}_z$ and $\beta\mathcal{H}_{ss}$, the density matrix is reduced to

$$\rho_s = \{1 - \alpha \mathcal{H}_z - \beta \mathcal{H}_{ss}\} / Sp\hat{1} \tag{26}$$

Here $\hat{1}$ is the unit matrix. Therefore these expressions describe the expectation value of the Zeeman energy and the expectation value of the energy of spin-spin interactions:

$$\begin{aligned} \langle \mathcal{H}_z \rangle &= Sp\rho_s \mathcal{H}_z \simeq -\alpha Sp \mathcal{H}_z^2 / Sp\hat{1} \\ \langle \mathcal{H}_{ss} \rangle &= Sp\rho_s \mathcal{H}_{ss} \simeq -\beta Sp \mathcal{H}_{ss}^2 / Sp\hat{1} \end{aligned} \tag{27}$$

where we have taken into account the orthogonality of operators $\mathcal{H}_z$ and $\mathcal{H}_{ss}$ following from Equations 23 and 24:

$$Sp \mathcal{H}_z \mathcal{H}_{ss} = 0. \tag{28}$$

One can see from Equations 27 that in the linear approximation in $\alpha\mathcal{H}_z$ and $\beta\mathcal{H}_{ss}$:

$$\frac{\partial \langle \mathcal{H}_z \rangle}{\partial \beta} = 0. \tag{29}$$

On the other hand we have the explicit expression for this partial derivative :

$$\frac{\partial \langle \mathcal{H}_z \rangle}{\partial \beta} = \frac{\partial}{\partial \beta} \frac{Sp \exp\{-\alpha \mathcal{H}_z - \beta \mathcal{H}_{ss}\} \mathcal{H}_z}{Sp \exp\{-\alpha \mathcal{H}_z - \beta \mathcal{H}_{ss}\}} = -\langle \mathcal{H}_z \mathcal{H}_{ss} \rangle + \langle \mathcal{H}_z \rangle \langle \mathcal{H}_{ss} \rangle \tag{30}$$

Comparing the right-hand sides of Equation 29 and Eqquation 30 we conclude that in the linear approximation in $\alpha\mathcal{H}_z$ and $\beta\mathcal{H}_{ss}$

$$\langle\mathcal{H}_z\mathcal{H}_{ss}\rangle \simeq \langle\mathcal{H}_z\rangle\langle\mathcal{H}_{ss}\rangle \tag{31}$$

The Equation 31 means the statistical independence of the first integrals. We remind the reader that two or more events are statistically independent if each individual event is not influenced by the occurrence of any other and Equation 31 corresponds to the well-known rule of probabilities multiplying.

An entropy of the quasi-equilibrium distribution defined by Equation 26 is equal in the linear approximation in $\alpha\mathcal{H}_z$ and $\beta\mathcal{H}_{ss}$ to

$$S = -k_B Sp\rho_s \ln\rho_s \simeq -k_B\{-\ln Sp\hat{1} + \alpha^2 Sp\mathcal{H}_z^2/Sp\hat{1} + \beta^2 Sp\mathcal{H}_{ss}^2/Sp\hat{1}\}. \tag{32}$$

We see that the entropy is just a sum of contributions coming from the Zeeman interaction and spin-spin interactions. Each of these contributions depends on α or β only. The obvious identities follow from Equation 32:

$$\begin{aligned}\alpha &= \frac{1}{k_B}\frac{\partial S}{\partial\langle\mathcal{H}_z\rangle} = \frac{1}{k_B T_z}\\ \beta &= \frac{1}{k_B}\frac{\partial S}{\partial\langle\mathcal{H}_{ss}\rangle} = \frac{1}{k_B T_{ss}}\end{aligned} \tag{33}$$

where the parameters T_z and T_{ss} have the physical meaning of temperature of the Zeeman subsystem and temperature of the reservoir of spin-spin interactions respectively. The linear approximation in $\alpha\mathcal{H}_z$ and $\beta\mathcal{H}_{ss}$ corresponds to the situation when the energy of spin in external magnetic field and one in internal fields are small compared to the heat energy. Therefore such a linearization corresponds to the high temperature approximation. Further according to the tradition we shall call the parameters α and β the temperatures of subsystems.

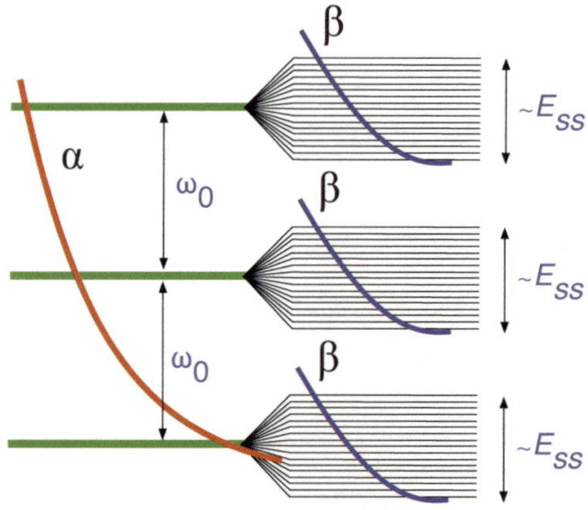

Fig. 3. The distribution of the spins over the energy levels of the spin system at high temperatures. It is possible to say about two spin temperatures - α and β in high-temperature approximation

In the thermal equilibrium with lattice the spin system is characterized by one single temperature (the temperatures α and β coincide). In the general case any thermodynamic state of spin system in high temperature approximation can be described by two temperatures: one for a Boltzmann distribution between the Zeeman levels, and a second one describing

the ordering of the spins in the local fields induced by spin-spin interactions. A pictorial illustration is given in Figure 3.
So at high temperatures we have two statistically independent subsystems - the Zeeman subsystem and the reservoir of spin-spin interactions - which are characterized by the corresponding independent temperatures. It is the approximation in which the well-known Provotorov's theory was built (Provotorov, 1962a;b; 1963),. The developed two spin temperature formalism approach was appeared to be very fruitful to understand and to describe many problems of magnetic resonance, like saturation, spin-lattice relaxation etc. (Abragam, 2004; Atsarkin & Rodak, 1972; Goldman, 1970; Wolf, 1979).

3.3 Low-temperature thermodynamics of spin system

At low temperatures the factorization condition (see Equation 31) is violated because in the expanding of exponent one has to take high-order terms into account and the right-hand side of Equation 29 will not be equal to zero. Consequently our subsystems became statistically dependent (for more details the reader should refer to (Philippot, 1964; Tayurskii, 1989; 1990)). The advantage of the above-mentioned choice of thermodynamic coordinates is lost. Besides the Zeeman temperature doesn't have the physical meaning of temperature, and the temperature of the reservoir of spin-spin interactions is the real temperature of the whole system (Philippot, 1964).
In order to be convinced of it we now note the following. As long as the Zeeman energy and the energy of spin-spin interactions are the first integrals then any linear combination of these energies with constant coefficients will be the first integral as well. In particular the total energy of spin system

$$E_s = Sp\rho_s\mathcal{H}_s = Sp\rho_s\mathcal{H}_z + Sp\rho_s\mathcal{H}_{ss} = \langle\mathcal{H}_{zz}\rangle + \langle\mathcal{H}_{ss}\rangle \tag{34}$$

and the expectation value of z-component of the total spin

$$\langle S\rangle = Sp\rho_s \sum_j S_j^z = \frac{1}{\omega_0} Sp\rho_s\mathcal{H}_z = \frac{1}{\omega_0}\langle\mathcal{H}_z\rangle \tag{35}$$

are the first integrals. If we choose these invariants as thermodynamic coordinates we should have an analogy of a thermodynamic system with fixed number of particles. The total number of particles and the z-component of total spin are the operators with discrete spectrum and these operators play a similar role. Therefore we can use a method analogous to that of the grand canonical ensemble and consider a Gibbs ensemble of systems characterized by the density matrix

$$\begin{aligned}\tilde{\rho}_s &= \tilde{Q}_s^{-1}\exp\{-\beta\mathcal{H}_s + \nu S_z\}\\ \tilde{Q}_s &= Sp\exp\{-\beta\mathcal{H}_s + \nu S_z\},\end{aligned} \tag{36}$$

where the multiplier ν is fixed by the condition that $\langle S_z\rangle$ is given (now $\langle\ldots\rangle$ denotes the averaging with the density matrix 36).
The entropy of the distribution in Eq.[19] may be written as

$$S = -k_B Sp\tilde{\rho}_s \ln\tilde{\rho}_s = -k_B\{-\ln\tilde{Q} - \beta\langle\mathcal{H}_s\rangle + \nu\langle S_z\rangle\}. \tag{37}$$

Then taking into account the identity

$$\langle\mathcal{H}_s\rangle = \omega_0\langle S_z\rangle + \langle\mathcal{H}_{ss}\rangle, \tag{38}$$

which follows from the Equations 22 and 23 the entropy may be written as

$$S = k_B \beta \langle \mathcal{H}_{ss} \rangle - k_B \langle S_z \rangle (\nu - \omega_0 \beta) + k_B \ln \tilde{Q} \tag{39}$$

One can use the definition of temperature for systems with fixed number of particles and obtain from Equations 37 and 39

$$\frac{1}{k_B T_s} = \left(\frac{\partial S}{\partial \langle \mathcal{H}_s \rangle} \right)_{\langle S^z \rangle} = \frac{S}{\partial \langle \mathcal{H}_{ss} \rangle} = \beta \tag{40}$$

This means that the temperature of the reservoir of spin-spin interactions is the real temperature of spin system (see Figure 4). On the other hand one can write from Equation 37 the expression for the total energy

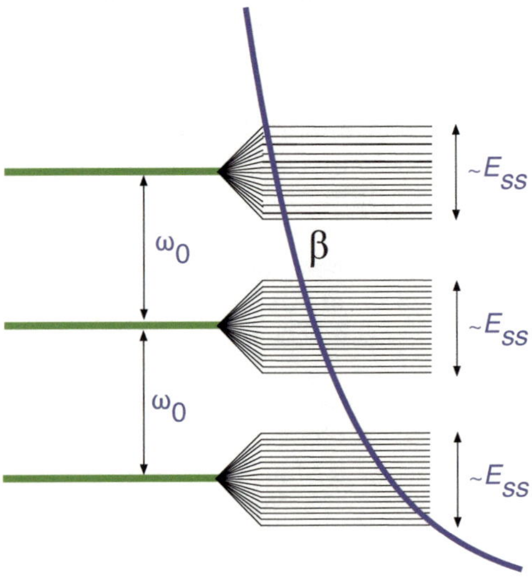

Fig. 4. The distribution of the spins over the energy levels of the spin system at low temperatures. Only the temperature of the reservoir of spin-spin interactions β is the real temperature of spin system

$$\langle \mathcal{H}_s \rangle = \frac{1}{k_B \beta} S + \frac{\nu}{\beta} - \frac{1}{\beta} \ln \tilde{Q} \tag{41}$$

whence a simple relation between multiplier ν and the chemical potential follows

$$\mu = \left(\frac{\partial \langle \mathcal{H}_s \rangle}{\partial \langle S^z \rangle} \right)_S = \frac{\nu}{\beta} \tag{42}$$

(here we used the definition of the chemical potential for the grand canonical ensemble). The state of the Zeeman subsystem is determined by the value of z-component of total spin or by the multiplier ν related to the chemical potential. Comparing the Equations 25, 36 and 42 one gets a simple relation between the chemical potential of spin system and the parameters α and β:

$$\mu = \omega_0 \left(1 - \frac{\alpha}{\beta} \right) \tag{43}$$

For further details concerning the using of chemical potential in spin thermodynamics the reader should refer to (Philippot, 1964). To calculate the expectation values of energies $\langle \mathcal{H}_Z \rangle$ and $\langle \mathcal{H}_{ss} \rangle$ at low temperatures one has to take into account the higher-order terms in the expansion of the density matrix in Equation 26. As a consequence the factorization condition 28 is violated and the Zeeman subsystem and the reservoir of spin-spin interactions cannot be considered as independent. So the advantage of the above-mentioned choice of thermodynamic coordinates is lost. Besides at low temperatures the entropy written in terms α and β

$$S = \alpha\langle \mathcal{H}_Z \rangle + \beta\langle \mathcal{H}_{ss} \rangle + k_B \ln Q_S \tag{44}$$

is not a sum of items each of which depends on α or β only. Therefore at low temperatures the parameter a has no physical meaning of the inverse temperature of the Zeeman subsystem. In such a situation it is more plausible to use the expectation value of jz-component of total spin (in other words the magnetization of spin system or the number of flipped spins) and the total energy as thermodynamic coordinates. Chemical potential and spin temperature are thermodynamically conjugated coordinates for them respectively. Such a choice of thermodynamic coordinates (with taking into account the Equation 43) has been made in (de Haas et al., 1980; 1981; Tayurskii, 1989; 1990). We note that one can also choose α or β as the thermodynamic coordinates at low temperatures but it is necessary to take into account the correlation between them and the fact that parameter a has no physical meaning of the inverse temperature of the Zeeman subsystem.

So, by use very simple, model system - systems of interacting spins in an external magnetic field - we have demonstrated that the definition of temperature depends crucially on the assumptions made for this procedure. And one has to be very careful doing so.

4. Conclusion

The physical quantity "temperature" is a cornerstone of thermodynamics and statistical physics. In the present paper the short introduction to the classical concept of temperature for macroscopic equilibrium systems was given. The concept of temperature was discussed regarding the nanoscale physics and non-extensive systems. It was shown forget that it is necessary to remember about the conditions to be satisfied in order to introduce "temperature" in macroscopic physics. The concept of "spin temperature" in condensed matter physics was reviewed and the advantages of thermodynamic approach to the problems of magnetism was illustrated. Partially, two temperatures spin thermodynamics was analyzed and the conditions when such approach is valid was studied.

5. References

Abe, S. & Okamoto, Y. (2001). *Nonextensive statistical mechanics and its applications*, Spinger, ISBN 3-540-41208-5.

Abe, S.; Beck, C.; Cohen, G.D. (2007). Superstatistics, thermodynamics and fluctuations. *Physical Review E*, Vol. 76, 031102(1-5).

Abragam, A. & Goldman, M. (1982). *Nuclear magnetism: Order and disorder*, Clarendon Press, ISBN 0198512945, 626 p.

Abragam, A. (1994). *The principles of nuclear magnetism*, Oxford University Press, ISBN 019852014X, 9780198520146, 599 p.

Atsarkin, V.A. & Rodak, M.I. (1972). Spin-Spin Interaction Temperature in the Electronic Paramagnetic Resonance. *Uspekhi Fiz. Nauk: Soviet Physics Uspekhi*, V. 107, 3-27.

Biró, T. S. (2011). *Is There a Temperature?: Conceptual Challenges at High Energy, Acceleration and Complexity (Fundamental Theories of Physics)*, Spinger, ISBN 978-1441980403, 323 p.

Bloembergen, N.; Pursell, E.M.; Pound, R.V. (1948). Relaxation Effects in Nuclear Magnetic Resonance Absorption. *Physical Review*, V. 73, 679-712.

Bloembergen, N.; Shapiro, S.; Pershan, P.S.; Artman, J.D. (1959). Cross-Relaxation in Spin-Systems. *Physical Review*, V. 114, 445-459.

Bonamour, P.; Bouffard, V.; Fermon, C.; Goldman, M.; Jacquinot, J.F.; Saux, G. (1991). Nuclear Ferromagnetism in a Perovskite Structure. *Physical Review Letters*, V. 66, 2810-1813.

Casimir, H.B.G. & Du Pré. F.K. (1938). Note on the Thermodynamic Interpretation of Paramagnetic Relaxation Phenomena. *Physica*, V. 5, 507-512.

Gell-Mann, M. & Tsallis, C. (2004). *Nonextensive Entropy: Interdisciplinary Applications*, Oxford University Press, ISBN 978-0195159776, 440 p.

Goldman, M. (1970). *Spin Temperature and Nuclear Magnetic Resonance in Solids*, Clarendon Press, 246 p.

de Haas, L.J.; Wenckenbach, W.Th.; Poulis, N.J. (1980). Magnetic resonance saturation in solids at finite temperatures. *Physica*, V. 103A, 295-315.

de Haas, L.J.; Wenckenbach, W.Th.; Poulis, N.J. (1981). Magnetic resonance saturation and cross-relaxation at low temperature. *Physica*, V. 111B, 219-230.

Hartmann, M.; Mahler, G.; Hess, O. (2004). Existence of Temperature on the Nanoscale. *Physical Review Letters*, Vol. 93, No. 8, 080402.

Lounasmaa, O.V. (1989). Nuclear Magnetic Ordering at Nanokelvin Temperatures. *[Physics Today]*, October 1989, 26-33.

Philippot, J. (1964). Spin-Spin Relaxation and Spin Temperature. *Physical Review*, V. 133A, 471-480.

Provotorov, B.N. (1962). About Magnetic Resonance Saturation in Crystals. *Soviet Physics JETP*, V. 14, 1126.

Provotorov, B.N. (1962). Theory of Double Magnetic Resonance in Solids *Physica Review*, V. 128, 75-76.

Provotorov, B.N. (1963). About Double Magnetic Resonance in Crystals. *Soviet Physics Solid State*, V. 4, 2155.

Redfield, A.G. (1955). Nuclear Magnetic Saturation and Roatry Saturation in Solids. *Physical Review*, V. 98, 1789-1809.

Shaposhnikov, I.G. On the Thermodynamic Theory of Paramagnetic Absorption in Weak Fields. *Zh. Eksp. Teor. Fiz.*, V. 17, 824-829.

Shaposhnikov, I.G. On the Thermodynamic Theory of Spin-Spin Relaxation in Paramagnets. *Zh. Eksp. Teor. Fiz.*, V. 18, 533-538.

Shaposhnikov, I.G. On the Thermodynamic Theory of Paramagnetic Absorption in Perpendicular Fields. *Zh. Eksp. Teor. Fiz.*, V. 19, 225-230.

Tayurskii, D.A. (1989). Spin Kinetics in Paramagnets at Low Temperatures. *physica status solidi (b)*, V. 152, 645-655.

Tayurskii, D.A. (1990). Cross-Relaxation in Paramagnetic Crystals at Low Temperatures. *Physica*, V. 165-166, 231-232.

Tsallis, C. (2009). Non-additive entropy and non-extensive statistical mechanics - An overview after 20 year. *Brasilian Journal of Physics*, Vol. 39, No. 2A, 337-356.

Wolf, D. (1979). *Spin-temperature and nuclear-spin relaxation in matter: basic principles and applications*, Clarendon Press, 462 p.

3

Photosynthetic Productivity: Can Plants do Better?

John B. Skillman[1], Kevin L. Griffin[2], Sonya Earll[1] and Mitsuru Kusama[1]
[1]*Department of Biology, California State University, San Bernardino, California*
[2]*Lamont-Doherty Earth Observatory, Columbia University, New York, New York*
USA

1. Introduction

By the time you finish this paragraph you will be a changed person. Change, of course, takes a variety of forms, from eroding mountains and melting snowflakes to the major changes we experience in our own lives - birth, growth, reproduction, and death. Less tangible, but changes nonetheless, are the small shifts in perspective that occur as we go through our days - new ideas, new feelings, and new understandings. But, as the Second Law of Thermodynamics reminds us, these varied and seemingly unrelated expressions of change do, in fact have a common source. All change arises from the decay of order - an increase in entropy. Sometimes the increased disarray that defines change is obvious, as when a crystalline cube of ice turns into a higgledy-piggledy puddle of water. Other times, the increased disorder is less obvious. Indeed, in living systems, change often seems to be associated with an *increase* in order. The gradual recovery of a lush and materially diverse living forest in the years following a catastrophic fire would seem to represent a net increase in order. The sequential nature of biological development, from zygote to adult, or from acorn to ancient oak, would seem to represent a net increase in order. The regulated changes in neuronal connectivity and brain architecture that occurs as the mind grasps new insights would seem to represent a net increase in order. Subtler still are the numerous changes in living systems where the initial and final states appear unaltered. The life of every cell depends upon the numbers, locations, and shapes of a multitude of complex molecules. These molecules themselves have a finite existence, a 'lifespan' that, in most cases, is considerably shorter than the life of the cell itself. The median half-life of various protein molecules sampled from various eukaryotes is only 1.8 hours (Hatfield & Vierstra, 1997). Consequently, even static living materials, such as the long-lived neurons in a human brain or the living cells in the trunk of an ancient oak tree, are constantly in the process of rebuilding themselves from within. Even with no obvious change, we are all in a state of constant material flux for the duration of our lives. We are different now from who we were at the start of this paragraph.

At first glance life seems to violate the Second Law of thermodynamics. The fact that all changes in the Universe lead to decreases in order and the fact that living matter acts to increase its own internal order is sometimes referred to as Schrödinger's paradox. But Schrödinger (1992) resolved this famous paradox. He acknowledged that the order inherent in living systems represented a thermodynamically off-balance condition. But Schrödinger also realized that a continuous input of 'high-quality' energy was required to maintain this off-balance or non-equilibrium state. Live organisms require a reliable source of energy in the form of food or chemically-reduced minerals or sunlight in order to postpone the descent into the disorder of thermodynamic equilibrium. At the same time, the production and maintenance of this non-equilibrium living state is coupled to energetic processes that do increase entropy in the Universe. For example, energy released in thermonuclear reactions on the sun simultaneously powers photosynthesis on Earth while adding to the overall entropy of the Universe. Likewise, energy released in the digestion of food powers the ordered changes in our bodies while also increasing the overall kinetic chaos of the matter around us thereby adding to the entropy of the Universe. Life exploits rather than violates the Second Law of thermodynamics.

Positioned at the interface of biology and physics, of life and non-life is photosynthesis. This biophysical process is the key to life's exploitation of the Second Law. Plants, algae, and many bacteria use solar energy to convert simple energy-poor substrates of water and carbon dioxide into complex, energy-rich, organic materials. More simply, plants and their photoautotrophic kin use sunlight to reverse the local flow of entropy. The carbohydrate products of photosynthesis are, in turn, used to energize and materialize the production, operation, and maintenance of the rest of the organism. These green organisms themselves serve as the energetic foundation of most biological communities. Indeed, photosynthesis is typically the central biological determinant of the overall productivity and species composition for any local community of organisms (Whittaker, 1975). Globally, photosynthesis also helps dictate the habitability of the planet. The soils that support our natural and agricultural ecosystems, the oxygen-rich atmosphere that permits our own existence, and the stratospheric ozone that shields us from hazardous ultraviolet sunlight are all byproducts of photosynthesis (Schlesinger, 1997). Our modern civilizations and economic systems are also beholden to this fundamental bioenergetic process. Food, fibers, pharmaceuticals, and fossil fuels are a few examples of economically important goods that are derived directly or indirectly from photoautotrophs (Roston, 2008). Today we have begun to ask even more of photosynthesis as we confront the tripartite challenge of overpopulation, dwindling fossil-fuels, and anthropogenic climate change. Provisioning food for a population of 10 billion in the face of finite water, deteriorating soils, and changing climates will be a daunting task. There is hope that improvements in the photosynthetic productivity of crop species can make a substantial contribution to this effort (Evans, 1993; Long et al., 2006). Meeting rising demands for global energy will also depend in part on replacing non-renewable fossil fuels with renewable plant-based resources. Presently there are many uncertainties associated with biofuels (Tilman et al., 2009). But, if properly executed, the use of renewable plant-based fuels, while leaving the remaining geological deposits of organic carbon in the ground, would have great economic, ecological, and climate-stabilizing benefits (Calvin, 1980; Robertson et al., 2011). Climate-stabilizing strategies also include long-term storage of photosynthetically-fixed CO_2 in wild and managed forests. Forest growth in the United States presently offsets approximately 15% of the annual U.S. emissions of CO_2 from fossil fuels (Ryan et al., 2010). A global increase in photosynthetic re-cycling of atmospheric CO_2 and the sequestration of that carbon into long-lived trees and soils should compensate for much of the CO2 emissions arising from humanities' profligate use of carbon-based fuels (Griffin & Seemann, 1996). The energetic, environmental and economic importance of photosynthesis cannot be overestimated.

One can imagine several strategies for squeezing more human benefit out of plant production including increased plant resistance to diseases and pests, increased plant tolerance to environmental stressors such as drought or heat, increased allocation of plant growth to the desired product such as grain in cereal crops or woody stems in forestry trees. Fig. 1 emphasizes how the overall rate of light-driven carbon fixation by the plant underpins all of these production-enhancement strategies. Thus, the most generally applicable approach to improving plant productivity, regardless of the species or the desired plant product, may be to find ways of increasing plant photosynthesis itself. Moreover, it has been argued that improving photosynthesis is the only viable production-enhancing strategy that remains for row crop species (Long et al., 2006; Zhu et al., 2010). Success at improving photosynthetic plant productivity requires an appreciation of how cellular photosynthesis translates to increased whole-plant production and an appreciation of where constraints are to be found that limit the flow of energy and carbon from the individual green cell up to the whole plant.

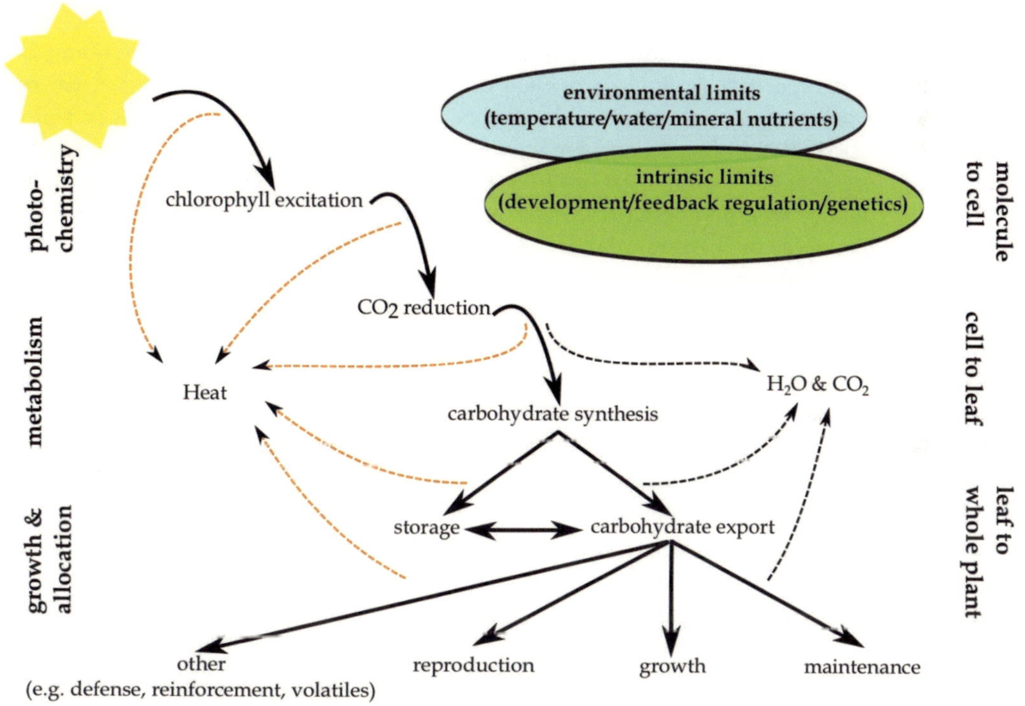

Fig. 1. A simplified energy flow diagram for pondering plant productivity possibilities. Moving down through the figure we go from quantum mechanical phenomena at sub-cellular scales (photochemistry) to biochemical and physiological phenomena at the cell to leaf scale (metabolism) to developmental phenomena at the leaf to whole-plant scale (growth and allocation). Solid black arrows represent energy transformation steps as light-energy is transduced into plant biomass and associated biological functions. Orange dashed arrows indicate the loss of useful energy to entropy ('heat') at each energy transformation step. Black dashed arrows indicate the loss of organic carbon to CO_2 at energy transformation steps powered by respiratory carbon oxidation. After Horton (2000).

Several key steps in photosynthetic production, defined as the net rate of light-driven plant growth, are depicted in Fig. 1. Briefly, light absorbed by photosynthetic pigments in the green cells of plants is used to power the conversion of CO_2 into a simple sugar product. This product is then used in the synthesis of more complex organic molecules. These photosynthates (i.e., the organic carbon products of photosynthesis) may be stored within the leaf or transported for use elsewhere in the plant. Photosynthates are considered as both energy and material resources for cellular metabolism and plant growth. Energetically, these photosynthates may be re-oxidized to CO_2 in mitochondrial respiration in order to produce cellular ATP, which, in turn, powers all other metabolic and transport processes in the plant. Materially, these photosynthates may serve as the carbon 'backbone' for the assembly of the rest of the cell, the leaf, and the whole plant. All essential whole-plant functions (e.g., growth, defense, maintenance, reproduction) depend upon this common and finite supply of photosynthate.

We define 'constraint' as a mechanism or process that limits the evolutionary or engineered response of a biological trait (or set of traits) to natural selection or genetic manipulations. For simplicity and clarity, this review, like the energy flow diagram in Fig. 1, tends to locate constraints on photosynthetic productivity at the level of the cell, the leaf, or the whole-plant. This simplifying perspective omits a great deal of important biology. In particular, we note that all of these processes are conditioned by the developmental and/or acclimatory state of the plant ('intrinsic limits' in Fig. 1), and are subject to the complex vagaries of environmental resources and stressors ('environmental limits' in Fig. 1). These dynamic processes also interact in complex regulatory web-like networks that are necessarily omitted from Fig. 1. This chapter presents some examples of these regulated interactions in order to illustrate how the system-like nature of plants can constrain their photosynthetic productivity.

This review examines the question of whether or not it is reasonable to expect success at squeezing more benefit out of plant photosynthesis. This is a broad question encompassing a multitude of disciplines and perspectives. We will limit ourselves to exploring some representative intrinsic biological constraints on the photosynthetic productivity of terrestrial C3 plants that emerge at the level the cell, the leaf, and the whole-plant. At the cellular level we will briefly describe the primary carbon metabolism of terrestrial C3 plants. This will make it possible to review results from a variety of molecular studies that have shown how targeted manipulations of carbon metabolism can improve on photosynthesis in ways that potentially enhance overall plant production. These kinds of studies imply that improved photosynthetic productivity, at least among plants grown in intensively managed agricultural and forest plantation settings, is possible. However, as Fig. 1 implies, complex leaf-level traits can limit plant productivity. There is broad variation among plant species in various aspects of leaf structure, function, and persistence or 'lifespan' and comparative studies can give insight into these limits. We will present results from a comparative leaf-level study that demonstrate how the photosynthetic potential of a leaf is constrained by traits associated with leaf persistence. Photosynthetic metabolism and foliar lifespan must be considered together in predicting the overall carbon-gain potential of an individual leaf. Next, moving up to the whole-plant level, we will present results from a study using plants that have been genetically modified to have longer-lived leaves in order to see how this affects aspects of whole-plant performance, including growth. We will conclude with some thoughts on the central question of whether or not plants can 'do better' as informed by our review of advances in the cellular-molecular genetics domain and the trade-offs that emerge as cellular processes scale up to the leaf- and whole-plant levels. We will suggest that analysis of results from studies designed to improve plant productivity will yield the greatest insight if considered from a 'systems' perspective. This perspective enforces a dialectic view of both reductionist and holistic understandings of plants and their energetic activities.

2. What constrains photosynthetic productivity?

The interest here is on the feasibility of increasing the intrinsic potential for photosynthesis and growth. The photosynthetic response of a leaf to light (Fig. 2) illustrates some important aspects of the intrinsic constraints on photosynthesis. Here we see that in complete darkness leaves are net producers of CO_2 as a result of mitochondrial respiration (R_m). Although R_m rates decrease in illuminated leaves, respiration does not stop entirely (Atkin et al., 1998). Thus, in the light, the rate of carbon assimilated by a cell or a leaf or a plant must be understood to be the net balance between ongoing carbon oxidation processes, including R_m, and chloroplast carbon reduction (i.e., gross photosynthesis). Mitochondrial densities and respiratory activity vary among species, among tissues in the same plant, and across growing conditions (Griffin et al., 2001). Wilson & Jones (1982, as cited in Long et al., 2006) were able to improve biomass production in rye grass by selecting for plants with reduced respiration rates. These observations emphasize the importance of R_m as a determinant of net carbon gain and implicate R_m as a target process for improving plant production.

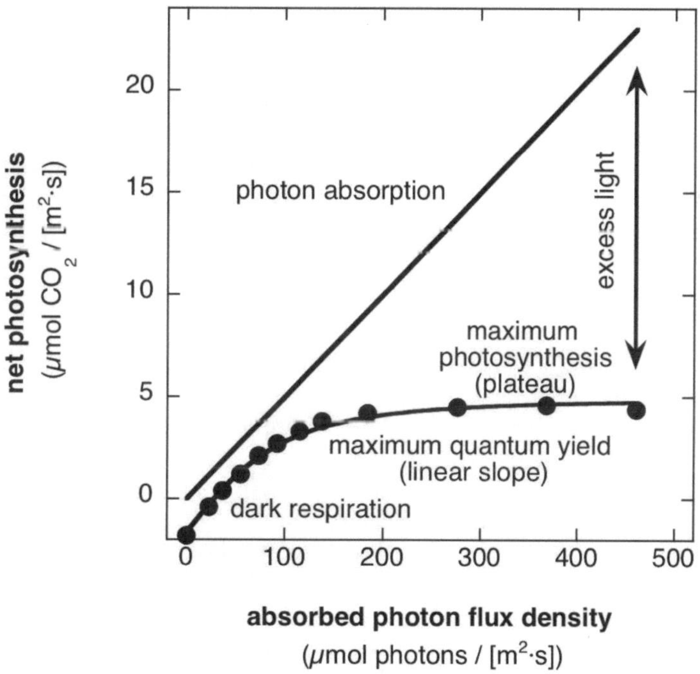

Fig. 2. Photosynthetic response of a healthy C3 leaf to variation in light under ambient CO_2, O_2, and moderate temperatures. The curvilinear scatter plot depicts the net rate of CO_2 uptake of a leaf as a function of light (photon flux density;PFD) absorbed by the leaf. Noteworthy parameters of the curve are discussed in the text.

The slope of the linear portion of the light-response curve (Fig. 2), the quantum yield (QY), is a measure of the maximum *realized* efficiency for the conversion of light energy into carbohydrate by the leaf. When measured under similar conditions, there is little variation in the maximum QY across the breadth of healthy, non-stressed C3 species (Ehleringer & Björkman, 1977). This suggests there is strong selection for plants to produce leaves that are as efficient as possible in capturing solar energy in carbohydrates. Interestingly, the maximum *realized* QY, measured under ambient CO_2 and O_2 concentrations, always falls well below the maximum *potential* QY determined in the lab under optimal, albeit artificial, controlled CO_2 and O_2 concentrations (Skillman, 2008). This 'real world' inefficiency is largely due to energy losses associated with photorespiration (discussed below). The observed difference between the maximum potential QY and the maximum realized QY in healthy C3 plants suggests photorespiration might be a suitable target for improving photosynthetic productivity.

The maximum capacity for net photosynthesis (Pmax) is quantified as the rate of CO_2 uptake (or O_2 production) at light-saturation (Fig. 2). Pmax, measured under identical conditions, varies considerably across species and varies for the same species grown under different conditions (e.g., Skillman et al., 2005). Much of our review focuses on efforts to increase Pmax as a means of improving photosynthetic productivity. But the assumption that changes in Pmax translate to changes in whole-plant growth is open to debate (Evans, 1993; Kruger & Volin, 2006; Poorter & Remkes, 1990).

The solid diagonal line in Fig. 2 takes its slope from the linear portion of the light response curve (QY). This shows that, in principle, if there were no upper saturation limit on Pmax, increasing absorbed light would continue to produce increasing amounts of carbohydrate all the way up to full-sun (~2000 μmol m^{-2} s^{-1} PFD). However, real leaves become light-saturated well below full-sun. The shade grown leaf in Fig. 2 is fully saturated at 200 μmol m-2 s-1 PFD or about 10% of full-sun. As photosynthesis becomes increasingly light-saturated, an increasingly greater portion of the light absorbed by the photosynthetic pigments is not used to drive carbon fixation and must therefore be considered as excess light. If the 'ceiling' on Pmax could be raised, leaves in high-light could theoretically improve plant production by using a greater portion of the available photo-energy for carbon fixation. In the context of trying to improve on photosynthetic production, the fate of excess absorbed light is intriguing and will be discussed below.

2.1 Cellular photosynthesis: molecular manipulations of carbon metabolism

At the cell-molecular level, net photosynthesis depends upon the integrated interactions of a set of interdependent biochemical processes. An abbreviated and simplified illustration of nine of these key interactive processes is given in Fig. 3: (i) ATP & NADPH-producing light- or photochemical-reactions on thylakoid membranes in the chloroplast (green stacked ovals in Fig. 3); (ii) CO_2 diffusion from the atmosphere into the leaf, the cell, and the chloroplast (dashed arrows from Ca to Ci to Cc at the top of Fig. 3); (iii) ATP and NADPH-dependent fixation and chemical reduction of CO_2 in the chloroplastic Calvin cycle (circular reaction sequence in the chloroplast in Fig. 3); (iv) chloroplastic starch biosynthesis from carbohydrate products of the Calvin cycle (linear reaction sequence in the chloroplast in Fig. 3); (v) cytosolic sucrose biosynthesis from Calvin cycle carbohydrates exported from the chloroplast (left-hand linear reaction sequence in the cytosol in Fig. 3); (vi) cytosolic glycolysis where hexoses (glucose or fructose) are oxidized to form pyruvate (right-hand linear reaction sequence in the cytosol in Fig. 3); (vii) mitochondrial citric acid cycle where pyruvate imported from the cytosol is oxidized to CO_2, producing chemical reducing equivalents $FADH_2$ and NADH (circular reaction sequence in the mitochondria in Fig. 3); (viii) respiratory electron transport on the inner mitochondrial membrane wherein electrons from NADH and $FADH_2$ are passed sequentially onto O_2, establishing a H^+ gradient which, in turn, drives mitochondrial ATP synthesis (membrane-bound reaction sequence near the bottom of the mitochondria in Fig. 3); and (xi) the photorespiration cycle where phosphoglycolate, a side-reaction product off the chloroplastic Calvin cycle, is modified and transported over a series of reactions spanning the chloroplast, the peroxisome, and the mitochondria before the final product, glycerate, feeds back into the Calvin cycle (cyclic sequence of reactions near top of figure occurring across all three organelles in Fig. 3). Below we will see that each of these interdependent cellular processes have been targeted for molecular manipulations of cellular carbon metabolism and we will note some cases where these modifications have improved photosynthesis.

2.1.1 The light-reactions and the fate of excess light

The photochemical- or light-reactions of photosynthesis involve light-driven electron and proton (H^+) movement at the inner set of chloroplast membranes (thylakoids) leading to the oxidation of H_2O to O_2 and the production of ATP and NADPH (Fig. 3; Blankenship, 2002). ATP and NADPH, the key light-reaction products, are needed for the subsequent fixation and chemical reduction of CO_2 in the Calvin cycle. The passage of electrons from H_2O to NADPH is mediated by a chain-like series of thylakoid-bound electron-carrier molecules including a special set of electron-carriers called photosystems. Photosystems (PS) are trans-membrane, multi-subunit, chlorophyll-binding protein complexes in the thylakoids where light-energy is transduced first to electrical- and then chemical-energy (Fig. 4). Two different classes of photosystems exist (PSII and PSI) that work in series in the light-reactions. In this reaction sequence, PSII precedes PSI (Fig. 4). Electron transport from H_2O to NADPH could not occur without the PS-mediated input of light energy. In particular, the light-dependent PSII-mediated oxidation of H_2O, releasing O_2 as a byproduct, is quite exceptional. The disassociation of H_2O into O_2 and H^+ and electrons does not normally happen under conditions present at the Earth's surface. This light-driven flow of electrons from H_2O to NADPH also results in the movement of protons (H^+) from the stroma space of the chloroplast into the inner thylakoid space (the lumen). This light-generated H^+ gradient is, in turn, used to drive ATP synthesis via another thylakoid multi-subunit protein complex called ATP-synthase, (not shown). The chemical energy held in the light-reaction products ATP and NADPH, represents a fraction of photo-energy initially absorbed by the PS pigments (Fig. 3). Indeed, a variable but substantial fraction of the absorbed light-energy is dissipated as thermal-energy from PS associated pigments ('heat' in Fig. 3 & 4) thereby lowering the energetic efficiency of the light-reactions.

Chida et al. (2007) showed the potential for increasing plant production by increasing electron transport rates. Cytochrome c6 is a photosynthetic electron transport carrier that operates between PSII and PSI in algae but which does not normally occur in land plants. *Arabidopsis thaliana* plants transformed to constitutively express the algal CytC6 gene sustained higher electron transport rates and had 30% higher Pmax rates and growth rates than wild-type plants (Table 1). Notably, these plants were grown under modest light levels (50 μmol photons/(m^2• s) PFD). It would be interesting to know how these transformed plants perform in brighter light because, as discussed below, rapid photosynthetic electron transport potential can actually be a liability in bright light.

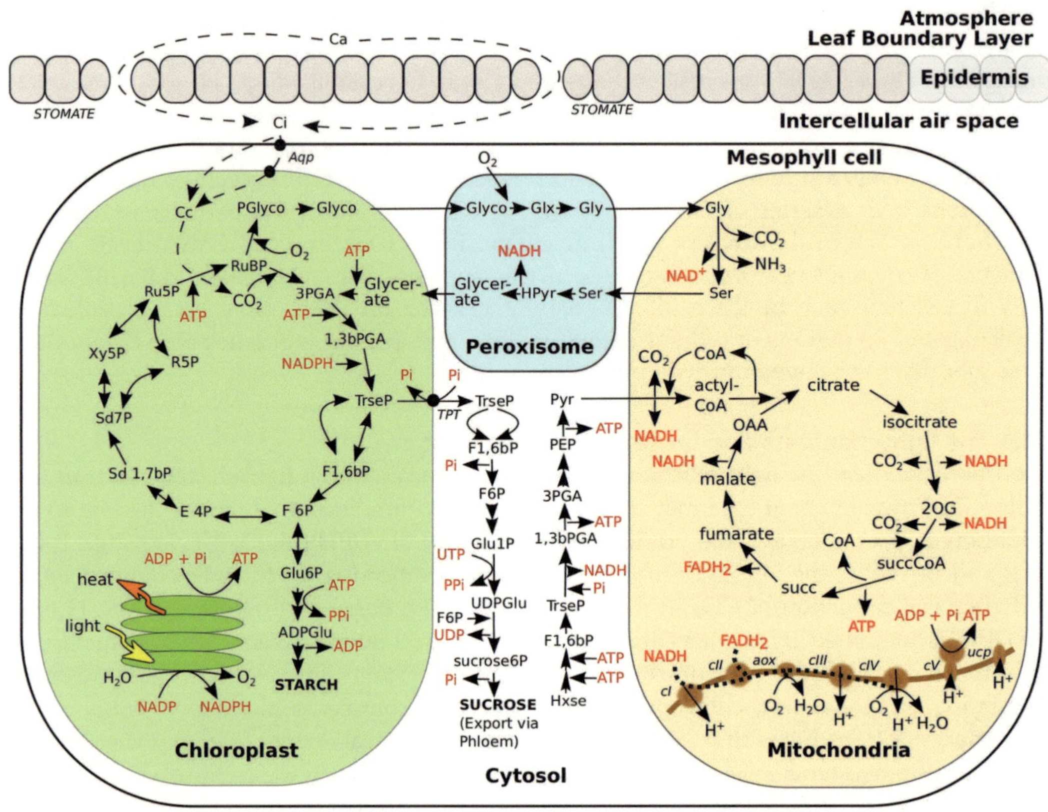

Fig. 3. Primary carbon metabolism in a photosynthetic C3 leaf. An abbreviated depiction of foliar CO2 uptake, chloroplastic light-reactions, chloroplastic carbon fixation (Calvin cycle), chloroplastic starch synthesis, cytosolic sucrose synthesis, cytosolic glycolysis, mitochondrial citric acid cycle, and mitochondrial electron transport. The photorespiration cycle spans reactions localized in the chloroplast, the peroxisome, and the mitochondria. Stacked green ovals (chloroplast) represent thylakoid membranes. Dashed arrows near figure top represent the CO2 diffusion path from the atmosphere (Ca), into the leaf intercellular airspace (Ci), and into the stroma of the chloroplast (Cc).Solid black arrows represent biochemical reactions. Enzyme names and some substrates and biochemical steps have been omitted for simplicity. The dotted line in the mitochondria represents the electron transport pathway. Energy equivalent intermediates (e.g., ADP, UTP, inorganic phosphate; Pi) and reducing equivalents (e.g., NADPH, $FADH_2$, NADH) are labeled in red. Membrane transporters Aqp (CO_2 conducting aquaporins) and TPT (triose phosphate transporter) are labeled in italics. Mitochondrial inner-membrane electron transport and proton transport proteins are labeled in small case italics.

Abbreviations (listed alphabetically); 1,3bPGA, 1,3-bisphosphoglyceric acid; 2OG, 2-oxoglutaric acid; 3PGA, 3-phosphoglyceric acid; acetylCoA, acetyl coenzyme A; ADP/ATP, adenosine diphosphate and triphosphate; ADPGlu, adenosine diphosphate glucose; AOX, alternative oxidase; Aqp, aquaporin; Ca, atmospheric CO_2; Cc, chloroplast CO_2; Ci, intercellular CO_2; cI, mitochondrial Complex I; cII, mitochondrial Complex II; cII, mitochondrial Complex III; cIV, mitochondrial Complex IV (cytochrome oxidase); cV, mitochondrial Complex V (ATP Synthase); citrate, citric acid; CoA, coenzyme A; E4P, erythrose 4-phosphate; F1,6bP, fructose 1,6-bisphosphate; F6P, fructose 6-phosphate; $FADH_2$, flavin adenine dinucleotide; Fumarate, fumaric acid; Glu1P, glucose 1-phosphate; Glu6P, glucose 6-phosphate; Glx, glyoxylic acid; Gly, glycine; Glycerate, glyceric acid; Glyco/PGlyco, glycolic acid/phosphoglycolic acid; Hpyr, hydroxypyruvic acid; Hxse, hexose (glucose and/or fructose); Isocitrate, isocitric acid; Malate, malic acid; NADH/NAD, oxidized and reduced forms of nicotinamide adenine dinucleotide; NADPH/NADP, oxidized and reduced forms of nicotinamide adenine dinucleotide phosphate; OAA, oxaloacetic acid; PEP, phosphoenol pyruvate; Pi, orthophosphate; PPi, pyrophosphate; Pyr, pyrivic acid; R5P, ribose 5-phosphate; Ru5P, ribulose 5-phosphate; RuBP, ribulose 1,5-bisphosphate; Sd1,7bP, sedoheptulose 1,7-bisphosphate; Sd7P, sedoheptulose 7-phosphate; Ser, serine; Starch, poly-glucan; Succ, succinic acid; succCoA, succinyl coenzyme A; sucrose, sucrose; sucrose6P, sucrose 6-phosphate; TPT, trisose phosphate translocator; TrseP, triose phosphate, collectively dihydroxyacetone phosphate and 3-phosphoglyceraldehyde; ucp, uncoupling factor; UDPGlu, uracil-diphosphate glucose; Xy5P, xylulose 5-phosphate.

Light is highly dynamic in time and space. The cellular photosynthetic apparatus must be able to balance the need to maximize photon absorption and use in the shade against the danger of excessive chlorophyll excitation in bright light (Anderson et al., 1988). Plants in the shade employ multiple cellular traits to maximize the efficient interception, absorption, and utilization of light for photosynthesis. But, under bright light, the challenge is in how to deal with a surplus of photo-energy. If light-driven electron transport exceeds chloroplastic capacity to utilize this chemical reducing power it can lead to the formation of singlet oxygen and other harmful reactive oxygen species (ROS). High-light stress can potentiate cell death when endogenous ROS are permitted to accumulate and damage cellular materials (Takahashi and Badger, 2010). Plants have evolved a suite of processes that lower the risk of broad scale cellular photo-oxidative damage under excess light (Demmig-Adams and Adams, 2006; Raven, 2011). These protective processes include features that (i) limit light absorption by chlorophyll, (ii) dissipate excess absorbed light as heat, (iii) divert light-driven electron transport away from an energy-saturated Calvin cycle towards alternative pathways, (iv) lower the number of functional PSII centers thereby impeding chloroplast electron transport, and (v) maintain high chloroplast complements of antioxidants to scavenge excess ROS. Paradoxically, these protective processes, to one extent or another, have the effect of lowering the energetic efficiency of photosynthesis because a portion of the available photo-energy is not available for carbon fixation. This low efficiency manifests as a light-induced reduction in QY, a phenomenon referred to as photoinhibition (e.g., Skillman et al., 1996).

Thermal dissipation of absorbed photo-energy from the PS pigments (before the initiation of photosynthetic electron transport) is one means for avoiding ROS production in excess light (Fig. 3). This process, termed feedback dissipation (FD), depends upon the conversion of xanthophyll pigments from one form to another in PS-associated proteins. Xanthophyll-dependent FD requires the activity of a number of gene products (Jung & Niyogi, 2009). The best-studied contributor to FD is the PsbS protein, a PSII-associated subunit. Plants lacking PsbS have restricted xanthophyll conversion, are more sensitive to photo-oxidative damage including persistent photoinhibition, and exhibit lower growth rates and reproduction under fluctuating light conditions (Krah & Logan, 2010; Külheim et al., 2002). Thus, xanthophyll-dependent FD, for its role in restricting chloroplast ROS formation, confers a strong fitness advantage for plants growing under natural fluctuating light conditions. High-light induced FD lowers the photosynthetic QY even when the cell is returned to low-light. This is because it takes several minutes for the xanthophyll pigments to return to the non-dissipating state.

Diversion of photosynthetic electron transport to non-productive pathways is another means of minimizing ROS production in excess light (Fig. 4). Alternative electron transport (AET) may be understood as a general strategy for dealing with an over-reduced electron transport chain which, in chloroplasts, manifests as an excessive $NADPH/NADP^+$ concentration ratio. Many metabolic processes fit this definition. The water-water cycle is an AET path that simultaneously acts as a sink for excess reductant and minimizes the accumulation of toxic ROS (Fig. 4). In excess light, when chloroplast NADPH oxidation rates are slower than light-dependent NADPH production rates, $NADP^+$ concentrations begin to restrict the photo-reduction of $NADP^+$ to NADPH. In this over-reduced state, PSI may pass electrons on to O_2 to form superoxide ($O_2{}^{-}$), a highly reactive and toxic ROS (i.e., the Mehler reaction). Superoxide is potentially hazardous to the cell but chloroplasts have a suite of antioxidant metabolites and enzymes that can usually scavenge it before it can do much damage (Foyer & Shigeoka, 2011). Chloroplast antioxidants (e.g. glutathione, ascorbic acid) can rapidly detoxify superoxide by reducing it sequentially back to H_2O (Fig. 4). This sequence of reactions from the PSII oxidation of water (yielding O_2 as a byproduct) through the sequential reduction of O_2 first to superoxide at PSI, and finally back to H_2O (i.e., the water-water cycle) is a energetically futile but it turns out to be an elegant solution to the problem of excess light (Asada, 1999). Photorespiration, another well-known AET example, will be discussed in a later section.

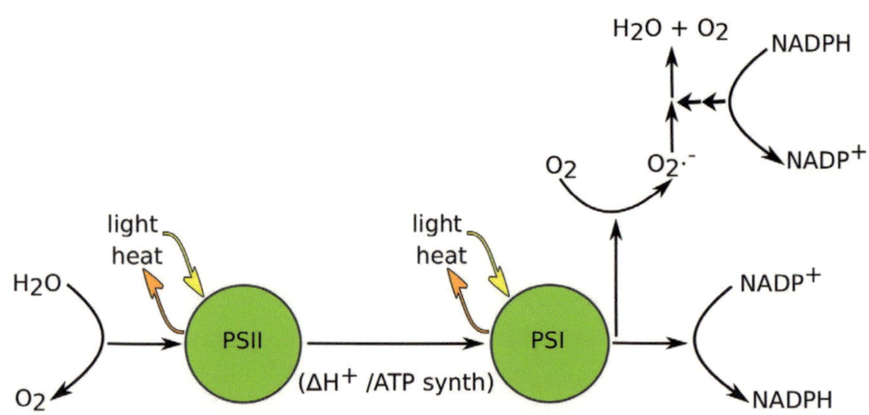

Fig. 4. Photosynthetic transport of electrons from PSII water oxidation is normally used to reduce $NADP^+$ to NADPH for use in Calvin cycle carbon fixation. But, under excess light, when the electron transport chain is over-reduced, electron flow may be diverted at PSI to the Mehler reaction, reducing O_2 to superoxide ($O_2{}^{\cdot -}$). Chloroplast antioxidant systems can further reduce $O_2{}^{\cdot}$ back to water, allowing the water-water cycle to function as a protective alternative electron transport path.

Finally, if other protective photoinhibitory processes (e.g., increased xanthophyll-dependent FD or increased AET diversions) are insufficient, photo-generated ROS can cause a net loss of functional PSII reaction centers (Takahashi & Badger, 2010). The ultimate threat of excess-light is that it can lead to unregulated photo-oxidative cellular damage. Paradoxically, the ROS-mediated net loss of functional PSII, is viewed as a last-ditch defense against excess endogenous ROS production and cell damage. Loss of active PSII centers will necessarily inhibit rates of thylakoid electron transport and therefore lower the rate of light-driven ROS production. But, the protective benefits of photoinhibition, whether arising from the net loss of functional PSII centers or from any other of a suite of ROS avoidance processes, come at a cost. All of these processes lower the efficiency of capturing light energy in plant organic carbon, and so reduce QY.

Several recent studies have made theoretical considerations of the costs of various aspects of photoinhibition (Murchie & Niyogi, 2011; Raven, 2011; Zhu et al. 2004; Zhu et al., 2010). Raven (2011) observes that photoinhibition can slow growth both because of the energetic costs of PSII repair/turnover and because of the foregone photosynthesis resulting from stress-induced QY reductions. Zhu et al. (2004) estimate that a speedier reversion of xanthophyll-dependent FD could improve daily whole-plant carbon gain as much as 25%. Our discussion so far holds a central lesson; photosynthesis requires a capacity for energetic flexibility. Photosynthesis must be both highly efficient and highly inefficient in its use of light, depending on the light level and the state of the plant. This capacity for regulated adjustments of light-use efficiency by the photosynthetic apparatus appears to be an important and conserved trait among plants. This complicates plant productivity improvement strategies that target gains in cellular photosynthetic efficiency.

2.1.2 Diffusion limitations on carbon acquisition

Products of the light-reactions, ATP and NADPH, are used primarily to energize CO_2 fixation and reduction (Fig. 3). But, along with ATP and NADPH, Calvin cycle carbon fixation also depends upon the stromal CO_2 concentration (Cc). Physical barriers (e.g. cell wall and membranes) and gas- to liquid-phase transitions between the external air and the enzymatic site of carbon fixation lowers the diffusional conductance between the atmosphere and the stroma (Terashima et al., 2011). Conductance limited diffusion from Ca to Ci to Cc varies among species and with environmental conditions and can strongly limit photosynthesis (Warren, 2008).

Stomata, in the leaf epidermis, are key sites of regulated control of carbon acquisition along this diffusion path (Fig. 3). Behaviorally, changes in stomatal aperture regulate the foliar rates of CO_2 uptake and transpirational water loss. Recently Araújo et al. (2011) studied respiratory and photosynthetic physiology in wild-type (WT) and antiSDH2-2 tomato (*Solanum lycopersicum*) plants grown under optimal greenhouse conditions. The SDH2-2 gene encodes a sub-unit of mitochondrial succinate dehydrogenase. This enzyme normally catalyzes the citric acid cycle conversion of succinate (succ) to fumarate (Fig. 3). Engineered SDH2-2 anti-sense plants had as much as 25% greater growth than WT plants (Table 1). Several differences in relevant primary carbon metabolism were observed between the two genotypes including a 30% enhancement in Pmax in antiSDH2-2 plants. Araújo et al. (2011) observed that antiSDH2-2 had lower tissue concentrations of malate, a citric acid cycle intermediate formed down-stream of the succinate dehydrogenase reaction. Malate is known to promote stomatal closure. They concluded that the Pmax and growth enhancements were pre-dominantly a result of greater stomatal conductances in the antiSDH2-2 plants arising from the reduced concentrations of malate.

The relative number of stomata in the leaf epidermis (stomatal density; SD) is subject to developmental control, depending upon the conditions under which the plant is grown (Beerling, 2007; Nadeau, 2009). Schülter et al. (2003) studied photosynthetic physiology in wild-type (WT) and sdd1-1 *Arabidopsis thaliana* plants. The sdd1-1 genotype has a point mutation that results in greater stomatal densities. Over a range of constant light conditions, sdd1-1 plants consistently had double the leaf SD of the WT plants. Under constant conditions the sdd1-1 plants also had higher rates of leaf transpiration but maximum carbon uptake rates (Pmax) were indistinguishable between genotypes. Thus, under constant light conditions, increased stomatal densities had no detectable effect on carbon gain and lowered the leaf-level water use efficiency (WUE; carbon gain per unit water lost). Interestingly, when low-light grown plants were transferred to high-light, leaf Pmax in the sdd1-1 plants was ~ 25% greater than in the transferred WT plants (Table 1). Apparently, upon transfer to bright light, stomatal density limited photosynthesis in WT but not sdd1-1 plants. The transfer had no effect on relative transpiration rates of the two genotypes and so leaf WUE increased with the change in light more for sdd1-1 than for WT plants. These studies illustrate the potential for bioengineering of stomatal behavior and/or density as means to increasing photosynthetic carbon gain. However, the inevitable trade-offs with water-use suggest the practical applications of these kinds of manipulations would ultimately be limited to plants grown under highly managed cultivation systems where water deficits can be minimized.

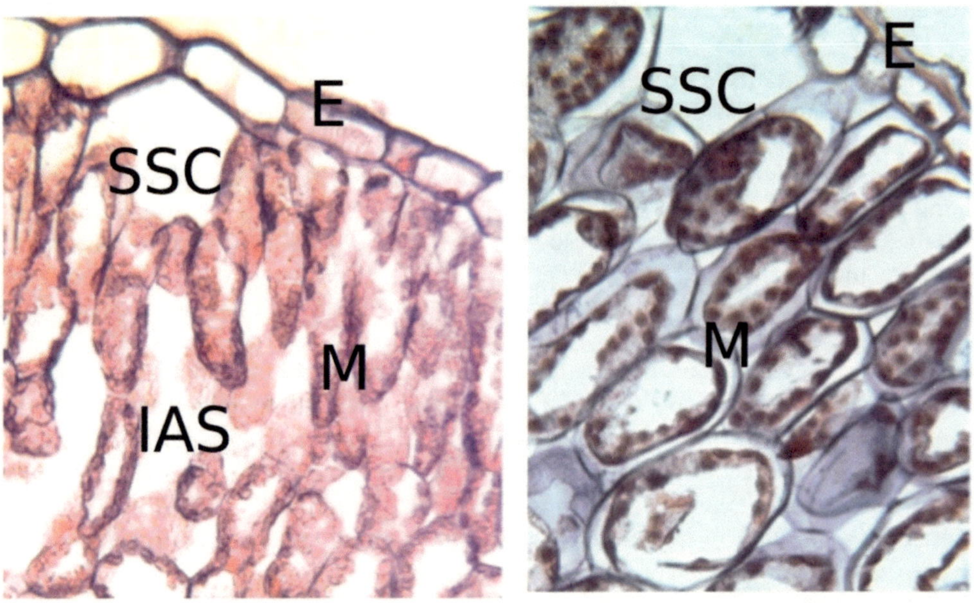

Fig. 5. Leaf anatomy differs among species in ways that affect the mesophyll conductance to CO_2 diffusion. Thin mesophytic *Nicotiana tabacum* leaves (left) have abundant intercellular air space, thin mesophyll cell walls, and, presumably a high mesophyll conductance that could sustain high rates of photosynthesis. Thick sclerophyllous *Agave schidigeri* leaves (right) have large, tightly packed, thick-walled cells, and, presumably a low mesophyll conductance that could restrict photosynthesis. E =epidermis, M=mesophyll cells, SSC=sub-stomatal cavity, IAS=intercellular airspace. (Micrographs by Bruce Campbell.)

After passage through the stomata, CO_2 diffusion from the intercellular space into the stroma where carboxylation occurs depends upon a series of conductances that are referred to collectively as the internal or mesophyll conductance (Terashima et al., 2011). As it turns out, these combined internal conductances substantially limit photosynthesis, and explain nearly half of the drawdown in CO_2 concentration between the atmosphere and the stroma

(Ca - Cc; Warren, 2008). Variation in mesophyll conductance depends upon structural features such as leaf morphology & anatomy, cell wall thickness and composition, cell packing, chloroplast position and density (Fig. 5). Mesophyll conductance is also affected by biochemical factors such as aquaporin membrane transport proteins (Aqp in Fig. 3).

Aquaporins (Aqps) are small trans-membrane proteins that facilitate the osmotic movement of water across membranes (Maurel et al., 2008). Some aquaporins will also transport other small uncharged molecules like CO_2 thereby potentially increasing mesophyll conductance (Fig. 3). Flexas et al. (2006) produced tobacco plants that were either deficient in or over-expressed aquaporin NtAQP1. Under optimal conditions, plants over-expressing NtAQP1 had mesophyll conductances 20% greater than wild-type (WT) plants and Pmax rates 20% greater than WT (Table 1). By contrast, the NtAQP1 deficient plants had mesophyll conductances and Pmax rates that were 30% and 13% lower, respectively, than the WT plants. Aquaporin density as a factor in mesophyll conductance CO_2 is complicated by its role in maintaining tissue water relations and by the fact that other determinants of mesophyll conductance are plastic and highly variable (Tholen et al., 2008; Tsuchihira et al., 2010). Nevertheless, Flexas et al. (2006) provide proof-of-concept evidence that bioengineered enhancements of mesophyll conductance can stimulate photosynthesis.

2.1.3 The carbon-reactions

The fate of stromal CO_2 and light-reaction products is followed here through the Calvin cycle, photorespiration, starch and sucrose synthesis, glycolysis and mitochondrial respiration in the so-called 'carbon-reactions' of a typical photosynthetic cell. In the chloroplast, the key reaction for Calvin cycle carbon fixation is the binding of CO_2 to ribulose 1,5-bisphosphate (RuBP), the five-carbon organic acceptor molecule (Fig. 3). This reaction yields two molecules of 3-phosphoglycerate (3PGA). This newly fixed organic carbon is re-arranged as it is shuttled through the early stages of the Calvin cycle before being diverted primarily to one of three different fates; (i) continuation on through the Calvin cycle for the regeneration of RuBP to sustain ongoing carbon fixation, (ii) departure from the Calvin cycle as six-carbon phosphorylated sugars (fructose 6-phosphate; F6P) for starch synthesis within the chloroplast, or (iii) departure from the Calvin cycle as three-carbon phosphorylated sugars (Triose phosphates; TrseP) for export to the cytosol. Carbohydrates exported to the cytosol are chiefly used for synthesis of sucrose, or re-oxidation in glycolysis and mitochondrial respiration. The Calvin cycle, and starch and sucrose synthesis are anabolic processes requiring the input of energy and reducing equivalents derived from either the chloroplastic light-reactions or from carbohydrate catabolism via respiration.

Efforts at enhancing photosynthesis include attempts at improving the Calvin cycle. Ribulose bisphosphate carboxylase/oxygenase (Rubisco), the crucial and enigmatic carbon-fixing enzyme that first introduces new carbon into the cycle has been a particular focus of these efforts (Fig. 3). This is because Rubisco's carboxylation reaction is slow and because it catalyzes two competing reactions: RuBP carboxylation and RuBP oxygenation. RuBP carboxylation products feed entirely into the Calvin cycle and grow the plant. RuBP oxygenation products partially divert carbon away from the Calvin cycle to the non-productive photorespiratory cycle resulting in losses of as much as 25% of the fixed carbon. Plants partially compensate for the inherent inefficiencies of this crucial enzyme by maintaining Rubisco at very high concentrations inside mesophyll chloroplasts. But this represents a resource cost. More than 25% of leaf nitrogen may be allocated to this one protein (Makino, 2011). The enigma of this enzyme is that it is the means by which virtually all biological organic carbon is produced from inorganic CO_2 and yet, even after >3 billion years of selection, it remains slow, confused, and costly (Tcherkez et al., 2006). To date, molecular engineering efforts have had little success at improving reaction rates or restricting RuBP oxygenase rates through modifying Rubsico kinetic properties (Whitney et al. 2011). Structure/function surveys of Rubsico from different taxonomic groups seem to indicate that the oxygenase reaction is an unavoidable feature of this crucial enzyme (Whitney et al., 2011). Somewhat surprisingly, targeting other Calvin cycle enzymes as a means of improving plant production have had more success (Raines, 2011). For example, tobacco TpS11 plants transformed by Tamoi et al. (2006) expressed Sedoheptulose bisphosphatase (SBPase) 60% greater than wild-type (WT) plants, had Pmax rates ~25% greater than WT, and had 50% greater biomass than WT (Table 1). These authors concluded that SBPase, which converts sedoheptulose 1,7 bisphosphate (Sd1,7bP) to sedoheptulose 7-phosphate (Sd7P), plays a critical and potentially limiting role in the Calvin cycle regeneration of the CO_2 (and O_2) acceptor molecule, RuBP (Fig. 3).

Starch and sucrose may be viewed as alternative and complementary end-products of cellular photosynthesis (Fig. 3). Typically, during the day, both carbohydrates are produced directly from new photosynthate (Stitt et al., 2010). Sucrose, in most species, is the major form by which carbon is transported elsewhere within the plant via the conducting cells of the phloem. Starch serves primarily as a stored carbohydrate reserve that may be used later to support growth, maintenance, reproduction and other carbon demanding functions (Fig. 1). Starch produced and stored in the chloroplast in the day is referred to as 'transitory starch' because it is generally broken down the following night and the sugar products exported to the cytosol. This continual sugar efflux from the chloroplast ensures a relatively constant source of substrate for cytosolic sucrose synthesis throughout the day/night cycle. Starch may also be stored in other tissues (e.g. stems or roots) as a long-term reserve. Both long-term and transitory starch storage represent diversions of carbohydrate away from immediate growth and thus a potential limit on plant production. Indeed, metabolite-profiles of 94 *Arabidopsis* accessions revealed that genotype variation in mesophyll starch content was negatively correlated with genotype differences in growth (Sulpice et al., 2009). Nevertheless, the advantages of carbohydrate reserves for plant resilience are clear, even at a cost of reduced allocation to growth and reproduction (Chapin et al., 1990). For example, studies of *Arbidopsis thaliana* starchless mutants show that the inability to accumulate transitory starch reduced growth and caused carbon starvation symptoms under day/night cycles when night length exceeds about 12 hours (reviewed in Stitt et al., 2010). These and other studies suggest growth is maintained at sub-maximal levels by diverting photosynthate to storage pools to enable plants to cope with periods unfavourable for photosynthesis.

Interestingly, overall plant demand for carbohydrate can feedback to affect the regulation of mesophyll photosynthetic capacity which, in turn, affects subsequent rates of starch and sucrose production (Paul & Pellny, 2003). Feedback regulation, the phenomenon where low carbohydrate demand feeds back to lower Pmax was elegantly demonstrated by Thomas & Strain (1991) who showed that cotton plants raised in small pots grew slower with lower Pmax rates than plants in larger pots. Further, they showed that as simple an act as transplanting plants from small to large pots stimulated root and whole-plant growth, reduced starch reserves, and increased Pmax. This illustrates how limits on plant growth sometimes control photosynthesis rather than the other way around. Carbohydrate storage and carbohydrate-mediated feedback regulation complicate efforts to increase plant production by enhancing photosynthesis.

Despite these interesting complications, withdrawal of carbon from the Calvin cycle for sucrose and starch synthesis is central to plant productivity (Fig. 1). By contrast, RuBP oxygenation results in the formation of phosphoglycolate (PGlyco) which represents a non-productive drain on the Calvin cycle (Fig. 3). Following the fate of PGlyco in Fig. 3, we see a series of reactions that form a biochemical cycle traversing the chloroplast, the peroxisome, and the mitochondria. This cycle behaves as a salvage pathway because it restores 75% of the carbon lost in the initial RuBP oxygenation reaction back into the Calvin cycle. Photorespiration largely explains why, in C3 plants, the maximum realized QY falls below the maximum potential QY (Fig. 2). The maximum potential QY can only be achieved when measuring photosynthesis under artificial atmospheric mixtures of CO_2 and O_2 that are sufficient for inhibiting the RuBP oxygenation reaction. Photorespiratory effects on QY become worse in C3 plants as Cc declines as happens, for instance, when stomata close to conserve water.

Photorespiration costs also include the resources allocated to the production and maintenance of the elaborate photorespiratory metabolic machinery (Foyer et al., 2009). Interestingly, genetic elimination of components of the photorespiratory cycle turns out to reduce plant production, sometimes to the point of lethality (Somerville, 2001). Thus, in spite of its obvious inefficiency, photorespiration plays various essential roles for C3 plants including service as an AET pathway (Osmond & Grace, 1995). For example, when stomata close and CO_2 becomes limiting for RuBP carboxylation, the coupled operation of the Calvin cycle and the photorespiratory cycle helps poise ADP/ATP and NADP/NADPH concentration ratios and minimize the over-production of ROS from the light-reactions. Kozaki and Takeba (1996) engineered tobacco plants that under-expressed chloroplastic glutamine synthetase (GS2), a necessary enzyme of the photorespiratory cycle (not shown in Fig. 3). As expected, the GS2 under-expressing plants exhibited less photorespiration. But these plants were also more susceptible to ROS-mediated loss of PSII function, presumably because the photorespiratory cycle was not available as a protective 'escape valve' for the flow of excess reducing power.

The carbon-concentrating mechanism found in C4 plants like corn and sugarcane represents an elaborately evolved solution to photorespiration. C4 plants engage additional upstream biochemistry to capture inorganic carbon and concentrate it in chloroplasts near Calvin cycle machinery (Sage, 2004). This high Cc sufficiently inhibits RuBP oxygenation reactions and virtually eliminates photorespiration in C4 plants. This would seem, at first glance, to be the perfect solution to the problem of photorespiration, and there is great interest in trying to engineer C4 physiology into C3 crop plants (Sheehy et al., 2008; Sage & Zhu, 2011). But C4 comes with its own set of trade-offs. For example, the maximum potential QY for C4 plants falls short of the maximum potential QY of C3 plants. This is because additional ATP is required to run the carbon-concentrating metabolism of C4 photosynthesis (Ehleringer & Björkman, 1977). The vast majority (~90%) of described plant species rely upon C3 photosynthesis, suggesting that across most growing conditions, the energetic penalty of C3 photorespiration does not outweigh the energetic cost of the C4 carbon-concentrating mechanism (Foyer et al., 2009).

Kebeish et al. (2007) took a different approach to minimizing the energetic penalty of photorespiration. *Arabidopsis thaliana* plants were transformed by inserting the glycolate degradation pathway genes from the bacterium *Escherichia coli*. The glycolate degradation enzymes were expressed in the chloroplast. This allowed chloroplastic glycolate to be converted directly to glycerate in the chloroplast, effectively bypassing the photorespiratory cycle (see chloroplastic Glyco and Glycerate in Fig. 3). Photorespiration was greatly diminished in transformed plants. Pmax was ~50% greater and shoot growth was ~66% greater in the transformed plants than in WT plants (Table 1). Although untested, this approach presumably permitted the continued operation of the photorespiratory cycle as a protective AET path thereby circumventing the problems reviewed above that arise with the elimination of the photorespiratory cycle.

Leaf respiration (R_m) - comprising glycolysis, the citric acid cycle, and mitochondrial electron transport (Fig. 3) - is coupled to, and coordinately regulated with, photosynthesis (and photorespiration) through multiple metabolic linkages (Nunes-Nesi et al., 2008). The study described above by Araújo et al. (2011) with reduced succinate dehydrogenase expression in antiSDH2-2 tomato plants demonstrates one of these linkages. These authors emphasized how down-regulation of this citric acid cycle enzyme promoted growth through indirect effects on stomatal conductance and photosynthesis. But the antiSDH2-2 plants also had leaf respiration rates that were 10-17% lower than WT plants. A diminished R_m can have major effects on whole-plant daily carbon gain (Amthor, 2010). We suggest that the growth enhancement observed by Araújo et al. (2011) in the antiSDH2-2 tomato plants was a consequence of both reduced respiratory carbon losses as well as the increased foliar carbon uptake associated with greater stomatal conductance emphasized by the authors (Table 1).

Plant mitochondria express a number of gene products that act to lower the energetic efficiency of R_m including the alternative oxidase (AOX) and uncoupling proteins (UCP). AOX reduces O_2 at an early step in the normal electron transport chain thereby reducing the ATP respiratory yield (Fig. 3). AOX activity varies with growth conditions (Searle et al., 2011), is required for heat-production in selected tissues (Miller et al., 2011), and is believed to function as a mitochondrial AET path thereby minimizing mitochondrial ROS production (Maxwell et al., 1999). Mitochondrial UCP also lowers the ATP respiratory yield because it permits H^+ passage across the inner mitochondrial membrane without driving ATP synthesis at Complex IV (Fig. 3). Sweetlove et al. (2006) observed that plants with reduced UCP levels had lower photorespiration rates, lower Pmax rates, and reduced growth. They interpret their findings to mean that, paradoxically, reduced R_m energetic-efficiency, as mediated by UCP, is essential for permitting high rates of coupled photosynthesis and photorespiration. It would be interesting to see what effect mitochondrial UCP over-expression has on plant productivity.

Our review of cellular primary carbon metabolism (Fig. 3) reveals three important points: First, these multiple processes are highly interactive, exhibiting elaborate, adaptive, system-level coordination and regulation (e.g., UCP-mediated support of high Pmax rates or the essential coupling of photorespiration to the Calvin cycle). Second, this coordinated system-level activity is highly variable/flexible and frequently effects low energetic efficiency. Controlled water-conserving stomatal closure, protective photoinhibition, and carbohydrate-mediated feedback are representative processes that down-regulate photosynthetic efficiency and remind us that natural selection acts on whole-organism lifetime fitness, not maximized momentary energetic efficiencies. Third, in spite of complex,

Changed trait (reference)	species	Molecular genetics	Pmax effect	Growth effect	comments
Added thylakoid electron carrier & increased electron transport rate. (Chiba et al., 2007)	*Arabidopsis thaliana*	Inserted CytC6 gene from red algae *Porphyra yezoensis*	~30% increase	~30% increase	Higher leaf levels of ATP and NADPH were also observed in transformed plants
Lower succinate dehydrogenase activity & higher stomatal conductance. (Araújo et al., 2011)	*Solanum lyco-persicum (tomato)*	Anti-sense lowered expression of succinate dehydro-genase gene SDH2-2	~30% increase	~20% increase	antiSDH2-2 plants also had lower respiration rates
Increased stomatal density & stomatal conductance. (Schlüter et al., 2003)	*Arabidopsis thaliana*	Point mutation in sdd1-1	~30% increase	not reported	Pmax enhancement realized for plants after transferring from low- to high-light
Increased aquaporin expression & mesophyll conductance. (Flexas et al., 2006)	*Nicotiana tabacum (tobacco)*	Over-expression of NtAQP	~20% increase	not reported	NtAQP was expressed in both plasma membrane and chloroplast envelope
Increased sedoheptulose bisphosphatase & greater Calvin cycle activity. (Tamoi et al., 2006)	*Arabidopsis thaliana*	Inserted SBPase gene from Chlamydo -monas reinhardtii	~30%	~50%	Leaf RuBP also increased ~25%
Addition of a glycolate catabolic pathway & minimized photorespiration. (Kebish et al., 2007)	*Arabidopsis thaliana*	Inserted five *E. coli* genes encoding glycolate catabolic enzymes	~50%	~66%	This bypass also increases Cc since a de-carboxylation step occurs in the chloroplast

Table 1. Selected genetic modifications at various sites in primary carbon metabolism that have yielded increased maximum photosynthesis.

fine-tuned, system-level co-regulation, targeted molecular manipulations have been able to enhance photosynthetic production for selected species when grown under optimally controlled environmental conditions (Table 1). Squeezing more out of photosynthesis, at least under suitable conditions, is clearly possible.

2.2 Leaf photosynthesis: Co-variation in leaf longevity and leaf photosynthetic capacity

The thesis of this review is that system-level regulation restricts overall plant productivity. New sets of limits to photosynthesis emerge as we move from the cell up to the leaf-level of organization (Fig. 1). In particular, leaf-level carbon gain will depend upon both the carbon costs and benefits of leaf photosynthetic performance integrated over the time the leaf is on the plant. Interestingly, the maximum possible duration of leaves on a plant (leaf lifespan; LL) and Pmax are known to vary inversely with each other across different species (Reich et al., 1997). This may seem counterintuitive because the cumulative contribution an individual leaf can make to the productivity of the plant would be predicted to depend upon both its maximum rate of carbon gain (Pmax) and the time period over which that rate is potentially realized (LL). As such, we might expect selection for various species to produce long-lived leaves capable of high Pmax rates. It seems that no such species exists!

2.2.1 Materials and methods

A comparative study was carried out across a broad range of tropical species to explore relationships between leaf longevity, photosynthetic capacity, leaf structure and nitrogen status, and the potential lifetime carbon gain for the individual leaf. Forty study species were selected from plants growing in separate distinct habitats or 'mesocosms' in Biosphere 2, a novel controlled-environment research facility in southern Arizona (Leigh et al., 1999). Selected species represented a broad range of growth forms and taxonomic groups. Leaf number per branch and leaf 'birth-rates' per branch were followed for ~1 year on three or more separate branches from one or more individual plants of each study species in order to get demography-based maximum LL estimates (Bazzaz & Harper, 1977). Pmax was measured on intact, healthy, fully-enlarged leaves of each species using a flow-through infrared gas analysis system (Li-Cor 6400, Li-Cor Instruments, Lincoln, Nebraska). Specific leaf area (SLA), the ratio of leaf laminar projected area per unit dry mass, was assessed on tissue samples from healthy, fully-enlarged leaves off the same plants. Leaf nitrogen concentration (N_L) was determined from Kjeldahl digests of tissue samples from healthy, fully-enlarged leaves off the same plants. N_L and leaf heat-of-combustion contents were used to calculate leaf energetic Construction Costs (CC) for leaf samples taken from a subset (18 out of 40) of the study species (Williams et al., 1987). Reported CC estimates assume all plants relied solely on nitrate as their nitrogen source. Reported Pmax, SLA, N_L, and CC values are the means of 3-5 measurements from independent leaves from each of the study species growing in Biosphere 2.

These data allowed the application of an empirical model to estimate the maximum potential lifetime net carbon gain that an individual leaf could make to the overall productivity of the plant. Zotz and Winter (1996) found a strong linear association between instantaneous *in situ* measures of Pmax and the total net 24 hour carbon gain (P_{24h}) for individual leaves from a broad range of different species and growth forms growing in a tropical forest in central Panama. For each Biosphere 2 study species, the average measure of

Pmax was input into the empirical formula from Zotz and Winter (1996) to get a best estimate of P_{24h}. This species-specific P_{24h} value was then multiplied over the estimated species-specific LL to get an estimated maximum potential leaf-lifetime carbon gain (P_{life}). One advantage to this empirical approach is that it incorporates the otherwise uncertain effects of day and night respiration into P_{24h} and P_{life} estimates. Likewise, it incorporates maintenance respiration into P_{24h} and P_{life} estimates. We note that the model holds the Pmax value constant over the projected life of the leaf. Leaves often show a linear decline in Pmax with age (Kitajima et al., 1997). An assumption of a linear decline in Pmax is sometimes incorporated into leaf lifetime carbon gain models (Hiremath, 2000; Kikuzawa 1991). As it turns out, making an assumption of a linear decline in Pmax has the simple effect of reducing estimates of P_{life} for all species by half. Because we have no actual measures of how Pmax varies with leaf age and because it has no qualitative effect on interspecific comparisons, a linear decline assumption was not factored into the Pmax-P_{life} model.

2.2.2 Results and discussion

Our study revealed a strong negative association between leaf lifespan and Pmax (Fig. 6A) and, to a lesser extent, negative associations between leaf lifespan and SLA (Fig. 6B) and N_L (Fig. 6C). Study species included various plants sampled from each of four simulated biome mesocosms (tropical rainforest, savannah-orchard, dry thorn-scrub, and sandy beach). There were, in some cases, significant differences in leaf-level characteristics between plants from different mesocosms which explains much of the plot scatter in Fig. 6 (not shown). However, the overall trends were quite robust even without accounting for these mesocosm differences.

This underscores the global nature of these leaf-level patterns. The results agree with observations made frequently on various C3 plant species from various terrestrial ecosystems (Reich et al., 1997). These relationships appear to be so robust that they are now referred to collectively as the worldwide leaf economics spectrum (WLES; Wright et al., 2004). One end of this spectrum represents species having short-lived, thin, high surface area/volume leaves (i.e. high SLA), with high protein or N_L contents and high photosynthetic rates. The other end of this spectrum represents species having long-lived, thick durable leaves (i.e., low SLA) with low protein or N_L contents and low photosynthetic rates. Among plants producing leaves that live ~1 year or less, there appears to be considerable scope for adjustments in Pmax as a means of increasing plant productivity. But, among plants producing leaves that potentially persist more than a year, there appears to be almost no scope for adjustments in Pmax.

The global nature of the patterns exhibited in Fig. 6A, B & C are interpreted as fundamental leaf structure/function trade-offs maintained by natural selection (Donovan et al., 2011; Reich et al., 1997). Thin leaves with low-density tissue can sustain high photosynthetic rates (high Pmax) in part because there is relatively little intra-leaf chloroplast shading and mesophyll conductances are large. But these same leaves will have low durability (short LL). Thick, high-density leaves are more durable (long LL) but tend to be photosynthetically limited (low Pmax) by intra-leaf chloroplast shading and low mesophyll conductances. This leaf-level pattern where Pmax varies inversely with leaf lifespan is not immediately obvious or explainable based on our previous analyses of cellular and metabolic limits on photosynthesis. These patterns illustrate how functional traits above the level of primary cellular carbon metabolism can place strong constraints on Pmax.

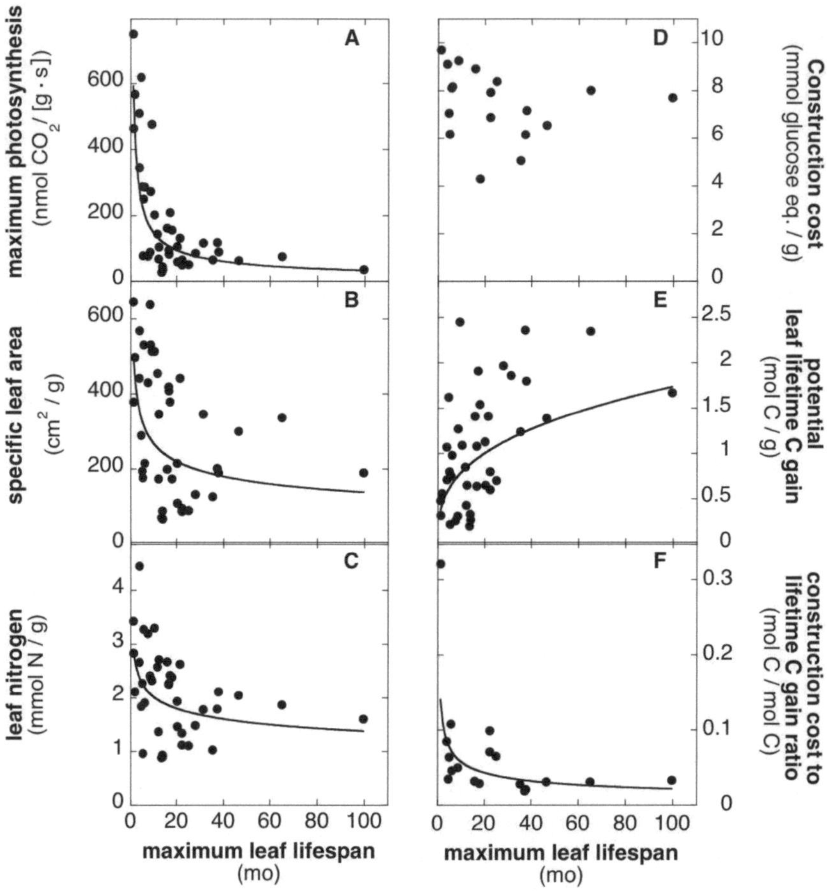

Fig. 6. Interspecific variation in leaf lifespan versus (A) Pmax, (correlation coefficient, r=0.82), (B) SLA, (r=0.51), (C) N_L, (r=0.47), (D) leaf CC, (Non-significant correlation), (E), estimated P_{life}; r=0.54), and (F) a carbon-based leaf cost/benefit ratio (CC/P_{life}; r=0.80). Each datapoint is the mean from 3-4 observations made on each of 40 (A, B, C and E) species or a subset of 18 species (D and F) in controlled environment mesocosms in Biosphere2 in Arizona, U.S.A. Lines of best-fit for power functions are plotted when the association between variables was significant ($P<0.05$) with correlation coefficients (r) as reported above.

Efforts to bioengineer improvements in photosynthetic productivity would generally focus on economically important plants that are grown in managed settings (e.g., commercial greenhouses, agricultural fields, orchards, forest plantations). Many, but not all, plants that might be subject to productivity enhancement efforts will be on the short LL end of the WLES. Genetically modified plants discussed above (e.g., *Arabidopsis* and tobacco) produce leaves that persist no more than a few weeks. Like these species, other economically important species with short-lived leaves may be relatively amenable to bioengineered improvements in Pmax. However, constraints implicit in the WLES suggest that species on the long LL end of the distribution may not be amenable to enhanced productivity through bioengineered improvements in Pmax. Many plant species that are sources of economically important commodities (e.g., coffee, rubber, olive oil, bananas, citrus), and thus possible targets for improved productivity efforts, are on the long LL end of this spectrum. Likewise, evergreen coniferous trees that produce much of the world's fuel, lumber, and wood pulp produce long-lived foliage with low Pmax rates (Reich et al., 1995). As we begin to extend our efforts at engineered improvements in plant productivity beyond typical lab species (e.g., *Arabidopsis thaliana*), consideration of these leaf-level constraints will become increasingly important.

Bioenergetic costs of producing leaves may be expected to differ among species as a function of foliar structure and chemical composition (Griffin, 1994). However, in the present study, variation in CC across the sampled species was modest and was not significantly correlated (Pearson Correlation Coefficient Test) with leaf lifespan (Fig. 6D). Leaf CC among the 18 species ranged from 4.30 mmol glucose to 9.71 mmol glucose equivalents/g dry leaf mass (0.77 to 1.75 g glucose equivalents/g dry mass), falling well within the range of other published CC values (Nagel et al., 2004; Poorter et al., 2006; Williams et al., 1989).

The potential contribution an individual leaf can make to the overall carbon budget of the plant, P_{life}, depends upon both leaf longevity and Pmax. Interestingly, modeled estimates of P_{life} tended to *increase* with leaf lifespan (Fig. 6E). Hiremath (2000) made a similar observation for a small number of early-successional tropical tree species in the field. Thus, even though Pmax declines as leaf longevity increases, it appears that increased time for photosynthetic operation associated with a prolonged leaf lifespan more than compensates for this. This finding is quite striking in the context of pondering plant productivity enhancements because it implies that targeting the molecular controls on delayed leaf senescence might yield greater carbon gain benefit than targeted enhancements of Pmax. This idea assumes that resource costs to the plant associated with producing more durable and longer-lived leaves is not prohibitive.

The dataset from our study permitted estimation of a carbon cost/benefit ratio for the 18 species for which both Pmax and CC data were available (Fig. 6F). This approach uses CC values as a measure of the carbon cost incurred to the plant for producing a gram of leaf tissue. In turn, P_{life} estimates are a measure of the maximum potential net carbon benefit a gram of leaf may provide back to the plant. A CC/P_{life} value of 1.0, expressed as mol C per mol C, can be considered a 'break even point' in this cost/benefit analysis. All species are expected to fall below this threshold value. Indeed, leaves from all species should fall substantially below 1.0 because many leaves operate much of the time under sub-optimal conditions (e.g., low-light, cold temperatures) and so perform well below Pmax, thereby reducing actual P_{life} below its potential. In addition, many leaves are damaged or abscised long before achieving their maximum leaf lifespan which would also reduce actual P_{life} below its potential. Inspecting the data reveals that the CC/P_{life} values for all species were well below 1.0 mol C/mol C, precisely as expected (Fig. 6F). It is noteworthy that the association between leaf lifespan and CC/P_{life} was quite strong ($r=0.80$) with short leaf lifespan species exhibiting relatively high carbon cost/benefit values and long leaf lifespan having relatively modest carbon cost/benefit values. This trend emphasizes the importance of having a high Pmax in short-lived leaves as a means of 'paying back' the construction cost before leaf death (see Poorter et al., 2006). This is important because it was suggested above that, in general, the best way to increase overall leaf lifetime carbon gain was through increasing LL. The pattern in Fig. 6F gives nuance to this view because it indicates that for species bearing short-lived leaves (e.g. annuals and deciduous perennials), there is great advantage to be had from increasing Pmax as a means to improve the overall carbon budget of the leaf and therefore the plant.

Interestingly, Williams et al (1989), using a similar approach, had different results. Where Fig 6F shows a negative association between the cost/benefit ratio and leaf lifespan, Williams et al. (1989) observed a positive association between their estimated carbon cost/benefit and leaf lifespan. In this prior study, the cost/benefit ratio was established as leaf CC divided by estimated A_{24h}. Williams et al., (1989) examined leaf traits in seven different rainforest successional shrub species, some specialized for the shaded understory and some specialized for sunny open sites. It seems that the main factor driving the positive association between CC/A_{24h} and leaf longevity in this earlier study was the segregation of shade and sun specialists. Species from open habitats had leaf lifespans of approximately 100 days and a lower overall CC/A_{24h} as a result of high photosynthetic rates in the bright sun. Species from understory habitats had leaf lifespans of approximately 700 days and a higher overall CC/A_{24h} as a result of limited photosynthetic activity in the deep shade. In contrast, the species selected for our study growing in Biosphere 2 occurred over a range of mostly intermediate light habitats (Cockell et al., 2000; Leigh et al., 2000). Consequently we believe that light effects on Pmax and leaf-lifespan in our study would be modest compared to the work reported by Williams et al., (1989).

Our comparative leaf-level study reveals two important points: First, across different plant species, foliar photosynthetic potential co-varies with leaf composition and longevity. This confirms the generality of the WLES and illustrates the emergence of system-level constraints on photosynthesis not predicted from our knowledge of cell metabolism. Second, an integrated leaf-lifetime cost/benefit analysis of net carbon gain suggests that direct manipulations of cellular photosynthesis may be a useful productivity-enhancing approach only in a limited set of plant species. At the same time, it suggests engineered alterations of other foliar traits such as leaf structure or leaf lifespan may be alternative or complementary strategies for enhancing photosynthetic productivity, depending upon the species.

2.3 Canopy photosynthesis: Does prolonged leaf lifespan enhance whole-plant production?

New sets of restrictions on photosynthesis emerge as we move from individual leaves up to the whole-plant level of organization (Fig. 1). At the whole-plant level, the canopy is the fundamental unit of photosynthesis. Various aspects of whole-plant structure, function, and development can limit canopy photosynthesis and whole-plant productivity. In the previous section we saw how leaf Pmax were constrained by differences in LL and associated structural and compositional traits. Here we examine the influence of leaf-lifespan on plant productivity further in order to illustrate how contrasts in canopy structure might differentially limit whole-plant productivity.

2.3.1 Materials and methods

Gan & Amasino (1995) produced tobacco plants carrying a genetic insert (P_{SAG12}:IPT) that effectively prolongs leaf lifespan over that of wild-type (WT) plants. The auto-regulating physiology underlying this prolonged leaf lifespan phenotype is that the cellular onset of leaf senescence in the transformed plants stimulates endogenous production of the anti-senescent plant hormone, cytokinin. The inhibition of senescence in turn, lowers the rate of cytokinin production. This elegant auto-regulated approach minimizes pleiotropic effects otherwise associated with constitutive over-production of cytokinins.

We quantified leaf and whole-plant characters in 3.5-mo old P_{SAG12}:IPT and WT tobacco plants to explore how variation in leaf lifespan affects whole-plant performance. Seven plants each of the P_{SAG12}:IPT and wild-type (WT) tobacco genotypes were grown 3.5 mos (to early flowering stage) at low density for maximum canopy light transmittance in controlled environment growth cabinets. Plants were grown under standard soil culture conditions. Plants were initially fertilized weekly with standard commercial nutrient solution (Miracle-Gro All Purpose, Scotts Company, Marysville, Ohio). Starting at the 6-8 leaf stage, plants were fertilized twice weekly. Plants were watered as needed early on and as they matured they were watered daily. Day/night temperatures were set at 20°C/15°C. Relative humidity was not controlled but generally was over 90% at night and dropped no lower than 40% during the day. Light incident at final canopy height was 700±80 µmol photons/($m^2 \cdot s$) PFD. Axial vegetative buds were excised, dried and weighed while the plants grew to prevent branch formation. This maintained monopodal stem architecture and maximized canopy light transmittance. Leaves were mapped to main stem node positions and tagged as the plants grew to follow growth and leaf demography. Dead leaves were collected, dried, and weighed at the time of abscission. Light incident on leaves at different canopy nodal positions was measured *in situ* using a pre-calibrated galium arsenide photodiode to assess self-shading within the canopy (Pearcy, 1991). The light response of photosynthetic oxygen production was measured on tissues from newly enlarged mature leaves near the top of the canopy prior to harvest using a leaf disc oxygen electrode (LD2; Hansatech Instruments, King's Lynn, UK). Light-saturated photosynthesis (Pmax) was also measured on leaves of different ages on representative plants of both genotypes. Standard gravimetric methods were used to assess whole-plant water use (McCulloh et al., 2007). Harvested plants were separated into component organs to quantify live leaf area and the dry weights of different component organs.

2.3.2 Results and discussion

The P_{SAG12}:IPT plants retained their leaves longer than WT plants as expected based upon previous observations (Boonman et al., 2006; Jordi et al, 2000; Gan & Amasino, 1995). The difference between the total number of nodes on the stem, a measure of all the leaves that had ever been produced, and the live leaves present at the time of the harvest is a measure of the number of leaves that had been lost in the canopy due to senescence and abscission (Table 2). The P_{SAG12}:IPT plants still had essentially all the leaves they had ever produced and these leaves were all still intact and green on the plant. The WT plants had lost approximately 12 leaves from the bottom of their canopy as a result of natural senescence and abscission. The lifespan of the seemingly immortal leaves of the P_{SAG12}:IPT plants could not be quantified. Leaf lifespan on the WT plants was approximately 8 weeks. At the time of the final harvest, the P_{SAG12}:IPT plants held about 50% more leaves than the WT plants. Older lower-canopy leaves retained by the P_{SAG12}:IPT plants were considerably larger than the younger mid- and upper-canopy leaves. When compared to WT plants, the P_{SAG12}:IPT plants had double the total photosynthetic leaf area. Total plant production (live + dead tissue) was ~15% greater for the P_{SAG12}:IPT plants than for the WT plants (Table 2). Enhanced leaf lifespan yielded no differences in plant height, live root mass, or allocation to reproduction (Table 2).

Photosynthetic carbon gain becomes light-limited in lower-canopy leaves due to self-shading. This represents an important constraint on photosynthetic productivity that only emerges at the whole-plant level (Valladares et al., 2002). Measures of leaf light-interception as a function of canopy position showed that the youngest vertically-oriented leaves at the shoot apex receive somewhat less overhead light than more-horizontally oriented fully-opened leaves a few nodes down from the apex (Fig. 7A). From here, light availability was attenuated linearly with leaf node position within the plant canopy. Light availability within plant canopies often declines exponentially (Hikosaka, 2005). The linear decline in light observed in the present study presumably results from the wide spacing among plants and the intentional monopodal canopy architecture. The spatial pattern of light availability was indistinguishable for the two tobacco genotypes except near the bottom of the canopy as a result of differences in foliar senescence and abscission.

Photosynthetic capacity (Pmax) tends to decline with canopy position both because of acclimation to the canopy light-gradient and because of age-dependent leaf senescence and associated resource re-mobilization (Hikosaka et al., 1994; Kitajima et al., 1997). To assess how photosynthetic potential varied through the plant canopy, Pmax was measured on selected fully-enlarged leaves at different canopy positions for which leaf age was known (Fig. 7B). Light-saturated photosynthetic O_2 production declined with leaf age in both genotypes but the decline rate was faster for WT than for P_{SAG12}:IPT plants. For example, 50-day old leaves in WT plants had Pmax rates that were approximately half that of similar aged leaves in the P_{SAG12}:IPT plants. Leaves of WT plants did not persist beyond about 60 days but leaves as old as 80 days in the P_{SAG12}:IPT plants were still present and photosynthetically competent, albeit at very low levels. Since intra-canopy light availability was the same for both genotypes, these contrasts represent differences in leaf senescence.

TRAIT	WILD-TYPE	P_{SAG12}:IPT
Live leaves present	19 ± 1 (A)	29 ± 2 (B)
Total live canopy leaf area (cm^2)	2,536 ± 93 (A)	5,222 ± 110 (B)
Stem height (cm)	67.0 ± 3.5 (A)	63.5 ± 5.5 (A)
Nodes present on stem	31 ± 3 (A)	31 ± 2 (A)
Live root mass (g)	30.3 ± 3.4 (A)	31.0 ± 2.2 (A)
Cumulative flower bud mass (g)	4.6 ± 0.4 (A)	6.0 ± 0.9 (A)
Cumulative whole plant mass (g)	95 ± 6 (A)	110 ± 5 (B)

Table 2. Live biomass allocation and cumulative production (living + abscised dead biomass) patterns for PSAG12:IPT & WT tobacco plants at 3.5 months after germination. All masses are dry weight. Values are means of 6-7 plants per genotype ± 1 S.E. Different letters (A or B) within a row indicate significant differences between genotypes (t-test, $p<0.05$).

Complete photosynthetic light-response curves were also measured on fully mature, upper-canopy leaves (2-3 weeks in age) where we expected to find no genotype differences in light-dependent acclimation or age-dependent senescence (Fig. 7C). Dark respiration rates (R_m) and QY values were similar in both genotypes. Likewise, the light level where photosynthesis and respiration are exactly balanced (photosynthetic light compensation point) was ~50 µmol photons m^{-2} s^{-1} PFD for WT and P_{SAG12}:IPT plants alike (Fig. 7C). Photosynthetic tissues below the light compensation point lose more carbon than they fix. We note that only the P_{SAG12}:IPT plants still had lower-canopy leaves growing in light levels below this critical level (Fig. 7A). Assuming that light compensation point changes little with leaf age and/or canopy position, it would appear that the lower leaves on the P_{SAG12}:IPT stems were a net carbon drain on the plant as a result of self-shading. Boonman et al. (2006) made similar observations for P_{SAG12}:IPT and WT tobacco plants grown at high densities where lower leaves were subject to both intra-canopy and inter-canopy shading. Boonman et al. found that growth rates were indistinguishable between the two genotypes at these high planting densities even though the P_{SAG12}:IPT plants had substantially more photosynthetic tissue than the WT plants. This uncoupling of total leaf area from plant growth was attributed to the older, deeply-shaded leaves in the P_{SAG12}:IPT plants acting as net respiratory tissues even during the day (Boonman et al., 2006). Unexpectedly, in the present study, light response curves for upper-canopy leaves indicated that the Pmax rates of the new, upper-canopy leaves tended to be lower ($p=0.07$; t-test) in the P_{SAG12}:IPT than in the WT plants (Fig. 7C). Given the canopy position and age of these leaves, this tendency for a genotype effect on upper-canopy Pmax rates cannot be explained directly by differences in light availability or leaf senescence.

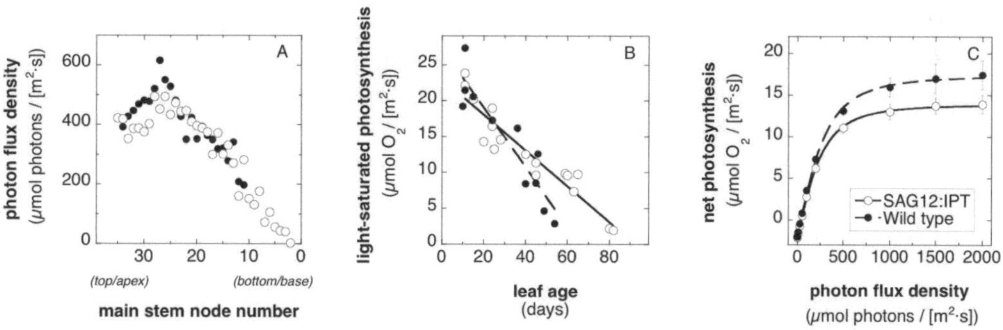

Fig. 7. Light interception by leaves at different positions in the canopy of WT and P_{SAG12}:IPT tobacco plants. Each data point is the mean from 3 different plants of each genotype (A). Pmax for leaves of different ages within the canopy of three WT and three P_{SAG12}:IPT tobacco plants (B). Photosynthetic light response of newly mature upper canopy leaves (2-3 weeks old) from three WT and three P_{SAG12}:IPT tobacco plants (C).

In general, plant canopies exhibit characteristic distribution and re-mobilization patterns of N_L that tend to maximize whole-canopy photosynthesis (Hikosaka, 2005). Upper-canopy leaves typically have higher N_L concentrations, higher complements of Rubisco, and sustain higher Pmax rates than shaded lower-canopy leaves. As older leaves senesce, degrading tissues release nutrient resources, including nitrogen, that may be transported for use in new growing tissues elsewhere in the plant, especially newly emerging, fully-illuminated, upper-canopy leaves. Canopy recycling of N_L is particularly important for plants grown in nitrogen poor soils. Jordi et al. (2000) demonstrated the detrimental effects of prolonged leaf lifespan on N_L and photosynthesis in upper canopy leaves of WT and P_{SAG12}:IPT plants grown under N-starvation conditions. As in the present study, Jordi et al. (2000) found that older leaves in WT plants dried, yellowed, and abscised whereas older leaves on the P_{SAG12}:IPT plants remained intact, green, and photosynthetically competent. However, newer upper-canopy leaves in the nitrogen-limited WT plants had more N_L, and supported higher Pmax rates than comparable leaves in nitrogen-limited P_{SAG12}:IPT plants. The forced retention of older P_{SAG12}:IPT leaves prevented canopy recycling of N_L to the upper canopy leaves. Under N-starvation conditions this prevented the production of new high N, high Pmax leaves at the top of the canopy. In the present study, N_L was not evaluated. However, the plants were well fertilized with complete nutrient solutions. All living leaves on both sets of plants were a rich green color implying no nitrogen limitations. Limited nitrogen availability seems an unlikely explanation for the marginal genotype differences in Pmax observed in Fig 7C.

All else being equal, a plant with greater canopy leaf area will lose more water through transpiration than a plant with less leaf area. We carried out gravimetric measures of 24 hour water-use for plants of both genotypes to see if whole-plant water use scaled linearly with the total surface area of the leaf canopy (Table 3). It did not. Despite having two-fold greater canopy leaf area, daily whole-plant water-use was ~13% lower in the P_{SAG12}:IPT plants than in the WT plants. When corrections for contrasts in total live leaf area were made, the difference in foliar transpiration was even more striking. The leaf-area based rate of water use was about 36% lower in the P_{SAG12}:IPT plants compared to the WT plants (Table 3). The marginally lower photosynthetic capacities observed for upper-canopy leaves of the P_{SAG12}:IPT plants (Fig. 7C) may be a consequence of stomatally-limited leaf gas-exchange rates presumably to compensate for the greater transpiring surface area in these plants. We speculate that the modest 15% difference in total plant production of the P_{SAG12}:IPT plants, despite having twice as much photosynthetic leaf area with minimal intra- and inter-canopy shading, may partially arise from reduced leaf gas-exchange rates for water-conservation.

	WILD-TYPE	**P_{SAG12}:IPT**
24-hour water use per plant (ml H_2O per plant)	349 ± 15 (A)	303 ± 17 (B)
24-hour water use per unit canopy leaf area (ml H_2O per m^2)	614 ± 53 (A)	391 ± 43 (B)

Table 3. Patterns of whole-plant water use for P_{SAG12}:IPT and wild-type tobacco plants at 3.5 months after germination. Values are means for 4 plants per genotype ± 1 S.E. Soil evaporative water losses have been factored out. Different letters (A or B) within a row indicate significant differences between genotypes (t-test, $p<0.05$)

This study, along with those of Gan & Amasino (1995), Jordi et al. (2000), and Boonman et al. (2006) indicate that longer lived leaves can help to increase overall plant production provided the plants are grown under optimal conditions. This result is qualitatively consistent with predictions made from leaf-level considerations (Fig. 6E). However, these various studies with the two tobacco genotypes also indicate some of the ways that whole-plant structure (e.g., canopy architecture and light gradients) and composition (e.g., canopy gradients in N_L concentration) can limit whole-plant productivity. Table 2 shows that a doubling of leaf area only yields a 15% gain in plant production indicating diminishing returns associated with prolonged leaf lifespan, even when grown under presumably optimal conditions. Prolonged retention of canopy-shaded leaves may slow growth if lower leaves act as a net carbon drain on the plant. Low canopy transpiration rates and low Pmax rates in fully illuminated upper-canopy leaves of P_{SAG12}:IPT plants suggest that whole-plant water conservation, even under well-watered conditions, may slow plant growth too. We note that unrecognized pleiotropic effects of the genetic transformation and/or unforeseen differential effects of axial bud removal in our study may also have contributed to the observed differences in photosynthesis, transpiration, and plant growth. The assessment of these possibilities awaits further study.

Comparative studies with P_{SAG12}:IPT plants also give insight into the basis of the WLES patterns shown in Fig 6. Leaf lifespan differences can allow for associated differences in leaf lifetime carbon gain if there are adequate resources, such as water, light, and nitrogen, for the leaf to sustain positive net carbon assimilation. But these comparative P_{SAG12}:IPT studies demonstrate how extended leaf lifespan is also associated with increased canopy-self shading, increased transpiring surface area, and reduced N_L re-mobilization rates due to increased nitrogen residence time within individual leaves. Consequently, a *low* Pmax should be favored in species producing longer-lived leaves because of the associated lower water and nitrogen costs and because the carbon gains otherwise associated with a high Pmax leaf would seldom be realized in long-lived leaves due to intra-canopy light-limitations. The global nature of the WLES patterns are re-interpreted as fundamental structure/function trade-offs arising at both the leaf *and* the whole-plant level.

3. Conclusions

The productive capture and use of sunlight by plants and their photoautotrophic kin makes the ordered changes of life on Earth thermodynamically possible. There is great interest in finding ways to increase plant production through different means including new approaches to enhanced photosynthesis. This is inspired, in part, by the need for practical solutions to various global problems of increasing urgency, and, in part, by advances in genetic engineering. Selected examples here illustrated how efforts at improving photosynthetic productivity must be considered from a systems perspective. A 'system' is a set of interacting and interdependent entities that function as a coherent whole (Lucas et al., 2011). Biological systems exhibit three properties; hierarchy, emergence, and resilience. The hierarchical nature of plant photosynthesis was emphasized here by focusing on carbon metabolism at the cellular level, CO_2 uptake at the leaf level, and plant growth at the whole-plant level (Fig. 1). At the cellular level there has been tremendous progress in our understanding of photosynthesis and related metabolic processes and in our ability to improve photosynthesis in selected species under carefully controlled cultivation conditions (Fig. 3; Table 1). These molecular studies demonstrate that plants can 'do better', giving a preliminary positive answer to the question posed by the title of this review. But emergent properties arising from the interactive nature of cellular carbon metabolism also demonstrated many ways in which photosynthetic efficiency is sacrificed for metabolic flexibility, a necessary condition for accommodating the variable environmental conditions plants normally experience. Selected studies at higher hierarchical levels were used to illustrate some ways that constraints on plant production can emerge that are otherwise unforeseen at more reductionist scales. The observed association between leaf lifespan and maximum photosynthesis (Fig. 6) is a leaf-level pattern that is not predictable from cellular biochemistry. A leaf-level view would attribute this trade-off to leaf structural differences associated with leaf 'toughness' that influence light transmission and gas diffusion. But our consideration of whole-plant performance in tobacco plants with contrasting leaf lifespans (Table 2 & 3; Fig. 7) also indicate that whole-plant water use, canopy light transmission, and canopy nitrogen distribution act to constrain leaf-level traits of photosynthesis and leaf lifespan. Thus, new properties that can limit plant growth continue to emerge as one proceeds up through additional hierarchical levels. Another theme that repeatedly arises as we consider limits on plant productivity is photosynthetic variation within and among species. This is related to the resilience property of complex living systems, where resilience is the ability to perform under a wide range of conditions by having the capacity to accommodate or recover from imposed changes in the state of the system. Stomatal closure for water-conservation, xanthophyll-mediated dissipation of absorbed light energy, divergence of chloroplastic (or mitochondrial) electron flow to non-productive processes, re-mobilization of leaf resources from old canopy-shaded leaves, storage of carbohydrates for future use, production of durable, long-lived leaves with low metabolic activity; these are all examples whereby diminished photosynthetic productivity may permit increased resilience and survivorship. It seems that natural selection favors resilient systems even if there is an associated marginal cost to energetic efficiency. This is an insight we must take into consideration as we strive to develop more productive plants. A more definitive answer to the question posed by the title of this review will emerge as we begin to take various species of photosynthetically-improved plants out of the lab and into the field. Systems-based analyses of results from such studies will give great insight into the extent to which bioengineering of photosynthesis can enhance >3 billion years of evolutionary innovation in photosynthetic productivity.

4. Acknowledgements

We thank Howard Fung for his assistance in measuring leaf construction costs. We thank Audrey Perkins for helping to reduce the entropy in this review.

5. References

Amthor, J.S. (2010). From Sunlight to Phytomass: On the Potential Efficiency of Converting Solar Radiation to Phyto-energy. *New Phytologist*, Vol.188, No.4, (December 2010), pp. 939-959.

Anderson, J.M., Chow, W.S. & Goodchild, D.J. (1988) Thylakoid Membrane Organization in Sun/Shade Acclimation. *Australian Journal of Plant Physiology*, Vol.15, No2. (March 1988), pp. 11-26.

Araújo, W.L., Nunes-Nesi, A., Osorio, S., Usadel, B., Fuentes, D., Nagy, R., Balbo, I., Lehmann, M., Studart-Witkowski, C., Tohge, T., Martinoia, E., Jordana, X., DaMatta, F.M. & Fernie, A.R. (2011). Antisense Inhibition of the Iron-Sulphur Subunit of Succinate Dehydrogenase Enhances Photosynthesis and Growth in Tomato via an Organic Acid-Mediated Effect on Stomatal Aperture. *Plant Cell*, Vol.23, No.2, (February 2011), pp. 600-627.

Asada, K. (1999). The Water-Water Cycle in Chloroplasts. *Annual Review of Plant Biology*, Vol.50, (June 1999), pp. 601-639.

Atkin, O.K., Evans, J.R. & Siebke, K. (1998) Relationship Between the Inhibition of Leaf Respiration by Light and Enhancement of Leaf Dark Respiration Following Light

Treatment. *Australian Journal of Plant Physiology*, Vol.24, No.4. (April 1998), pp. 437-443.

Bazzaz, F.A. & Harper, J.L. (1977). Demographic Analysis of the Growth of *Linum usitatissimum*. *New Phytologist*, Vol.78, No.1, (January 1977), pp. 193-208.

Blankenship, R.E. (2002). *Molecular Mechanisms of Photosynthesis*. (First Edition), Blackwell Science, ISBN 0632043210, Oxford.

Beerling, D. (2007). *The Emerald Planet: How Plants Changed Earth's History*. (First Edition), Oxford University Press, ISBN 9780199548149, Oxford.

Boonman, A., Anten, N.P.R., Dueck, T.A., Jordi, W.J.R.M., van der Werf, A., Voesenek, L.A.C.J. & Pons, T.L. (2006). Functional Significance of Shade-Induced Leaf Senescence in Dense Canopies: An Experimental Test Using Transgenic Tobacco. *American Naturalist*, Vol.168, No.5, (November 2006), pp. 597-607.

Calvin, M. (1980). Hydrocarbons from Plants: Analytical Methods and Observations. *Die Naturwissenschaften*, Vol.67, No.11, (November 1980), pp. 525-533.

Chapin F.S. III, Schulze, E.D. & Mooney, H.A. (1990). The Ecology and Economics of Storage in Plants. *Annual Review of Ecology and Systematics*, Vol.21, (June 1990), pp. 423-447.

Chida, H., Nakazawa, A. & 15 others. (2007). Expression of the Algal Cytochrome c6 Gene in *Arabidopsis* Enhances Photosynthesis and Growth. *Plant Cell Physiology*. Vol.47, No.7, (July 2007), pp. 948-957.

Cockell, C.S., Southern, A. & Herrera, A. (2000). Lack of UV Radiation in Biosphere 2-Practical and Theoretical Effects. *Ecological Engineering*, Vol.16, No.2, (November 2000), pp. 293-229.

Demmig-Adams, B. & Adams, W.W. III. (2006). Photoprotection in an Ecological Context: the Remarkable Complexity of Thermal Energy Dissipation. *New Phytologist*, Vol.172, No.1, (October 2006), pp. 11-21.

Donovan, L.A., Maherali, H., Caruso, C.M., Huber, H. & de Kroon, H. (2011). The Evolution of the Worldwide Leaf Economics Spectrum. *Trends in Ecology and Evolution*, Vol.26, No.2, (February 2011), pp. 88-95.

Ehleringer, J. & Björkman, O. (1977). Quantum Yields for CO_2 Uptake in C3 and C4 Plants. *Plant Physiology*, Vol.59, No.1, (January 1977), pp. 86-90.

Evans, L.T. (1993). *Crop Evolution, Adaptation and Yield* (First Edition), Cambridge University Press, ISBN 52129558, Cambridge.

Flexas, J., Ribas-Carbo, M., Hanson, D.T., Bota, J., Otto, B., Cifre, J., McDowell, N., Medrano, H. & Kaldenhoff, R. (2006). Tobacco aquaporin NtAQP1 is Involved in Mesophyll Conductance to CO_2 *in vivo*. *The Plant Journal*, Vol.48, No.3, (November 2006), pp 427-439.

Foyer, C.H., Bloom, A.J., Queval, G. & Noctor, G. (2009). Photorespiratory Metabolism: Genes, Mutants, Energetics, and Redox Signaling. *Annual Review of Plant Biology*, Vol.60, (June 2009), pp. 455-484.

Foyer, C.H. & Shigeoka, S. (2011). Understanding Oxidative Stress and Antioxidant Functions to Enhance Photosynthesis. *Plant Physiology*, Vol.155, No.1, (January 2011), pp. 93-100.

Gan, S. & Amasino, R.M. (1995). Inhibition of Leaf Senescence by Autoregulated Production of Cytokinin. *Science*, Vol.270, No.5244, (December 1995), pp. 1986-1988.

Griffin, K.L. (1994). Calorimetric Estimates of Construction Cost and Their Use in Ecological Studies. *Functional Ecology*, Vol.8, No.5, (October 1994), pp. 551-562.

Griffin, K.L. & Seemann, J.R. (1996). Plants, CO_2 and Photosynthesis in the 21st Century. *Chemistry & Biology*, Vol.3, No.4, (April 1996), pp. 245-254.

Griffin, K.L., Anderson, O.R., Gastrich, M.D., Lewis, J.D., Lin, G.H., Schuster, W., Seemann, J.R., Tissue, D.T., Turnbull, M.H. & Whitehead, D. (2001). Plant Growth in Elevated CO2 Alters Mitochondrial Number and Chloroplast Fine Structure. *Proceedings of the National Academy of Sciences*, Vol.98, No.5, (February, 2001). pp. 2473–2478.

Hatfield, P.M. & Vierstra, R.D. (1997). Protein Degradation. In: *Plant Metabolism*, D.T. Dennis, D.H. Turpin, D.D. Lefebvre & D.B. Layzell (Eds), pp. 26-36, Addison Wesley Longman, ISBN 0582259061, Essex.

Hikosaka, K., Terashima, I. & Katoh, S. (1994). Effects of Leaf Age, Nitrogen Nutrition and Photon Flux Density on the Distribution of Nitrogen Among Leaves of a Vine (*Ipomoea tricolor*) Grown Horizontally to Avoid Mutual Shading of Leaves. *Oecologia*, Vol.97, No.4, (April 1994), pp. 451-457.

Hikosaka, K. (2005). Leaf Canopy as a Dynamic System: Ecophysiology and Optimality in Leaf Turnover. *Annal of Botany*, Vol.95, No.3, (February 2005), pp. 521-533.

Hiremath, A.J. (2000). Photosynthetic nutrient-use efficiency in three fast-growing tropical trees with differing leaf longevities. *Tree Physiology*, Vol.20, No.14, (August 2000), pp. 937-944.

Horton, P. (2000). Prospects for Crop Improvement Through the Genetic Manipulation of Photosynthesis: morphological and Biochemical Aspects of Light Capture. *Journal of Experimental Botany*, Vol.51, No.2, (February 2000), pp. 475-485.

Jordi, W., Schapendonk, E., Davelaar, E., Stoopen, G.M., Pot, C.S., De Visser, R., Van Rhijn, J.A., Gan, S. & Amasino, R. (2000). Increased Cytokinin Levels in Transgenic P_{SAG12}:IPT Tobacco Plants Have Large Direct and Indirect Effects on Leaf Senescence, Photosynthesis and N Partitioning. *Plant, Cell and Environment*, Vol.23, No. 3, (March 2000), pp. 279-289.

Jung, H.S. & Niyogi, K.K. (2009). Quantitative Genetic Analysis of Thermal Dissipation in *Arabidopsis*. *Plant Physiology*, Vol.150, No.2, (June 2009), pp. 977-986.

Kebish, R., Niessen, M., Thiruveedhi, K., Bari, R., Hirsch, H.J., Rosenkranz, R., Stäbler, N., Schönfeld, B., Kreuzaler, F. & Peterhänsel, C. (2007). Chloroplastic Photorespiratory Bypass Increases Photosynthesis and Biomass Production in *Arabidopsis thaliana*. *Nature Biotechnology*, Vol.25, No.5, (May 2007), pp. 593-599.

Kikuzawa, K. (1991). A Cost-Benefit Analysis of Leaf Habit and Leaf Longevity of Trees and Their Geographical Pattern. *American Naturalist*, Vol.138, No.5, (November 1991), pp. 1250-1263.

Kitajima, K., Mulkey, S.S. & Wright, S.J. (1997). Decline of Photosynthetic Capacity with Leaf Age in Relation to Leaf Longevities for Five Tropical Canopy Tree Species. *American Journal of Botany*, Vol.85, No.5, (May 1997), pp. 702-708.

Kozaki, A. & Takeba, G. (1996). Photorespiration Protects C3 Plants from Photoinhibition. *Nature*. Vol. 384, ((December 1996), pp. 557-560.

Krah N.M. & Logan, B.A. (2010). Loss of *PSBS* Expression Reduces Vegetative Growth, Reproductive Output, and Light-Limited but not Light-Saturated Photosynthesis in *Arabidopsis thaliana* (Brassicaceae) Grown in Temperate Light Environments. *American Journal of Botany*, Vol.97, No.4, (April 2010), pp. 644-649.

Kruger, E.L. & Volin, J.C. (2006). Re-examining the Empirical Relationship Between Plant Growth and Leaf Photosynthesis. *Functional Plant Biology*, Vol.33, No.5, (May 2005), pp. 421-429.

Külheim, C., Agren, J. & Jansson, S. (2010). Rapid Regulation of Light Harvesting and Plant Fitness in the Field. *Science*, Vol.297, No.5578, (July 2002), pp. 91-93.

Leigh, L.S., Burgess, T., Marino, B.D.V. & Wei, Y.D. (1999). Tropical Rainforest Biome of Biosphere 2: Structure, Composition and Results of the First Years of Operation. *Ecological Engineering*, Vol.13, No.1, (June 1999), pp. 65-93.

Long, S.P., Zhu, X.G., Naidu, S.L. Ort, D.R. (2006). Can Improvement in Photosynthesis Increase Crop Yields? *Plant, Cell and Environment*, Vol.29, No.3, (March 2006), pp. 315-330.

Lucas, M., Laplaze, L. & Bennett, M. (2011). Plant System Biology: Network Matters. *Plant, Cell and Environment*, Vol.34, No.11, (November 2011), pp. 535-553.

Makino, A. (2011). Photosynthesis, Grain Yield, and Nitrogen Utilization in Rice and Wheat. *Plant Physiology*, Vol.155, No.1, (January 2011), pp. 125-129.

Maurel, C., Verdoucq, L., Luu, D.T. & Santoni, V. (2008). Plant Aquaporins: Membrane Channels with Multiple Integrated Functions. *Annual Review of Plant Biology*, Vol.59, No.1, (2008), pp. 595-624.

Maxwell, D.P., Wang, Y. & McIntosh, L. (1999). The Alternative Oxidase Lowers Mitochondrial Reactive Oxygen Production in Plant Cells. *Proceedings of the National Academy of Science*, Vol.96, No.14, (July 1999), pp. 8271-8276.

McCulloh, K.A., Winter, K., Meinzer, F.C., Garcia, M., Aranda, J. & Lachenbruch, B. (2007). A Comparision of Daily Water Use Estimates Derived From Constant-Heat Sap-Flow Probe Values and Gravimetric Measurments in Pot-Grown Saplings. *Tree Physiology*, Vol.27, No.9, (September 2007), pp. 1355-1360.

Miller, R.E., Grant, N.M., Giles, L., Ribas-Carlos, M., Berry, J.A., Watling, J.R. & Robinson, S.A. (2011). In the Heat of the Night-Alternative Pathway Respiration Drives Thermogenesis in *Philodendron bipinnatifidum*. *New Phytologist*, Vol.189, No.4, (March 2011), pp. 1013-1026.

Murchie, E.H. & Niyogi, K.K. (2011). Manipulation of Photoprotection to Improve Plant Photosynthesis. *Plant Physiology*, Vol.155, No.1, (January 2011), pp. 86-92.

Nadeau, J.A. (2009). Stomatal Development: New Signals and Fate Determinants. *Current Opinion in Plant Biology*, Vol.12, No.1, (February 2009), pp. 29-35.

Nagel, J.M., Huxman, T.E., Griffin, K.L. & Smith, S.D. (2004). CO_2 Enrichment Reduces the Energetic Cost of Biomass Construction in an Invasive Desert Grass. *Ecology*, Vol.85, No.1, (January 2004), pp. 100-106.

Nunes-Nesi, A., Sulpice, R., Gibon, Y. & Fernie, A.R. (2008). The Enigmatic Contribution of Mitochondrial Function in Photosynthesis. *Journal of Experimental Botany*, Vol.59, No.7, (March 2008), pp. 1675-1684.

Osmond, C.B. & Grace, S.C. (1995). Perspectives on Photoinhibition and Photorespiration in the Field: Quintessential Inefficiencies of the Light and Dark Reactions of Photosynthesis. *Journal of Experimental Botany*, Vol.46, No.9, (September 1995), pp. 1351-1362.

Paul, M.J. & Pellny, T.K. (2003). Carbon Metabolite Feedback Regulation of Leaf Photosynthesis and Development. *Journal of Experimental Botany*, Vol.54, No.382, (February 2003), pp. 539-547.

Pearcy, R.W. (1991). Radiation and Light Measurements. In: *Plant Physiologica Ecology Field Methods and Instrumentation*, R.W. Pearcy J Ehleringer, H.A. Mooney & P.W. Rundel (Eds), pp. 97-116, Chapman & Hall, ISBN 0412407302, London.

Poorter, H. & Remkes, C. (1990). Leaf Area Ratio and Net Assimilation Rate of 24 Wild Species Differing in Relative Growth Rate. *Oecologia*, Vol.83, No.4, (July 1990), pp. 553-559.

Poorter, H., Pepin, S., Rijkers, T., de Jon, Y., Evans, J.R. & Körner, C. (2006). Construction Costs, Chemical Composition and Payback Time of High- and Low-Irradiance Leaves. *Journal of Experimental Botany*, Vol.57, No.2, (February 2006), pp. 355-371.

Raines, C.A. (2011). Increasing Photosynthetic Carbon Assimilation in C3 Plants to Improve Crop Yield: Current and Future Strategies. *Plant Physiology*, Vol.155, No.1, (January 2011), pp. 36-42.

Raven, J.A. (2011). The Cost of Photoinhibition. *Physiologia Plantarum*, Vol.142, No.1, (May 2011), pp. 87-104.

Reich, P.B., Koike, T., Gower, S.T. & Schoettle, A.W. (1995). Causes and Consequences of Variation in Conifer Leaf Life-Span, In: *Ecophysiology of Coniferous Forests*, W.K. Smith & T.M. Hinckley (Eds), pp. 225-254, Academic Press, ISBN 0126528756, San Diego.

Reich, P.B., Walters, M.B. & Ellsworth, D.S. (1997). From tropics to tundra: Global convergence in plant functioning. *Proceedings of the National Academy of Science*, Vol.94, No.25, (December 1997), pp. 13730-13734.

Robertson, D.E., Jacobson, S.A., Morgan, F., Berry, D., Church, G.M. & Afeyan, N.B. (2011). A New Dawn for Industrial Photosynthesis. *Photosynthesis Research*, Vol.107, No.3, (March 2011), pp. 269-277.

Roston, E. (2008). *The Carbon Age.* (First Edition), Walker and Company, ISBN 9780802715579, New York.

Ryan, M.G., Harmon, M.E., Birdsey, R.A., Giardina, C.P., Heath, L.S., Houghton, R.A., Jackson, R.B., McKinley, D.C., Morrison, J.F., Murray, B.C., Pataki, D.E. & Skog, K.E. (2010). A Synthesis of the Science on Forests and Carbon for U.S. Forests. *Issues in Ecology*, Vol.13, (Spring 2010), pp. 1-16.

Sage, R.F. (2004). The Evolution of C4 Photosynthesis. *New Phytologist*, Vol.161, No.2, (February 2004), pp. 341-370.

Sage, R.F. & Zhu, X.G. (2011). Exploiting the Engine of C4 Photosynthesis. *Journal of Experimental Botany*, Vol.62, No.9, (May 2011), pp. 2989-3000.

Schlesinger, W.H. (1997). *Biogeochemistry: An Analysis of Global Change.* (Second Edition), Academic Press, ISBN 012625155, San Diego.

Schlüter, U., Muschak, M., Berger, D. & Altmann, T. (2003). Photosynthetic Performance of an *Arabidopsis* Mutant with Elevated Stomatal Density (sdd1-1) Under Different Light Regimes. *Journal of Experimental Botany*, Vol.54, No.383, (February 2003), pp. 867-874.

Schrödinger, E. (1992). *What is life?* (first published in 1944), Cambridge University Press, ISBN 9780521427081, Cambridge.

Searle, S.Y., Thomas, S., Griffin, K.L., Horton, T., Kornfield, A., Yakir, D., Hurry, V. & Turnbull, M.H. (2011). Leaf Respiration and Alternative Oxidase in Field-Grown Alpine Graases Respond to Natural Changes in Temperature and Light. *New Phytologist*, Vol.189, No.4, (March 2011), pp. 1027-1039.

Sheehy, J.E., Mitchell, P.L. & Hardy, B. (2008). *Charting New Pathways to C4 Rice*. (First Edition), World Scientific Publishing Company, ISBN 139789812708516, Singapore.

Skillman, J.B., Strain, B.R. & Osmond, C.B. (1996). Contrasting Patterns of Photosynthetic Acclimation and Photoinhibition in Two Evergreen Herbs from a Winter Deciduous Forest. *Oecologia*, Vol 107. No.4, (September 1996), pp. 446-455.

Skillman, J.B., Garcia, M., Virgo, A. & Winter, K. (2005). Growth Irradiance Effects on Photosynthesis and Growth in Two Co-occurring Shade-tolerant Neotropical Perennials of Contrasting Photosynthetic Pathways. *American Journal of Botany*, Vol.92, No.11, (November 2005), pp. 1811-1819.

Skillman, J.B. (2008). Quantum Yield Variation Across the Three Pathways of Photosynthesis. *Journal of Experimental Botany*, Vol.59, No.7, (March 2008), pp. 1647-1661.

Somerville, C.R. (2001). An Early Arabidopsis Demonstration. Resolving a Few Issues Concerning Photorespiration. *Plant Physiology*, Vol.125, No.1, (January 2011), pp. 20-24.

Stitt, M., Lunn, J. & Usadel, B. (2010). *Arabidopsis* and Primary Photosynthetic Metabolism-More than the Icing on the Cake. *The Plant Journal*, Vol.61, No.6, (March 2010), pp. 1067-1091.

Sulpice, R., Pyl, E.T. & 17 others. (2009). Starch as a Major Integrator in the Regulation of Plant Growth. *Proceedings of the National Academy of Science*, Vol.106, No.25, (June 2009), pp. 10348-10353.

Sweetlove,L.J., Lytovchenko, A., Morgan, M., Nunes-Nesi, A., Taylor, N.L., Baxter, C.J., Eickmeier, I. & Fernie, A.R. (2006). Mitochondrial Uncoupling Protein is Required for Efficient Photosynthesis. *Proceedings of the National Academy of Science*, Vol.103, No.51, (December 2006), pp. 19587-19592.

Takahashi, S. & Badger, M.R. (2010). Photoprotection in Plants: A New Light on Photosystem II Damage. *Trends in Plant Sciences*, Vol.16, No.1, (January 2010), pp. 53-60.

Tamoi, M., Nagaoka, M., Miyagawa, Y. & Shigeoka, S. (2006). Contribution of Fructose-1,6-bisphosphatase and Sedoheptulose-1,7-bisphosphatase to the Photosynthetic Rate and Carbon Flow in the Calvin Cycle in Transgenic Plants. *Plant and Cell Physiology*, Vol.47, No.3, (March 2006), pp. 380-390.

Tcherkez, G.G.B., Farquhar, G.D. & Andrews, T.J. (2006). Despite Slow Catalysis and Confused Substrate Specificity, all Ribulose bisphosphate carboxylase may be Nearly Perfectly Optimized. *Proceedings of the National Academy of Science*, Vol.103, No.19, (May 2006), pp. 7246-7251.

Terashima, I., Hanba, Y.T., Tholen, D. & Niiemets, Ü. (2011). Leaf Functional Anatomy in Relation to Photosynthesis. *Plant Physiology*, Vol.155, No.1, (January 2011), pp. 108-116.

Tholen, D., Boom, C., Noguchi, K., Ueda, S., Katase, T. & Terashima, I. (2008). The Chloroplast Avoidance Response Decreases Internal Conductance to CO_2 Diffusion in *Arabidopsis thaliana* Leaves. *Plant, Cell and Environment*, Vol.31, No.11, (November 2008), pp. 1688-1700.

Thomas, R.B. & Strain, B.R. (1991). Root Restriction as a Factor in Photosynthetic Acclimation of Cotton Seedlings Grown in Elevated Carbon Dioxide. *Plant Physiology*, Vol.96 No.2, (June 1991), pp. 627-634.

Tilman, D., Socolow, R., Foley, J.A., Hill, J., Larson, E., Lynd, L., Pacala, S., Reily, J., Searchinger, T., Somerville, C. & Williams, R. (2009). Beneficial Biofuels-The Food, Energy, and Environment Trilemma. *Science,* Vol.325, No.5938, (July 2009), pp. 270-271.

Tsuchihira, A., Hanba, Y.T., Kato, N., Doi, T., Kawazu, T. & Maeshima, M. (2010) Effects of Overexpression of Radish Plasma Membrane Aquaporins on Water-Use Efficiency, Photosynthesis and Growth of *Eucalyptus* Trees. *Tree Physiology,* Vol.30, No.3, (March 2010), pp. 417-430.

Valladares, F., Skillman, J.B. & Pearcy, R.W. (2002). Convergence in Light Capture Efficiencies Among Tropical Forest Understory Plants: A Case of Morphological Compensation. *American Journal of Botany,* Vol.89, No.8, (August 2002), pp. 1275-1284.

Warren, C.R. (2008). Stand Aside Stomata, Another Actor Deserves Centre Stage: the Forgotten Role of the Internal Conductance to CO_2 Transfer. *Journal of Experimental Botany,* Vol.59, No.7, (March 2008), pp. 1475-1487.

Whitney, S.P., Houtz, R.L. & Alonso, H. (2011). Advancing Our Understanding and Capacity to Engineer Nature's CO_2-sequestering Enzyme, Rubisco. *Plant Physiology,* Vol.155, No.1, (January 2011), pp. 27-35.

Whittaker, R.H. (1975). *Communities and Ecosystems.* (Second Edition), Macmillan, New York.

Wiliams, K., Percival, F., Merino, J. & Mooney, H.A. (1987). Estimation of Tissue Construction Cost from Heat of Combustion and Organic Nitrogen Content. *Plant, Cell and Environment,* Vol.10, No.9, (December 1987), pp. 725-734.

Williams, K., Field, C.B. & Mooney, H.A. (1989). Relationships Among Leaf Construction Cost, Leaf Longevity, and Light Environments in Rain-Forest Plants of the Genus *Piper. American Naturalist,* Vol.133, No.2, (February 1989), pp. 198-211.

Wright, I.J., Reich, P.B. & 31 others. (2004). The Worldwide Leaf Economics Spectrum. *Nature,* Vol.428 (April 2004), pp. 821-827.

Zhu, X.G., Ort, D.R., Whitmarsh, J. & Long, S.P. (2004) The Slow Reversibility of PSII Thermal Energy Dissipation on Transfer from High to Low Light May Cause Large Losses in Carbon Gain by Crop Canopies: A Theoretical Analysis. *Journal of Experimental Botany,* Vol.55, No.400, (March 2004), pp. 1167-1175.

Zhu, X.G., Long S.P. & Ort D.R. (2010). Improving Photosynthetic Efficiency for Greater Yield. *Annual Review of Plant Biology,* Vol61, (June 2010), pp. 235-261.

Zotz, H & Winter, K. (1996). Diel Patterns of CO_2 Exchange in Rainforest Canopy Plants. In: *Tropical Forest Plant Ecophysiology,* S.S. Mulkey, R.L. Chazdon & A.P. Smith, (Eds), 89-113, Chapman & Hall, ISBN 0412035715, New York.

4

Thermodynamics in Mono and Biphasic Continuum Mechanics

Henry Wong[1], Chin J. Leo[2] and Natalie Dufour[1]
[1]*Ecole Nationale des Travaux Publics de l'Etat,*
[2]*University of Western Sydney,*
[1]*France*
[2]*Australia*

1. Introduction

This chapter applies the laws of thermodynamics to problems in continuum mechanics. Initially these are applied to a monophasic medium. The case of a biphasic porous medium is then treated with the aim of illustrating how a framework may be established for capturing possible couplings in the pertinent constitutive relationships. This approach is founded on the two laws of Thermodynamics. The first law expresses the conservation of energy when considering all possible forms while the second law postulates that the quality of energy must inevitably deteriorate in relation to its transformability into efficient mechanical work.

2. The principles of thermodynamics in the case of monophasic media

In order to simplify matters so that the reader can have a good intuitive understanding on the fundamental principles, in particular their physical contents, we begin with the simplest case of a monophasic continuous media.

Consider a solid body in movement, with mass density ρ and a velocity field $\boldsymbol{v}$ (figure 1). Our attention will be focused on an arbitrarily chosen part of this body, which occupies a volume Ω_t at time t. For ordinary problems of solid mechanics, we are concerned with mechanical and thermal energies. We therefore suppose that the body inside Ω_t is subject to a distributed body force $\boldsymbol{f}$ (for example gravity) and surface tractions $\boldsymbol{t}$ on its boundary surface, noted $\partial\Omega_t$. At the same time, the body is subject to a heat flux $\boldsymbol{q}$ on $\partial\Omega_t$ and an internal heat source r_q.

To begin with, we consider the energy and entropy balance of all the matter inside the volume Ω_t, using the two principles of thermodynamics.

3. The first principle of thermodynamics

The first principle stipulates that energy must be conserved under its different forms. Limiting our study here to thermal and mechanical energies, we can write:

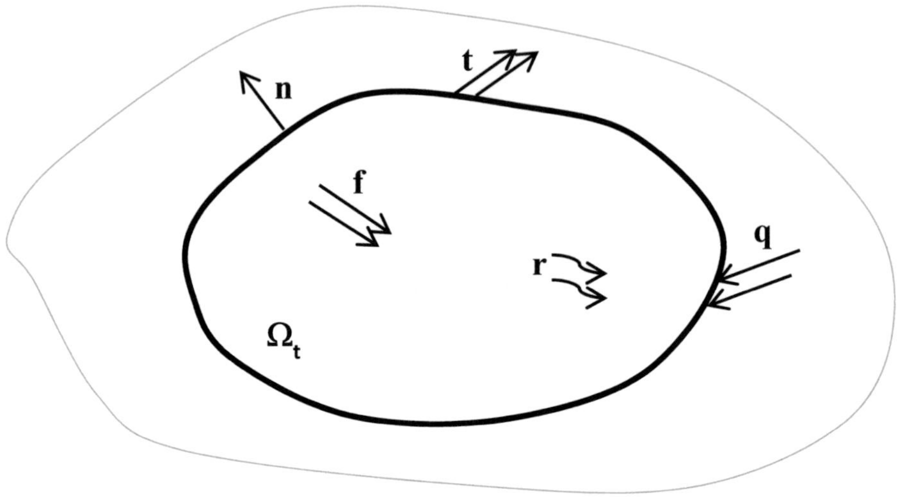

Fig. 1. Ω_t is a generic part of a body in movement, with distributed body forces $\boldsymbol{f}$, surface tractions t inward heat flux $\boldsymbol{q}$, and distributed heat source r_q, n is the outward unit normal.

$$\frac{d}{dt}(E+K) = P_x + Q \tag{1}$$

In the above equation, E and K are the global internal and kinetic energies, while P_x and Q are the total external supply of mechanical and thermal power for all matters inside Ω_t. The time derivative refers to the rate of increase of the energy content by following the same ensemble of material particles in their movement. This equation simply states that heat and mechanical energies received by a body which are not converted into kinetic energy become the internal energy. In continuum mechanics, physical quantities vary spatially from one point to another. The global quantities can be expressed in terms of the sum of local quantities:

$$\begin{aligned} E &= \int_{\Omega_t} \rho e \, d\Omega_t \\ K &= \int_{\Omega_t} \frac{1}{2} \rho \boldsymbol{v} \cdot \boldsymbol{v} \, d\Omega_t \\ P_x &= \int_{\Omega_t} \boldsymbol{f} \cdot \boldsymbol{v} \, d\Omega_t + \int_{\partial\Omega} \boldsymbol{t} \cdot \boldsymbol{v} \, dS \\ Q &= \int_{\Omega_t} r_q \, d\Omega_t - \int_{\partial\Omega} \boldsymbol{q} \cdot \boldsymbol{n} \, dS \end{aligned} \tag{2}$$

where e, the specific internal energy is defined as the internal energy per unit of mass. Substitution of equation (2) into (1) and on account of the classic equation $\boldsymbol{t} = \boldsymbol{\sigma} \cdot \boldsymbol{n}$ relating the surface traction $\boldsymbol{t}$ to the second order symmetric stress tensor $\boldsymbol{\sigma}$, we get after some simplifications:

$$\rho \dot{e} = \boldsymbol{\sigma} : \dot{\boldsymbol{\varepsilon}} + r_q - div\boldsymbol{q} \tag{3}$$

where $\boldsymbol{\varepsilon}$ denotes the strain tensor and a dot above a variable denotes the material derivative (i.e. total derivative with respect to time) by following the movement of an elementary solid particle. Internal energy is the energy content within a given mass of material. This includes the (A) kinetic energy due to the *disordered* thermal agitation and the (B) interaction, or potential, energy between molecules due to their relative positions (for example the elastic strain energy). It is the macroscopic description of (A) that leads to the introduction of the absolute temperature. The internal energy can also be the energy stored due to concentration of solutes (osmotic potential), but is outside the scope of this presentation. However, it should be noted that the following energies are not counted as internal energy:

1. Kinetic energy due to the macroscopic (ordered) movement of a material body
2. Potential energy due to the position of a body relative to an external field such as gravity

The last form of energy, namely the macroscopic potential energy, is accounted for by considering conservative body forces derivable from a potential, such as the gravity force per unit volume $\rho \boldsymbol{g}$, in the term $\boldsymbol{f}$ in the definition of P_x. Note that relative to the first principle, all forms of energy have an equal status.

4. The second principle of thermodynamics

The second principle of Thermodynamics confers a special status to heat, and distinguishes it from all other forms of energy, in that:

1. Once a particular form of energy is transformed into heat, it is impossible to back transform the entire amount to its original form without compensation.
2. To convert an amount of heat energy ΔQ into useful work, a necessary condition is to have at least two reservoirs with two different (absolute) temperatures T_1 and T_2 (suppose $T_1 > T_2$ to fix ideas).
3. Moreover, the above conversion can at best be partial in that the amount of work ΔW extractable from a given quantity of heat ΔQ admits a theoretical upper bound depending on the two temperatures:

$$\frac{\Delta W}{\Delta Q} \leq \frac{T_1 - T_2}{T_2} \tag{4}$$

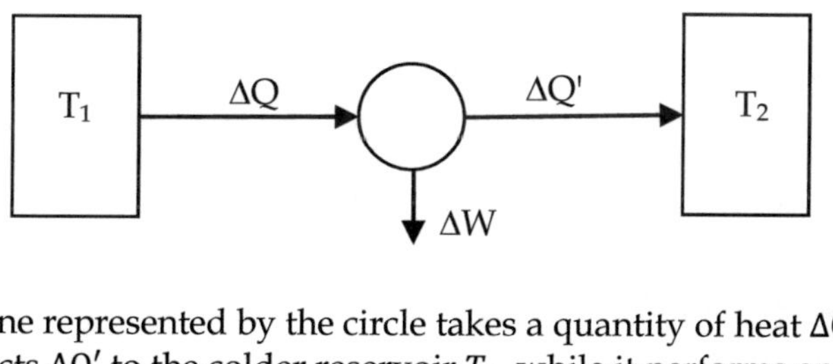

Fig. 2. The heat engine represented by the circle takes a quantity of heat ΔQ from the hotter reservoir T_1 and rejects $\Delta Q'$ to the colder reservoir T_2, while it performs an amount of useful work ΔW. The first principle requires $\Delta W = \Delta Q - \Delta Q'$ and the second principle sets a theoretical upper bound on the *efficiency* $\Delta W / \Delta Q$ attainable by heat engines.

Note that real efficiencies obtainable in practical cases are far less than that suggested by equation (4) due to unavoidable frictional losses. In the limit when the temperature becomes uniform, no mechanical work can be extracted anymore and this corresponds to some kind of *thermal-death*. In technical terms, when a particular form of energy is transformed into heat, the energy is *degraded* and becomes less *available* to perform *useful* work. The second principle gives a systematic and consistent account of why heat engines have theoretical upper limits of efficiency, and why certain phenomena can never occur spontaneously. For example, **we** cannot extract sea water at 20°C, cool it down to 0°C by extracting heat from it, and use that heat to drive the turbine and advance a ship! The theoretical formulation of the second principle via the concept of entropy derives its basis from a very large quantity of observations. The counter-part of the generality of its validity is the high level of abstraction, making it difficult to understand. Classical irreversible thermodynamics formulated directly at the macroscopic scale has an axiomatic appearance. The entropy change is defined axiomatically with respect to heat exchange and production. To understand its molecular original requires investigations at the microscopic scale. This is not necessary if the objective is to apply thermodynamic principles to build phenomenological models, although such investigations do contribute to a better understanding of the physical origin of the phenomena. Clausius (1850) invented the thermodynamic potential - the entropy - to describe this uni-directional and irreversible degradation of energy. Formulated in terms of entropy, the second principle of thermodynamics says that whenever some form of energy is transformed into heat, the global entropy increases. It can at best stay constant for reversible processes but can never decrease. If we denote s the specific entropy (per unit mass), the second principle writes:

$$\frac{d}{dt}\int_{\Omega_t} \rho s \, d\Omega_t \geq \int_{\Omega_t} \frac{r_q}{T} \, d\Omega_t - \int_{\partial\Omega_t} \frac{\boldsymbol{q}\cdot\boldsymbol{n}}{T} \, dS \tag{5}$$

In other words, for a fixed quantity of matter, the entropy increase must be greater than (resp. equal to) external heat supply divided by the absolute temperature in irreversible (resp. reversible) processes. The difference is due to other forms of energy being transformed into heat via *dissipative* processes. In our study here, this corresponds to internal frictional processes transforming mechanical energy into heat. Once this occurs, the process becomes irreversible. The previous inequality can be simplified to the following local form using Gauss' theorem:

$$\rho\dot{s} + div\frac{\boldsymbol{q}}{T} - \frac{r_q}{T} \geq 0 \tag{6}$$

As a macroscopic theory, irreversible thermodynamics does not give any explanation on the origin of entropy. Similarly to the case of plastic strains, the manipulation of entropy and other thermodynamic potentials will rely on postulated functions, valid over finite domains and containing coefficients to be determined by experiments.

5. Clausius-Duheim (CD) inequality

Combining the first and the second principle, we obtain the classic Clausius-Duhem (CD) inequality in the context of solid mechanics (electric, magnetic, chemical or osmotic terms etc. can appear in more general problems):

$$\Phi = \boldsymbol{\sigma}:\dot{\boldsymbol{\varepsilon}} + \rho(T\dot{s} - \dot{e}) - \frac{1}{T}\boldsymbol{q}\cdot\boldsymbol{grad}T \geq 0 \tag{7}$$

In the limiting case when the temperature field is uniform and the process is reversible, the above inequality becomes equality:

$$\frac{1}{\rho}\boldsymbol{\sigma}:\dot{\boldsymbol{\varepsilon}} + T\dot{s} - \dot{e} = 0 \quad \text{or} \quad de = \frac{1}{\rho}\boldsymbol{\sigma}:\mathrm{d}\boldsymbol{\varepsilon} + Tds \tag{8}$$

Since the specific internal energy is a state function and is supposed to be entirely determined by the state variables, we conclude from the differential form in (8) that e depends naturally on ε and s (i.e. $e = e(\boldsymbol{\varepsilon}, s)$) and that the following state equations hold:

$$\boldsymbol{\sigma} = \rho\frac{\partial e}{\partial \boldsymbol{\varepsilon}} \;\; ; \;\; T = \frac{\partial e}{\partial s} \tag{9}$$

However, the specific entropy s is not a convenient independent variable as it is intuitively difficult to comprehend and practically difficult to control. The classical approach consists of introducing another state function, the specific Helmholtz's free energy, via the Legendre transform:

$$\psi = e - Ts \tag{10}$$

to recast inequality (7) to the following form:

$$\Phi = \boldsymbol{\sigma}:\dot{\boldsymbol{\varepsilon}} - \rho\left(\dot{\psi} + s\dot{T}\right) - \frac{1}{T}\boldsymbol{q}\cdot\boldsymbol{grad}T \geq 0 \tag{11}$$

Again, in the absence of dissipative phenomena and a uniform temperature field, we have:

$$\frac{1}{\rho}\boldsymbol{\sigma}:\dot{\boldsymbol{\varepsilon}} - s\dot{T} - \dot{\psi} = 0 \quad \text{or} \quad \mathrm{d}\psi = \frac{1}{\rho}\boldsymbol{\sigma}:\mathrm{d}\boldsymbol{\varepsilon} - s\mathrm{d}T \tag{12}$$

via the same reasoning as previously, we deduce that the specific free energy ψ depends naturally on ε and T and satisfies the following state equations:

$$\boldsymbol{\sigma} = \rho\frac{\partial \psi}{\partial \varepsilon} \;\; ; \;\; s = -\frac{\partial \psi}{\partial T} \tag{13}$$

The Legendre transform (10) thus allows one to define a thermodynamic potential with natural independent variables which are more *accessible* (T instead of s in the present case). The quantity Φ, having the unit of energy per unit volume per unit time, is called *total dissipation*. It represents the transformation of non-thermal energy into heat via frictional processes, which then becomes *less available*.

6. How to use the second principle

There are two ways to make use of the second thermodynamic principle. We can first of all verify the consistency or the inconsistency of a given model with respect to the 2nd principle, in an *a posteriori* manner, in the sense that the construction of the model does not rely in any way on the 2nd principle. On the other hand, we can actually construct a model, starting from the Clausius-Duhem inequality, by specifying appropriate functional forms for the Helmholtz's free energy and the dissipation. Naturally, there is no unique way to achieve this goal since thermodynamics does not supply any information on the specific behavior of a particular material under study. This process must therefore integrate experimental data so that the model predictions are consistent with the reality. Among different representations (or models) consistent with thermodynamic principles, the best is the one with a clear logical structure and comprising a minimum number of parameters (simplicity). This last criterion allows to minimise the amount of experimental work necessary to identify these parameters, which is always a very time-consuming task.

7. Implicit but essential assumptions

All classic developments based on irreversible thermodynamics assume implicitly that the process does not deviate significantly from thermodynamic equilibrium. In consequence, despite the fact the system is in evolution therefore in *non-equilibrium*, the state equation expressing the condition of *thermodynamic equilibrium* can still be used to reduce the number of independent state parameters by one in complex problems (for example, the density, pressure and temperature of the pore fluid transiting a porous solid is related by a state equation). This is strictly speaking an approximation. Its efficiency can only be assessed *a posteriori* by the results.

In a heterogeneous system, the thermodynamic state hence the state parameters are position-dependent. This heterogeneity (hence non-equilibrium) is the driving force which tends to restore the system back to thermodynamic equilibrium. However, it is assumed that the (spatial) variation is sufficiently mild so that every elementary particle can be considered as under thermodynamic equilibrium. Its state parameters are therefore linked by the state equation expressing this equilibrium requirement. This assumption is called the "*hypothesis of local equilibrium*". This assumption excludes the treatment of fast processes (for example explosions) under the framework of classic irreversible thermodynamics.

8. Applications to plasticity and viscoplasticity: General equations

To illustrate how thermodynamic principles can be used to formulate physical laws, let us consider the particular case of the inelastic behaviour of solids. The classic partition:

$$\boldsymbol{\varepsilon} = \boldsymbol{\varepsilon}^e + \boldsymbol{\varepsilon}^p$$

Is assumed, where $\boldsymbol{\varepsilon}^e$ is the elastic strain and $\boldsymbol{\varepsilon}^p$ denotes for the time being all forms of irreversible (i.e. inelastic) strains. In order to satisfy the CD inequality (11), a common practice is to assume that $\psi = \psi(\boldsymbol{\varepsilon}^e, T, \boldsymbol{V}_k)$, so that $\dot{\psi} = \frac{\partial \psi}{\partial \varepsilon}\dot{\boldsymbol{\varepsilon}}^e + \frac{\partial \psi}{\partial T}\dot{T} + \frac{\partial \psi}{\partial \boldsymbol{V}_k}\dot{\boldsymbol{V}}_\mathrm{k}$. The scalar variables grouped into a tensor $\boldsymbol{V}_k$ are internal variables introduced to account for the state-dependent non-linear inelastic behaviour. In practice, this is often the irreversible strains or their scalar invariants. The CD inequality then becomes:

$$\Phi = \left(\boldsymbol{\sigma} - \rho\frac{\partial \psi}{\partial \boldsymbol{\varepsilon}}\right) : \dot{\boldsymbol{\varepsilon}}^e + \boldsymbol{\sigma} : \dot{\boldsymbol{\varepsilon}}^p - \rho\left(s + \frac{\partial \psi}{\partial T}\right)\dot{T} - \rho\frac{\partial \psi}{\partial \boldsymbol{V}_k} \cdot \dot{\boldsymbol{V}}_\mathrm{k} - \frac{\boldsymbol{q} \cdot \boldsymbol{grad}T}{T} \geq 0 \tag{14}$$

Consider the particular case of elastic (reversible) evolution corresponding to stationary values of the internal variables $\boldsymbol{V}$ and plastic strains, with uniform temperatures. We then have zero dissipation, retrieving the classic state equations (13). In the sequel it will be assumed that these state equations remain valid even under irreversible inelastic evolutions, so that the CD inequality becomes:

$$\Phi = \Phi_\mathrm{M} + \Phi_\mathrm{T} = \boldsymbol{\sigma} : \dot{\boldsymbol{\varepsilon}}^p - \boldsymbol{A}_k \cdot \dot{\boldsymbol{V}}_\mathrm{k} - \frac{\boldsymbol{q} \cdot \boldsymbol{grad}T}{T} \geq 0 \tag{15}$$

Under a simplified framework, we require the mechanical and thermal dissipations to be separately non-negative (this reduces the amount of coupling to account for in the model):

$$\Phi_\mathrm{M} = \boldsymbol{\sigma} : \dot{\boldsymbol{\varepsilon}}^p - \boldsymbol{A}_k \cdot \dot{\boldsymbol{V}}_\mathrm{k} \geq 0 \quad ; \quad \Phi_\mathrm{T} = -\frac{\boldsymbol{q} \cdot \boldsymbol{grad}T}{T} \geq 0 \tag{16}$$

The *thermodynamic force* $\boldsymbol{A}_k$, the conjugate variable to the *thermodynamic flux* $\boldsymbol{V}_k$, is "*defined*" as:

$$\boldsymbol{A_k} = \rho \frac{\partial \psi}{\partial \boldsymbol{V_k}} \tag{17}$$

In practice, $\boldsymbol{A}_k$ is often the variable which determines the size (isotropic hardening) or the amount of translation (kinematic hardening) of the yield surface and represents in a simplified manner all the effects of the loading history. One particular example is the pre-consolidation pressure which determines the current yield envelope of clays (as in Camclay model).
The non-negativity of the thermal dissipation can be satisfied by the classic Fourier Law:

$$\boldsymbol{q} = -\boldsymbol{K} \cdot \boldsymbol{grad}T \tag{18}$$

where the thermal conductivity tensor $\boldsymbol{K}$ must be symmetric and strictly positive, so that:

$$\Phi_{\mathrm{T}} = \frac{\boldsymbol{grad}T \cdot \boldsymbol{K} \cdot \boldsymbol{grad}T}{T} \geq 0 \tag{19}$$

It remains to satisfy the non-negativity of the mechanical (or intrinsic) dissipation:

$$\Phi_{\mathrm{M}} = \boldsymbol{\sigma} : \dot{\boldsymbol{\varepsilon}}^p - \boldsymbol{A}_k \cdot \dot{\boldsymbol{V}}_{\mathrm{k}} \geq 0 \tag{20}$$

The non-negativity of the mechanical dissipation forms the basis for the construction of the material behavioral laws. Note that the equation $\boldsymbol{A}_k = \rho \frac{\partial \psi}{\partial \boldsymbol{V}_k}$ only "defines" the variable $\boldsymbol{A}_k$ but does not contain any rule to calculate its evolution. Similarly, we need a rule to calculate the plastic strain rate $\dot{\boldsymbol{\varepsilon}}^p$.

9. Onsager's principle

In many physical problems, the total dissipation can be written as the sum of the products between a set of thermodynamic forces X and theirs conjugates, the thermodynamic flux x:

$$\Phi = X \cdot x = X_i x_i \geq 0 \tag{21}$$

Onsager, based on theoretical studies at molecular scales where all phenomena are reversible, suggested when the physical process only deviates slightly from the thermodynamic equilibrium, the thermodynamic forces and flux can be related by a set of phenomenological coefficients:

$$X_i = L_{ij} x_j \tag{22}$$

Onsager showed theoretically that the coefficients L_{ij} must be symmetrical. To ensure the non-negativity of the dissipation, it suffices to require L_{ij} to be definite positive, other than being symmetrical. The off-diagonal coefficients allow to account for cross-couplings. This formulation seems to be better suited to moderately non-linear problems. For example, it cannot lead to the classical plastic flow rule in solids.

10. Dissipation potentials

Another, more general, way to satisfy automatically the non-negativity of Φ_{M} is to introduce dissipation potentials. This can also handle more general non linear behaviours.

In the case of inelastic behaviour, we define a scalar function called the dissipation potential $\varphi(\dot{\boldsymbol{\varepsilon}}^p, \dot{\boldsymbol{V}}_{\mathrm{k}})$, convex and continuously differentiable with respect to both its arguments, positive everywhere and null at the origin, such that:

$$\boldsymbol{\sigma} = \frac{\partial \varphi}{\partial \dot{\boldsymbol{\varepsilon}}^p} \; ; \; \boldsymbol{A_k} = -\frac{\partial \varphi}{\partial \dot{\boldsymbol{V}}_{\mathrm{k}}} \tag{23}$$

We get immediately:

$$\Phi_{\mathrm{M}} = \boldsymbol{\sigma} : \dot{\boldsymbol{\varepsilon}}^p - \boldsymbol{A_k} \cdot \dot{\boldsymbol{V}}_{\mathrm{k}} = \frac{\partial \varphi}{\partial \dot{\boldsymbol{\varepsilon}}^p} : \dot{\boldsymbol{\varepsilon}}^p + \frac{\partial \varphi}{\partial \dot{\boldsymbol{V}}_{\mathrm{k}}} \cdot \dot{\boldsymbol{V}}_{\mathrm{k}} \geq \varphi \geq 0 \tag{24}$$

In general, it is more convenient to work with $\varphi^*(\boldsymbol{\sigma}, \boldsymbol{A}_k)$, the Legendre transform of φ, also convex and positive definite with respect to its arguments, zero at origin, with:

$$\dot{\boldsymbol{\varepsilon}}^p = \frac{\partial \varphi^*}{\partial \boldsymbol{\sigma}} \; ; \; \dot{\boldsymbol{V}}_{\mathrm{k}} = -\frac{\partial \varphi^*}{\partial \boldsymbol{A}_k} \tag{25}$$

So that:

$$\Phi_{\mathrm{M}} = \boldsymbol{\sigma} : \dot{\boldsymbol{\varepsilon}}^p - \boldsymbol{A}_k \cdot \dot{\boldsymbol{V}}_{\mathrm{k}} = \boldsymbol{\sigma} : \frac{\partial \varphi^*}{\partial \boldsymbol{\sigma}} + \boldsymbol{A}_k \cdot \frac{\partial \varphi^*}{\partial \boldsymbol{A}_k} \geq \varphi^* \geq 0 \tag{26}$$

Theoretically, once the free energy and the dissipation function are specified, the stress-strain relation is fully defined. This is therefore one possible way to construct a constitutive model. However the above reasoning does not work for plasticity.

11. Hardening plasticity for "standard" materials

In plasticity, the dissipation potential is not differentiable. Classically, the usual way to satisfy the dissipation inequality is to define a yield function:

$$F = F(\boldsymbol{\sigma}, \boldsymbol{A_k}) \tag{27}$$

(1) convex with respect to its arguments
(2) the "*elastic domain*" $F(\boldsymbol{\sigma}, \boldsymbol{A}_k) \leq 0$ contains the origin, and that:

$$\dot{\boldsymbol{\varepsilon}}^p = \dot{\lambda} \frac{\partial F}{\partial \boldsymbol{\sigma}} \; ; \; \dot{\boldsymbol{V}}_{\mathrm{k}} = -\dot{\lambda} \frac{\partial F}{\partial \boldsymbol{A}_k} \; ; \; \dot{\lambda} \geq 0 \tag{28}$$

where λ is the classic plastic multiplier, which obeys the conditions that:

$$\dot{\lambda} = 0 \text{ if } F < 0 \text{ or } \dot{F} < 0 \; ; \; \dot{\lambda} \geq 0 \text{ if } F = 0 \text{ and } \dot{F} = 0 \tag{29}$$

The first condition says if either the stress point is strictly inside the yield surface or if it is currently on the yield surface but moves inwards, the plastic multiplier, hence the plastic strain rate is null. The second condition expresses the condition of plastic loading when the current stress point is on the yield surface and it moves outwards. In this latter case, we have:

$$\Phi_{\mathrm{M}} = \dot{\lambda} \left(\boldsymbol{\sigma} : \frac{\partial F}{\partial \boldsymbol{\sigma}} + \boldsymbol{A_k} \cdot \frac{\partial F}{\partial \boldsymbol{A_k}} \right) \geq 0 \tag{30}$$

The non-negativity of the term between the parenthesis, namely:

$$\begin{Bmatrix} \boldsymbol{\sigma} \\ \boldsymbol{A_k} \end{Bmatrix} \cdot \begin{Bmatrix} \frac{\partial F}{\partial \sigma} \\ \frac{\partial F}{\partial A_k} \end{Bmatrix} \geq 0 \tag{31}$$

stems from geometric arguments (figure 3). This, together with $\dot{\lambda} \geq 0$, allows to ensure the non-negativity of Φ_M.

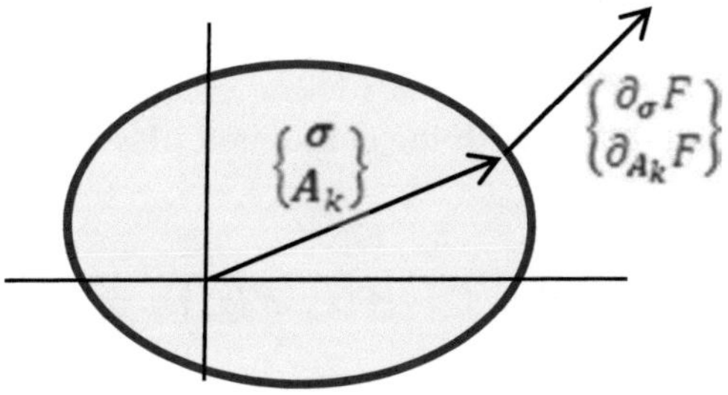

Fig. 3. The convex elastic domain contains the origin. Hence the position vector of a point on the boundary $\{\boldsymbol{\sigma}, \boldsymbol{A}_k\}$ and the normal vector at the same point $\{\boldsymbol{\partial}_\sigma F, \boldsymbol{\partial}_{A_k} F\}$ give a positive scalar product.

To construct an elastoplastic model, we need to define a hardening rule:

$$\boldsymbol{A_k} = \boldsymbol{A_k}(\boldsymbol{V_k}) \tag{32}$$

The plastic multiplier $\dot{\lambda}$ can then be determined by the classic consistency condition:

$$\dot{F} = \frac{\partial F}{\partial \sigma} \cdot \dot{\boldsymbol{\sigma}} + \frac{\partial F}{\partial A_k} \cdot \dot{\boldsymbol{A}}_{\mathrm{k}} = 0 \tag{33}$$

For stress-controlled evolutions, this yields, after a little substitution:

$$\dot{\lambda} = \frac{\frac{\partial F}{\partial \sigma} \cdot \dot{\boldsymbol{\sigma}}}{H} \quad ; \quad H = \frac{\partial F}{\partial \boldsymbol{A_k}} \cdot \frac{\partial \boldsymbol{A_k}}{\partial \boldsymbol{V_k}} \cdot \frac{\partial F}{\partial \boldsymbol{A_k}} \tag{34}$$

H is known as the hardening or plastic modulus. To relate the stress increment directly to the strain increment via the tangent stiffness tensor, we substitute:

$$\dot{\boldsymbol{\sigma}} = \boldsymbol{D_e} \cdot (\dot{\boldsymbol{\varepsilon}} - \dot{\boldsymbol{\varepsilon}}^p) \quad ; \quad \dot{\boldsymbol{\varepsilon}}^p = \dot{\lambda} \frac{\partial F}{\partial \sigma} \tag{35}$$

in the above to get:

$$\dot{\lambda} = \frac{\frac{\partial F}{\partial \sigma} \cdot \boldsymbol{D_e} \cdot \dot{\boldsymbol{\varepsilon}}}{H + \frac{\partial F}{\partial \sigma} \cdot \boldsymbol{D_e} \cdot \frac{\partial F}{\partial \sigma}} \tag{36}$$

Restarting with $\dot{\boldsymbol{\sigma}} = \boldsymbol{D_e} \cdot (\dot{\boldsymbol{\varepsilon}} - \dot{\boldsymbol{\varepsilon}}^p) = \boldsymbol{D_e} \cdot \left(\dot{\boldsymbol{\varepsilon}} - \dot{\lambda} \frac{\partial F}{\partial \sigma}\right)$ and after some manipulation leads to:

$$\dot{\sigma} = \boldsymbol{D_{ep}} \cdot \dot{\boldsymbol{\varepsilon}} \quad ; \quad \boldsymbol{D_{ep}} = \left(\boldsymbol{D_e} - \frac{\boldsymbol{D_e} \cdot \frac{\partial F}{\partial \sigma} \otimes \frac{\partial F}{\partial \sigma} \cdot \boldsymbol{D_e}}{H + \frac{\partial F}{\partial \sigma} \cdot \boldsymbol{D_e} \cdot \frac{\partial F}{\partial \sigma}} \right) \tag{37}$$

Note that the associative flow rule $\dot{\boldsymbol{\varepsilon}}^p = \dot{\lambda} \frac{\partial F}{\partial \sigma}$ renders the tangent matrix $\boldsymbol{D_{ep}}$ symmetric. This relation is also essential in the model construction to ensure the non-negativity of the dissipation. If we replace $\dot{\boldsymbol{\varepsilon}}^p = \dot{\lambda}\frac{\partial F}{\partial \boldsymbol{\sigma}}$ by $\dot{\boldsymbol{\varepsilon}}^p = \dot{\lambda}\frac{\partial g}{\partial \boldsymbol{\sigma}}$ with $g \neq F$ (non-associative flow rule), the CD inequality will no longer be automatically verified. This means that thermodynamic principles may then be violated in some evolutions. Note that in order to describe isotropic and kinematic hardening, the thermodynamic flux $\boldsymbol{V}_k$ is often decomposed into a tensor $\boldsymbol{\alpha}$ and a scalar r, associated with thermodynamic forces $\boldsymbol{X}$ and R. We would then have to write:

$$F = F(\boldsymbol{\sigma}, \boldsymbol{X}, R) \quad ; \quad \dot{\boldsymbol{\varepsilon}}^p = \dot{\lambda}\frac{\partial F}{\partial \boldsymbol{\sigma}} \quad ; \quad \dot{\boldsymbol{\alpha}} = -\dot{\lambda}\frac{\partial F}{\partial \boldsymbol{X}} \quad ; \quad \dot{r} = -\dot{\lambda}\frac{\partial F}{\partial R} \tag{38}$$

A common example is to identify r with the cumulated plastic deviatoric strain γ^p, defined as:

$$r = \gamma^p = \int_0^t \left(\frac{2}{3}\dot{\boldsymbol{e}}^p(\tau) : \dot{\boldsymbol{e}}^p(\tau)\right)^{1/2} d\tau \tag{39}$$

where $\dot{\boldsymbol{e}}^p = \mathrm{dev}(\dot{\boldsymbol{\varepsilon}}^p)$.

12. Viscoplasticity

We start with:

$$\boldsymbol{\varepsilon} = \boldsymbol{\varepsilon}^e + \boldsymbol{\varepsilon}^{vp} \tag{40}$$

then go through the same procedure as for plasticity:

$$\Phi = \boldsymbol{\sigma} : \dot{\boldsymbol{\varepsilon}} - \rho(\dot{\psi} + s\dot{T}) - \frac{\boldsymbol{q} \cdot \boldsymbol{grad} T}{T} \geq 0 \tag{41}$$

and:

$$\psi = \psi(\boldsymbol{\varepsilon}^e, T, \boldsymbol{V}_k) \tag{42}$$

We end up with the same dissipation inequality:

$$\Phi = \left(\boldsymbol{\sigma} - \rho\frac{\partial \psi}{\partial \boldsymbol{\varepsilon}}\right) : \dot{\boldsymbol{\varepsilon}}^e + \boldsymbol{\sigma} : \dot{\boldsymbol{\varepsilon}}^{vp} - \rho\left(s + \frac{\partial \psi}{\partial T}\right)\dot{T} - \rho\frac{\partial \psi}{\partial \boldsymbol{V}_k} \cdot \dot{\boldsymbol{V}}_k - \frac{\boldsymbol{q} \cdot \boldsymbol{grad} T}{T} \geq 0 \tag{43}$$

the same state equations:

$$\boldsymbol{\sigma} = \rho\frac{\partial \psi}{\partial \boldsymbol{\varepsilon}} \quad ; \quad s = -\frac{\partial \psi}{\partial T} \tag{44}$$

the same intrinsic dissipation (we discard the thermal part here):

$$\Phi_M = \boldsymbol{\sigma} : \dot{\boldsymbol{\varepsilon}}^{vp} - \boldsymbol{A}_k \cdot \dot{\boldsymbol{V}}_k \geq 0 \tag{45}$$

the same definition for the *thermodynamic force* $\boldsymbol{A}_k$ conjugate to the *thermodynamic flux* $\boldsymbol{V}_k$:

$$\boldsymbol{A}_k = \rho\frac{\partial \psi}{\partial \boldsymbol{V}_k} \tag{46}$$

However, a fundamental difference with plasticity intervenes here. In viscoplasticity, a continuously differentiable dissipation potential, definite positive, convex and contains the origin, can be defined:

$$\varphi^* = \varphi^*(\boldsymbol{\sigma}, \boldsymbol{A}_k)\ ;\ \dot{\boldsymbol{\varepsilon}}^{vp} = \frac{\partial \varphi^*}{\partial \boldsymbol{\sigma}}\ \ ;\ \dot{\boldsymbol{V}}_\mathrm{k} = -\frac{\partial \varphi^*}{\partial \boldsymbol{A}_k} \tag{47}$$

so that the non-negativity condition can be *a priori* satisfied:

$$\Phi_\mathrm{M} = \boldsymbol{\sigma}: \dot{\boldsymbol{\varepsilon}}^{vp} - \boldsymbol{A}_k \cdot \dot{\boldsymbol{V}}_\mathrm{k} = \boldsymbol{\sigma}: \frac{\partial \varphi^*}{\partial \boldsymbol{\sigma}} + \boldsymbol{A}_k \cdot \frac{\partial \varphi^*}{\partial \boldsymbol{A}_k} \geq \varphi^* \geq 0 \tag{48}$$

As for plasticity, in order to describe isotropic and kinematic hardening, the internal variable $\boldsymbol{V}_k$ is often decomposed into a tensor $\boldsymbol{\alpha}$ and a scalar r, associated with thermodynamic forces $\boldsymbol{X}$ and R:

$$\psi = \psi(\boldsymbol{\varepsilon}^e, T, \boldsymbol{\alpha}, r)\ ;\ \boldsymbol{X} = \frac{\partial \psi}{\partial \boldsymbol{\alpha}}\ \ ;\ R = \frac{\partial \psi}{\partial r} \tag{49}$$

The mechanical dissipation inequality then becomes:

$$\Phi_\mathrm{M} = \boldsymbol{\sigma}: \dot{\boldsymbol{\varepsilon}}^{vp} - \boldsymbol{X} \cdot \dot{\boldsymbol{\alpha}} - R\dot{r} \geq 0 \tag{50}$$

with the corresponding dissipation potential :

$$\varphi^* = \varphi^*(\boldsymbol{\sigma}, \boldsymbol{X}, R\,)\ ;\ \dot{\boldsymbol{\varepsilon}}^{vp} = \frac{\partial \varphi^*}{\partial \boldsymbol{\sigma}}\ \ ;\ \dot{\boldsymbol{\alpha}} = -\frac{\partial \varphi^*}{\partial \boldsymbol{X}}\ \ ;\ \dot{r} = -\frac{\partial \varphi^*}{\partial R} \tag{51}$$

We and up with:

$$\Phi_\mathrm{M} = \boldsymbol{\sigma}: \dot{\boldsymbol{\varepsilon}}^{vp} - \boldsymbol{X} \cdot \dot{\boldsymbol{\alpha}} - R\dot{r} = \boldsymbol{\sigma}: \frac{\partial \varphi^*}{\partial \boldsymbol{\sigma}} + \boldsymbol{X} \cdot \frac{\partial \varphi^*}{\partial \boldsymbol{X}} + R\frac{\partial \varphi^*}{\partial R} \geq 0 \tag{52}$$

For example, Lemaitre's model with isotropic hardening is based on the following dissipation potential:

$$\varphi^*(\boldsymbol{\sigma}, R\,) = \frac{K}{N+1}\left(\frac{\sigma_{eq} - R}{K}\right)^{\boldsymbol{N+1}} \frac{1}{\zeta} \tag{53}$$

Where ζ is considered as a parameter independent of the stress tensor, with:

$$\sigma_{eq} = \sqrt{\frac{3}{2}\boldsymbol{s}:\boldsymbol{s}}\ \ ;\ \ \boldsymbol{s} = \mathrm{dev}(\boldsymbol{\sigma}) = \boldsymbol{\sigma} - \frac{1}{3}tr(\boldsymbol{\sigma})\boldsymbol{I} \tag{54}$$

A differentiation gives:

$$\dot{\boldsymbol{\varepsilon}}^{vp} = \frac{\partial \varphi^*}{\partial \boldsymbol{\sigma}} = \left(\frac{\sigma_{eq} - R}{K}\right)^{\boldsymbol{N}} \frac{1}{\zeta}\left(\frac{3}{2}\frac{\boldsymbol{s}}{\sigma_{eq}}\right) \tag{55}$$

and:

$$\dot{r} = -\frac{\partial \varphi^*}{\partial R} = \left(\frac{\sigma_{eq} - R}{K}\right)^{\boldsymbol{N}} \frac{1}{\zeta} \tag{56}$$

where we have used the identity $\frac{\partial \sigma_{eq}}{\partial \boldsymbol{\sigma}} = \frac{3}{2}\frac{\boldsymbol{s}}{\sigma_{eq}}$. Note that the viscoplastic strain rate is purely deviatoric, in other words $tr(\dot{\boldsymbol{\varepsilon}}^{vp}) = 0$. Using the classic definition of the equivalent deviatoric viscoplastic strain rate:

$$\dot{\gamma}_{vp} = \sqrt{\frac{2}{3}\dot{\boldsymbol{\varepsilon}}^{vp}:\dot{\boldsymbol{\varepsilon}}^{vp}} \tag{57}$$

It can easily be verified that:

$$\dot{\gamma}_{vp} = \dot{r} = \left(\frac{\sigma_{eq}-R}{K}\right)^N \frac{1}{\zeta} \tag{58}$$

ζ is an intermediate variable to ensure the consistency of the relations. A particular choice of ζ can be $\zeta = r^{N/M}$ which is consistent with the text of Lemaitre & Chabouche (1990). In view of the above identity on $\dot{\gamma}_{vp}$ and $\dot{r}$, we can also write:

$$\dot{\boldsymbol{\varepsilon}}^{vp} = \dot{\gamma}_{vp}\left(\frac{3}{2}\frac{\boldsymbol{s}}{\sigma_{eq}}\right) = \dot{r}\left(\frac{3}{2}\frac{\boldsymbol{s}}{\sigma_{eq}}\right) \tag{59}$$

To define completely the model, we still need a (hardening) relation between R et r. This can either be defined explicitly $R = R(r)$ or by specifying a specific Helmholtz free energy ψ and then uses $R = \frac{\partial \psi}{\partial r}$.

13. Case of biphasic porous media

13.1 Fundamental hypotheses and definitions

In a macroscopic description, a biphasic medium is considered as the superposition of 2 continua. At a given time t and at a given position $\boldsymbol{x}$, 2 particles, one representing the solid and the other, the fluid, occupy simultaneously the same spatial region $d\Omega_t$ around the geometric point $\boldsymbol{x}$. In order to access separately the mass of each phase, we define the Eulerian porosity n (resp. the Lagrangian porosity ϕ) so that $n\, d\Omega_t$ (resp. $\phi\, d\Omega_0$) represents the current volume of fluid inside $d\Omega_t$. We have to deal with the macroscopic strain and porosity variations of the solid skeleton. Following Coussy (2004), we split the strain and porosity variation into a elastic and a plastic part:

$$\boldsymbol{\varepsilon} = \boldsymbol{\varepsilon}^e + \boldsymbol{\varepsilon}^p \; ; \; \dot{\phi} = \dot{\phi}_e + \dot{\phi}_p \quad ; \quad \Delta\phi = \phi - \phi_0 = \phi_e + \phi_p \tag{60}$$

We denote by ϵ and ϵ_s the volumetric component of the skeleton strain and that of the solid matrix (i.e. $\epsilon = tr(\boldsymbol{\varepsilon})$, etc.), which admit the same decomposition:

$$\epsilon = \epsilon^e + \epsilon^p \; ; \; \epsilon_s = \epsilon_s^e + \epsilon_s^p \tag{61}$$

The global volume change comes from those of the solid matrix and of the porous space. It can be proved that:

$$\epsilon = (1-\phi_0)\epsilon_s + \phi - \phi_0 \quad ; \; \epsilon^e = (1-\phi_0)\epsilon_s^e + \phi_e \; ; \quad \epsilon^p = (1-\phi_0)\epsilon_s^p + \phi_p \tag{62}$$

Extending equation (5) to include the contributions of the fluid, we write:

$$\frac{d^s}{dt}\int_{\Omega_t}(1-n)\rho_s s_s \, d\Omega_t + \frac{d^f}{dt}\int_{\Omega_t} n\rho_f s_f \, d\Omega_t \geq \int_{\Omega_t}\frac{r_q}{T}\, d\Omega_t - \int_{\partial\Omega_t}\frac{\boldsymbol{q}\cdot\boldsymbol{n}}{T}\, dS \tag{63}$$

where $\frac{d^s}{dt}(\cdot), \frac{d^f}{dt}(\cdot)$ express the kinematics of the solid skeleton and fluid phases respectively while ρ_s, s_s, ρ_f, s_f denote the respective density and entropy of the solid and fluid phases. The Clausius-Duhem inequality corresponding to deformable porous thus admits the following:

$$\Phi = \Phi_M + \Phi_F + \Phi_T \geq 0$$

where Φ_M, Φ_T are as before the intrinsic mechanical and thermal dissipations while Φ_F is the fluid dissipation. Going through the same procedure as in the case of monophasic media, but considering the contributions of both the solid and fluid phases, each with an independent kinematic field, the Clausius-Duhem inequality can be derived:

$$\Phi_{\mathrm{M}} = \boldsymbol{\sigma} : \dot{\boldsymbol{\varepsilon}} + p\dot{\phi} - \dot{\Psi}_{\mathrm{s}} \geq 0 \tag{64}$$

$$\Phi_F = \left(-\boldsymbol{grad}p + \rho_f\left(\boldsymbol{f} \quad - \boldsymbol{\gamma}^f\right)\right) \cdot \mathcal{V} \geq 0 \tag{65}$$

where $\rho_f\left(\boldsymbol{f} \quad - \boldsymbol{\gamma}^f\right)$ represents the body and inertia forces of the fluid; $\mathcal{V} = n\left(\boldsymbol{V}^f - \boldsymbol{V}^s\right)$ is the filtration vector and $\left(\boldsymbol{V}^f - \boldsymbol{V}^s\right)$ is the velocity of the fluid phase relative to the solid phase. Introduce the Gibb's free energy $G_s = \Psi_s - p(\phi - \phi_0) = \Psi_s - p(\phi_e + \phi_p)$ leads to:

$$\Phi_{\mathrm{M}} = \boldsymbol{\sigma} : \dot{\boldsymbol{\varepsilon}} - (\phi_e + \phi_p)\dot{p} - \dot{\mathrm{G}}_{\mathrm{s}} \geq 0 \tag{66}$$

Restricting to the case of reversible behaviour where the plastic components and the intrinsic dissipation Φ_{M} vanish, so that the above inequality becomes an equality, we deduce that $G_s = G_s(\varepsilon^e, p)$, and get the state equations:

$$\boldsymbol{\sigma} = \frac{\partial G_s}{\partial \varepsilon^e} \; ; \; \phi_e = -\frac{\partial G_s}{\partial p} \tag{67}$$

Differentiating the above leads to the following constitutive equations:

$$d\sigma_{ij} = C_{ijmn} d\varepsilon_{mn}^e - b_{ij} dp \; ; \; d\phi_e = b_{ij} d\varepsilon_{ij}^e + \frac{1}{N} dp \tag{68}$$

with:

$$C_{ijmn} = \frac{\partial^2 G_s}{\partial \varepsilon_{ij}^e \partial \varepsilon_{mn}^e} \; ; \; b_{ij} = -\frac{\partial^2 G_s}{\partial \varepsilon_{ij}^e \partial p} \; ; \; \frac{1}{N} = -\frac{\partial^2 G_s}{\partial p^2} \tag{69}$$

For isotropic behaviour, we have:

$$d\sigma_{ij} = \left(K - \tfrac{2}{3}G\right) d\varepsilon_{kk}^e \delta_{ij} + 2G d\varepsilon_{ij}^e - b\, dp\, \delta_{ij} \; ; \; d\phi_e = b\, d\epsilon^e + \frac{1}{N} dp \tag{70}$$

The first of the above equations can be rewritten to introduce an *elastic effective stress* σ_{ij}' which determines entirely the strain increments under elastic behaviour:

$$d\sigma_{ij}' = \left(K - \tfrac{2}{3}G\right) d\varepsilon_{kk}^e \delta_{ij} + 2G d\varepsilon_{ij}^e \; ; \; \sigma_{ij}' = \sigma_{ij} + b\, p\, \delta_{ij} \tag{71}$$

Recalling the following relation resulting from fluid mass conservation and the definition of fluid bulk modulus K_f:

$$\mathrm{d}\phi = \frac{dm_f}{\rho_f} - \phi \frac{dp}{K_f} \tag{72}$$

Recalling the definition of fluid volume content (neglecting 2nd order terms) $dv_f = \frac{dm_f}{\rho_f}$ and combining with the 2nd state equation, we obtain:

$$dv_f = d\phi_p + b\, d\epsilon^e + \frac{1}{M} dp \; ; \; \frac{1}{M} = \frac{1}{N} + \frac{\phi}{K_f} \tag{73}$$

To introduce a simple non linear skeleton behaviour, we restart with $\Phi_{\mathrm{M}} = \boldsymbol{\sigma}:\dot{\boldsymbol{\varepsilon}} + p\dot{\phi} - \dot{\Psi}_{\mathrm{s}} \geq 0$, and postulates that:

$$\Psi_s(\boldsymbol{\varepsilon}^e, \phi_e, \boldsymbol{V_k}) = W_s(\boldsymbol{\varepsilon}^e, \phi_e) + U(\boldsymbol{V_k}) \tag{74}$$

Where $U(\boldsymbol{V}_k)$ represents the trapped energy due to hardening, depending only on the internal state parameters $\boldsymbol{V}_k$. Substituting this into the Clausius-Duhem inequality and simplifying leads to:

$$\Phi_{\mathrm{M}} = \boldsymbol{\sigma}:\dot{\boldsymbol{\varepsilon}}^p + p\dot{\phi}_p + \boldsymbol{A_k}\cdot\dot{\boldsymbol{V}}_{\boldsymbol{k}} \geq 0 \tag{75}$$

with:

$$\boldsymbol{\sigma} = \frac{\partial \Psi_s}{\partial \boldsymbol{\varepsilon}^e} = \frac{\partial \mathrm{W}_s}{\partial \boldsymbol{\varepsilon}^e} \;;\; p = \frac{\partial \Psi_s}{\partial \phi_e} = \frac{\partial \mathrm{W}_s}{\partial \phi_e} \;;\; \boldsymbol{A_k} = -\frac{\partial \Psi_s}{\partial \boldsymbol{V_k}} = -\frac{\partial U}{\partial \boldsymbol{V_k}} \tag{76}$$

The above inequality can also be rewritten as:

$$\delta\Phi_{\mathrm{M}} = \delta W^p - dU \geq 0 \;;\; \delta W^p = \boldsymbol{\sigma}:d\boldsymbol{\varepsilon}^p + pd\phi_p \;;\; dU = \frac{\partial U}{\partial \boldsymbol{V_k}} d\boldsymbol{V_k} \tag{77}$$

Hence dU represents that part of the plastic work which is not dissipated into heat. Returning to (65), it is observed that the non-negativity of the dissipation Φ_F leads to Darcy's law as the constitutive equation of flow, which is defined for the isotropic case as:

$$\mathrm{n}\left(\boldsymbol{V}^f - \boldsymbol{V}^s\right) = \lambda_{\mathrm{h}}\left(-grad\boldsymbol{p} + \rho_f\left(\boldsymbol{f} - \boldsymbol{\gamma}^f\right)\right) \tag{78}$$

where λ_h is the hydraulic conductivity or coefficient of permeability of the medium. It is interesting to note that the thermodynamic approach confirms Darcy's law governs fluid flow relative to the solid matrix, and not with respect to a stationary observer.

13.2 Poroplastic behaviour

As for monophasic media, the dissipation potential is not differentiable in plasticity. To satisfy the non-negativity of the intrinsic dissipation, we postulate an elastic domain defined by a convex function f:

$$F(\boldsymbol{\sigma}, p, \boldsymbol{A_k}) \leq 0 \tag{79}$$

The domain contains the origin, in other words:

$$F(0,0,0) < 0 \tag{80}$$

Introducing the classic standard material behavioural law:

$$d\boldsymbol{\varepsilon}^p = d\lambda\frac{\partial F}{\partial \boldsymbol{\sigma}} \;;\; d\phi_p = d\lambda\frac{\partial F}{\partial p} \;;\; d\boldsymbol{V_k} = d\lambda\frac{\partial F}{\partial \boldsymbol{A_k}} \;;\; d\lambda \geq 0 \;;\; F \leq 0 \tag{81}$$

we have:

$$\Phi_{\mathrm{M}} = d\lambda\left[\boldsymbol{\sigma}:\frac{\partial F}{\partial \boldsymbol{\sigma}} + p\frac{\partial F}{\partial p} + \boldsymbol{A}_k\frac{\partial F}{\partial \boldsymbol{A}_k}\right] \geq 0 \tag{82}$$

The quantity between square brackets represents the scalar product between the position vector $(\boldsymbol{\sigma}, p, \boldsymbol{A}_k)$ and the outward normal vector $\left(\frac{\partial F}{\partial \boldsymbol{\sigma}}, \frac{\partial F}{\partial p}, \frac{\partial F}{\partial \boldsymbol{A}_k}\right)$ which is perpendicular to the boundary of the elastic domain $f = 0$. Its positivity comes from the geometric convexity of the domain $F \leq 0$ and the fact that the domain contains the origin. In the above formulation, the yield criterion is supposed to depend both on the total stress and the fluid pressure. This can be simplified if the plastic porosity change is related to the plastic volumetric strain:

$$\dot{\phi}_p = b'\dot{\epsilon}^p = b'\dot{\boldsymbol{\varepsilon}}^p\text{: I} \tag{83}$$

so that:

$$\Phi_\text{M} = \boldsymbol{\sigma}''\text{: }\dot{\boldsymbol{\varepsilon}}^p - \dot{\Psi}_s \geq 0 \;\; ; \;\; \boldsymbol{\sigma}'' = \boldsymbol{\sigma} + b'p\,\text{I} \tag{84}$$

Mechanical stress and fluid pressure then intervene in the yield function only via a *plastic effective stress* $\boldsymbol{\sigma}''$:

$$F(\boldsymbol{\sigma}'', \boldsymbol{A}_k) \leq 0 \tag{85}$$

with:

$$d\boldsymbol{\varepsilon}^p = d\lambda\frac{\partial F}{\partial \sigma''} \;\; ; \;\; d\boldsymbol{V}_k = d\lambda\frac{\partial F}{\partial \boldsymbol{A}_k} \;\; ; \;\; d\lambda \geq 0 \;\; ; \;\; F \leq 0 \tag{86}$$

However, there are two effective stresses $\boldsymbol{\sigma}'$ and $\boldsymbol{\sigma}''$, which is confusing. The situation will be optimum if we can assume either $b' = b$, hence $\boldsymbol{\sigma}' = \boldsymbol{\sigma}''$, or matrix incompressibility which implies $b' = b = 1$ and that $\boldsymbol{\sigma}' = \boldsymbol{\sigma}'' = \boldsymbol{\sigma} + p\mathbf{I}$. The last case is of particular importance and corresponds to the majority of cases in soils.The above flow rule is known as associative since the strain rate is normal to the yield surface, with the advantage that the non-negativity of the dissipation is always satisfied. Geomaterials exhibit complex volumetric behaviours and sometimes call for non associative flow rules:

$$d\boldsymbol{\varepsilon}^p = d\lambda\frac{\partial g}{\partial \sigma''} \;\; ; \;\; d\boldsymbol{V}_k = d\lambda\frac{\partial g}{\partial \boldsymbol{A}_k} \;\; ; \;\; d\lambda \geq 0 \;\; ; \;\; F \leq 0 \tag{87}$$

However, the non-negativity of the dissipation is not always satisfied in this last case.

13.3 Poroviscoplastic behaviour

Recall that we have to satisfy:

$$\Phi_\text{M} = \boldsymbol{\sigma}\text{: }\dot{\boldsymbol{\varepsilon}}^p + p\dot{\phi}_p + \boldsymbol{A}_k \cdot \dot{\boldsymbol{V}}_k \geq 0 \tag{88}$$

The dissipation potential is in this case differentiable so that we can write:

$$\varphi^* = \varphi^*(\boldsymbol{\sigma}, p, \boldsymbol{A}_k) \;\; ; \;\; \dot{\boldsymbol{\varepsilon}}^p = \frac{\partial \varphi^*}{\partial \sigma} ; \;\; \dot{\phi}_p = \frac{\partial \varphi^*}{\partial p} \;\; ; \;\; \dot{\boldsymbol{V}}_k = \frac{\partial \varphi^*}{\partial \boldsymbol{A}_k} \tag{89}$$

Hence:

$$\Phi_\text{M} = \boldsymbol{\sigma}\text{: }\frac{\partial \varphi^*}{\partial \sigma} + p\frac{\partial \varphi^*}{\partial p} + \boldsymbol{A}_k \cdot \frac{\partial \varphi^*}{\partial \boldsymbol{A}_k} \geq 0 \tag{90}$$

Similar to the case of plasticity, we can simplify by supposing $\dot{\phi}_p = b\dot{\epsilon}^p = b\dot{\boldsymbol{\varepsilon}}^p\text{: }\mathbf{I}$ and $\boldsymbol{\sigma}' = \boldsymbol{\sigma} + bp\,\mathbf{I}$. We then require the dissipative potential to satisfy:

$$\varphi^* = \varphi^*(\boldsymbol{\sigma}', \boldsymbol{A}_k) \;\; ; \;\; \dot{\boldsymbol{\varepsilon}}^p = \frac{\partial \varphi^*}{\partial \sigma'} \;\; ; \;\; \dot{\boldsymbol{V}}_k = \frac{\partial \varphi^*}{\partial \boldsymbol{A}_k} \tag{91}$$

For example, if we take:

$$\varphi^* = \frac{1}{2\eta}\langle F(\boldsymbol{\sigma}', \boldsymbol{A}_k)\rangle^2 \tag{92}$$

We get:

$$\dot{\boldsymbol{\varepsilon}}^{\boldsymbol{p}} = \frac{\partial\varphi^*}{\partial\sigma'} = \frac{1}{\eta}\langle F\rangle\frac{\partial F}{\partial\sigma'} \quad ; \quad \dot{\boldsymbol{V}}_{\boldsymbol{k}} = \frac{\partial\varphi^*}{\partial \boldsymbol{A}_k} = \frac{1}{\eta}\langle F\rangle\frac{\partial F}{\partial \boldsymbol{A}_k} \tag{93}$$

14. Applications

14.1 Example 1 – Hardening plasticity – EPS geofoam

In the following example we illustrate the first type of use of the second thermodynamic principle discussed in Section 6, namely, by verifying a constitutive model of EPS geofoam *a posteriori* for thermodynamic consistency. This model was developed by the authors (Wong and Leo, 2006) based on experimental results from a series of standard "drained" triaxial tests. It initially adopted the Mohr-Coulomb yield function used widely in soil mechanics but upon further testing with a true triaxial apparatus (Leo et al., 2008), a Drucker-Prager type yield function was subsequently preferred. This is written as:

$$F(\boldsymbol{\sigma}, a) = \sqrt{3\boldsymbol{J}_2} - bI_1 - a = 0 \tag{94}$$

i.e.

$$dF = \frac{\partial \boldsymbol{F}}{\partial \boldsymbol{\sigma}} \cdot \boldsymbol{\sigma} + \frac{\partial \boldsymbol{F}}{\partial \boldsymbol{a}} \cdot da = 0 \tag{95}$$

where $\boldsymbol{I}_1 = tr(\boldsymbol{\sigma})$ is the first stress invariant, $\boldsymbol{J}_2 = \frac{1}{2}\boldsymbol{s}:\boldsymbol{s}$ is the second stress invariant and b is a material constant. Here $a(r) = a_0 + \beta r$ is the hardening law accounting for the isotropic hardening effects; a_0, β are material constants and r is an internal variable chosen as the equivalent deviatoric plastic strain defined by:

$$r = \int_0^t \sqrt{\frac{1}{2}\dot{\boldsymbol{e}}^{\boldsymbol{p}}:\dot{\boldsymbol{e}}^{\boldsymbol{p}}}\, d\tau \tag{96}$$

Referring to the discussion in Section 11, we observe that equation (94) is a particular form of (27), $a(r)$ of $\boldsymbol{A}_k(\boldsymbol{V}_k)$, and (96) is the equivalent of (39). Geometrically, the surface of equation (94) corresponds to a conical surface, with the symmetry axis coinciding with the hydrostatic axis. The apex angle is governed entirely by the constant b, whereas a, together with b, determines the distance separating the cone tip from the origin. According to the laws of thermodynamics, an associative flow rule should have been adopted for the plastic strain (i.e. $\dot{\boldsymbol{\varepsilon}}^{p} = \dot{\lambda}\frac{\partial F}{\partial \sigma}$ in equation (28)) for this constitutive model, but we chose a non-associative flow rule instead where,

$$\dot{\boldsymbol{\varepsilon}}^{\boldsymbol{p}} = \dot{\lambda}\frac{\partial G}{\partial \sigma}; G(\boldsymbol{\sigma}) = \sqrt{3\boldsymbol{J}_2} - \boldsymbol{c}I_1 \tag{97}$$

c is a rheological parameter which depends on the initial stress. This is because experimental measurements suggest that the plastic volumetric strain is better represented by the plastic potential given in (97) rather than the yield function of (94). As discussed earlier, this means that the thermodynamics principle in terms of the non-negativity of the dissipation may possibly be violated in some evolutions since the normality rule (plastic strain increment being normal to the yield function) is not being followed. The associative flow rule, however, has been a problem with some geomaterials such as soils and rocks in that it tends to erroneously predict plastic volumetric strain. This is one instance where the insight provided by thermodynamics into post yielding volumetric behavior is seemingly at odds with experimental evidence. In these cases it is widely accepted that the plastic volumetric behavior would be better captured using a non-associative flow rule. These cases also demonstrate that while thermodynamics insights provide useful guidance to help engineers focus on important aspects of the constitutive relationships in continuum mechanics, it is necessary that these insights should ultimately be supported by experimental evidence.

14.2 Example 2 – Poroelasticity: closure of a spherical cavity

This example dealing with the closure of a deeply embedded cavity in poroelastic medium was previously studied by the authors (Wong et al. 2008). Here we illustrate the second type of use of the second thermodynamic principle discussed in Section 6, where the thermodynamics concepts from Section 13.1 are applied to formulate the constitutive relationships that lead, importantly, to the analytical solutions for the closure of a spherical cavity. The closure constitutes part of a life cycle of an underground mining cavity idealised by four stages. Initially, the ground is in a state of hydro-mechanical equilibrium. The cavity is then excavated and an internal support is provided to maintain its stability. Various techniques of support exist. For example, it can be evenly spaced steel bolts or a layer of shotcrete or a combination of them. For modelling purposes, this support can be assimilated to a layer of elastic material lining the cavity walls. At the end of its service life, the cavity is backfilled with a poro-elastic material before being abandoned. We were interested in the long term evolution of the hydro-mechanical fields in the surrounding medium and in the backfill after the its abandonment, when the support starts to deteriorate. This problem deals with a special case of the reversible behaviour where the intrinsic dissipation vanishes, namely $\Phi_M = 0$ (as opposed to the more general case of irreversible behaviour for materials with plasticity and/or viscosity), leading to the state equation (67) and the constitutive equations (70) for isotropic poroelastic material. Limiting ourselves to small strains, we define:

$$d\sigma_{ij} = \sigma_{ij} - \sigma_{ij}^0 \quad ; \quad d\varepsilon_{ij}^e = \varepsilon_{ij}^e - \varepsilon_{ij}^0 \quad ; \quad dp = p - p^0 \tag{98}$$

where $\sigma_{ij}^0, \varepsilon_{ij}^0, p^0$ denotes the initial stress, strain, pore pressure respectively. We make further assumptions that the solid grains of the medium are incompressible, and it thus holds that the skeletal volumetric change $d\epsilon^e = d\varepsilon_{kk}^e$ must be the same as the change in the porosity $d\phi^e$, that is:

$$d\epsilon^e = d\phi^e \tag{99}$$

By comparing (99) to the second equation of (70), it is evident that the values of Biot coefficients must be: $b = 1$ and $1/N = 0$. Taking initial strain $\varepsilon_{ij}^0 = 0$, equation (70) thus yields the following constitutive relationships for a linear isotropic poroelastic material:

$$\sigma_{ij} - \sigma_{ij}^0 = \left(K - \tfrac{2}{3}G\right)\epsilon^e\,\delta_{ij} + 2G\varepsilon_{ij}^e - (p - p^0)\,\delta_{ij} \quad ; \quad \phi - \phi_0 = \epsilon^e \tag{100}$$

Since we are clearly dealing with a poroelastic medium, the superscript 'e' denoting elastic strain shall be omitted in Example 2 without ambiguity, for the sake of brevity. For the fluid phase of the porous material, the constitutive equation follows from the thermodynamically consistent Darcy's law, equation (78). Here, after neglecting inertia effects but not body forces due to gravity g, the fluid mass flux, $\boldsymbol{w}_f = \rho_f n(\boldsymbol{V}^f - \boldsymbol{V}^s)$ is related to the thermodynamic forces as: $\boldsymbol{w}_f/\rho_f = \lambda_h(-\boldsymbol{grad}p + \rho_f\boldsymbol{g})$. At $t = 0$, the fluid is assumed to be in hydraulic equilibrium, implying that: $0 = \lambda_h(-\boldsymbol{grad}p_0 + \rho_f\boldsymbol{g})$. The difference between these two equations yields:

$$\boldsymbol{w}_f/\rho_f = -\lambda_h \boldsymbol{grad}(p - p_0) \tag{101}$$

As shown above, insights from thermodynamics principles have lead to constitutive equations (100) and (101). These equations thus allow us to develop a set of governing equations which is applicable to the cavity closure problem. These equations are then solved with respect to the initial and boundary conditions for a spherical cavity to obtain a set of analytic solutions, of which a detailed discussion is given in Wong et al. (2008).

14.3 Example 3 – Poroviscoelasticity: closure of long cylindrical tunnel

Example 3 illustrates the use of thermodynamics principles in formulating constitutive equations for a poro-viscoelastic medium. The ultimate purpose here is also to develop solutions for a long horizontally aligned tunnel with a circular cross-section embedded in a poro-viscoelastic massif. The setting of the problem is similar to Example 2 discussed above except that the spherical cavity is replaced by a long lined tunnel (Dufour et al. 2009). We start by restricting to small strain problems where the strain tensor of a viscoelastic material can be decomposed into an elastic part (denoted by superscript 'e') and a viscoelastic part (superscript 'ν'):

$$\varepsilon_{ij} = \varepsilon_{ij}^e + \varepsilon_{ij}^\nu \tag{102}$$

The strain and stress tensors are separated into isotropic and deviatoric parts as follow:

$$\varepsilon_{ij} = \frac{1}{3}\epsilon\delta_{ij} + e_{ij} \quad ; \quad \sigma_{ij} = \sigma\delta_{ij} + s_{ij} \tag{103}$$

where ϵ, e_{ij} are the mean and deviatoric strains defined previously; $\sigma = \sigma_{ij}/3$ is the mean stress and $s_{ij} = \sigma_{ij} - \sigma\delta_{ij}$ is the deviatoric stress tensor. It is noted that the decomposition into elastic and viscoelastic parts in (102) apply separately to ϵ, e_{ij} and the porosity as well such that:

$$\epsilon = \epsilon^e + \epsilon^\nu \quad ; \quad e_{ij} = e_{ij}^e + e_{ij}^\nu; \quad \phi - \phi_0 = \phi_e + \phi_\nu \tag{104}$$

Correspondence between volumetric strain and porosity change holds for each of the elastic and viscoelastic components:

$$\epsilon = \phi - \phi_0 \quad ; \quad \epsilon^e = \phi_e \quad ; \quad \epsilon^\nu = \phi_\nu \tag{105}$$

14.3.1 Poroviscoelastic constitutive equations

Following (74), we postulate the existence of trapped energy due to viscosity that depends on viscous strains only and write the free energy of the skeleton as:

$$\Psi_s(\epsilon^e, e_{ij}^e, \phi_e, \epsilon^v, e_{ij}^v) = W_s(\epsilon^e, e_{ij}^e, \phi_e) + U(\epsilon^v, e_{ij}^v) \tag{106}$$

where the following relationships are considered for the functions W_s, U:

$$W_s(\epsilon^e, e_{ij}^e, \phi_e) = \frac{1}{2}K_0\epsilon^{e2} + \mu_0 e_{ij}^{e\,2} + \frac{1}{2}N_0(\epsilon^e - \phi_e)^2 \tag{107}$$

$$U(\epsilon^v, e_{ij}^v) = \frac{1}{2}\xi\epsilon^{v2} + \chi e_{ij}^{v\,2} \tag{108}$$

Specialising to a linear isotropic porous material, after substituting (107) into (76) and taking into consideration the decomposition into the mean and deviatoric parts, and the initial stresses we get:

$$(\sigma - \sigma_0) + (p - p_0) = K_0(\epsilon - \epsilon^v) \tag{109}$$

$$s_{ij} - s_{ij}^0 = 2\mu_0(e_{ij} - e_{ij}^v) \tag{110}$$

$$p - p_0 = -N_0(\epsilon^e - \phi_e) \tag{111}$$

K_0, μ_0, N_0 are the initial or "short term" analogues of K, μ, N respectively. Further substitution of (106) – (111) into (64) yields:

$$\{(\sigma - \sigma_0) + (p - p_0) - \xi\epsilon^v\}\dot{\epsilon}^v + \{(s_{ij} - s_{ij}^0) - 2\chi e_{ij}^v\}\dot{e}_{ij}^v \geq 0 \tag{112}$$

In the next step, a convex dissipative potential $\mathcal{D}(\dot{\epsilon}^v, \dot{e}_{ij}^v)$ is introduced so that based on (112):

$$\frac{\partial\mathcal{D}}{\partial\dot{\epsilon}^v} = (\sigma - \sigma_0) + (p - p_0) - \xi\epsilon^v \quad ; \quad \frac{\partial\mathcal{D}}{\partial\dot{e}_{ij}^v} = (s_{ij} - s_{ij}^0) - 2\chi e_{ij}^v \tag{113}$$

which leads to:

$$\mathcal{D}(\dot{\epsilon}^v, \dot{e}_{ij}^v) = \frac{1}{2}\zeta\dot{\epsilon}^{v2} + \eta\dot{e}_{ij}^{v\,2} \tag{114}$$

where positivity of $\zeta \geq 0; \eta \geq 0$ ensures the convexity of $\mathcal{D}(\dot{\epsilon}^v, \dot{e}_{ij}^v)$. From the above developments, the constitutive equations relating stresses to strains for an isotropic poro-viscoelastic material may thus be defined by equations (109)-(111) as well as by the following equations.

$$(\sigma - \sigma_0) + (p - p_0) = \xi\epsilon^v + \zeta\dot{\epsilon}^v \tag{115}$$

$$s_{ij} - s_{ij}^0 = 2\chi e_{ij}^v + 2\eta\dot{e}_{ij}^v \tag{116}$$

where ξ, ζ, χ, η are rheological constants. Note that these equations have been formulated based on the thermodynamics approach while adopting the convex dissipative potential, $\mathcal{D}(\dot{\epsilon}^v, \dot{e}_{ij}^v)$, in equation (114). Before proceeding further, we will now introduce the Laplace transform, defined for a typical function $f(r,t)$ as follows:

$$\bar{f}(r,s) = L\{f(r,t)\} = \int_0^\infty f(r,t)e^{-st}\,dt; \bar{f}(r,s) = L^{-1}\{\bar{f}(r,t)\} = \frac{1}{2\pi i}\int_{\Gamma-i\infty}^{\Gamma+i\infty}\bar{f}(r,t)e^{st}\,ds \tag{117}$$

where s is the Laplace transform parameter and $i^2 = -1$. In the notations adopted here, the bar over the symbol denotes the transformed function represented by the symbol. The value

Γ is chosen such that all poles in the s-plane lie to the left of the vertical line Re(s) = Γ. Taking the Laplace transform of (109), (110), (113) and (116) and solving for the viscous volumetric and deviatoric strains give,

$$\overline{\epsilon^{v}} = \frac{K_0}{K_0+\xi+\zeta s}\bar{\epsilon} \quad ; \quad \bar{e}_{ij}^{v} = \frac{\mu_0}{\mu_0+\chi+\eta s}\bar{e}_{ij} \tag{118}$$

The constitutive equations (115), (116), (118) are then used to developed governing equations for the closure of a long cylindrical tunnel in poroviscoelastic massif. Laplace transform solutions have been developed and discussed in detail in Dufour et al. (2009) to which interested readers may refer.

15. References

Biot, M.A.: General theory of three-dimensional consolidation. Journal of Applied Physics, 12, pp155-164, 1941.

Clausius, Rudolf (1850). On the Motive Power of Heat, and on the Laws which can be deduced from it for the Theory of Heat. Poggendorff's *Annalen der Physik*, (Dover Reprint). ISBN 0-486-59065-8.

Coussy O. Poromechanics. John Wiley & Sons Ltd.; 2004.

Dufour N., Leo C. J., Deleruyelle F., Wong H. Hydromechanical responses of a decommissioned backfilled tunnel drilled into a poro-viscoelastic medium. Soils and Foundations 2009;49(4):495-507.

Lemaitre, J. and Chaboche, J. Mech of solid materials, Cambridge University Press, 1990

Leo, C.J., *Kumruzzaman, M.,* Wong, K, *Yin, J.H. Behaviour of EPS geofoam in true triaxial compression tests', Geotextiles and Geomembranes, 2008, 26(2), pp175-180.*

Wong, K. and Leo, C.J. A simple elastoplastic hardening constitutive model for EPS geofoam, *Geotextiles and Geomembranes*, 2006, 24, pp299-310.

Wong, H., Morvan, M.,Deleruyelle, F. and Leo, C.J. Analytical study of mine closure behaviour in a poro-elastic medium, Computers and Geotechnics, 2008, 35(5), pp645-654.

5

Heat – Mechanically Induced Structure Development in Undrawn Polyester Fibers

Valentin Velev[1], Anton Popov[2] and Bogdan Bogdanov[2]
[1]Konstantin Preslavsky University, 9712, Shumen,
[2]University "Prof. Dr. Assen Zlatarov", 8000, Burgas,
Bulgaria

1. Introduction

The performances of the non-isotropic polymer systems strongly depend on their super molecular structure (Wu et al., 2001; Shabana, 2004; Keum & Song, 2005; Ziabicki & Jarecki, 2007; Sulong et al., 2011).

The wide application and consequently higher production of flexible chain fiber forming polymers, in particular poly (ethylene terephthalate) (PET) is due to the possibility of the heat mechanical modification to obtain highly modular and high strength materials from them (Llana & Boyce, 1999; Bai et al., 2000; Dupaix & Boyce, 2005; Guzzato et al., 2009). PET is an essential engineering polymer with properties strongly depending of the degree of crystallinity and the perfection of crystal phase, too. The effects of some basic parameters of the heat mechanically treatment such as strain force extension rate and temperature on the structure development of PET have been studied using different methods as differential scanning calorimetry (DSC), wide angle X-ray scattering (WAXS) (Kong & Hay, 2003; Zhang et al., 2004; Karagiannidis et al., 2008; Raabe et al. 2004), dynamic mechanical analysis (Ma et al., 2003), laser irradiation (Wijayathunga et al., 2007) and other.

The optimal performance of the high-temperature orientation modification is a complicated and still not sufficiently well studied process. The simultaneous mechanical and thermal modification however is extremely complex phenomenon occurring on the basis of statistical probabilistic processes, as are also the possible results from it. In this sense the results from variations of heat mechanical modification are unpredictable not unique and often very different, contradictory and unexpected. Moreover for each specific object and purpose exist additional conditions, and therefore needed special study of orientation thermal treatment for the obtaining of best mechanical performance. If the samples simultaneous heat mechanical modification (SHMM) is carried out without accounting and control of a number of events, processes and parameters the results can easily prove contrary to the expectations. And to make the results from STMM easily predictable, susceptible to control and allowing obtaining of materials with improved predefined wanted properties it is necessary in depth study on the nature, mechanism and kinetics of the justifying processes and the relationship between them. Therefore the study of these processes is a permanent "ever green" interest in the polymer physics. One of the most interesting from this point of view objects are polyethylene terephthalate fibers. There are varieties of investigations of the affects of the thermal and mechanical treatments on the

relaxation and phase transitions in PET fibers. In some of them as-spun filaments are subjected to thermal treatment at constant temperatures without stress (Betchev, 1995; Bai et al., 2000) as well with application of tensile force (Zhang et al., 2004; Sharma et al., 1997).
Important is the answer of the question on what schemes and under what conditions should be conducted SHMM to maximize the orientation effect at the expense of minimal object destruction at high temperature uni-axial deformation. To obtain definite answer to a similar question is necessary a multifactorial planning and carrying out of massive diverse experiment. The preliminary suggestive for a range of the possible conditions of withdrawal experiments are impressive much.
For initial approbation of the behavior of the specific object to the complex SHMM we accepted the technologically real (and maximum possible) temperature interval from 20 ^0C to 200 ^0C and sufficient as a beginning, a range of orientation tensions from 0 *MPa* to 1.7 *MPa* with enough good resolution of 0.1 *MPa.* The experiment was carried out in combination of gravitational loading of the samples at a linear heating in line with the coefficient of fibers thermal conductivity average heating rate of 3.5 $^0C/min$.
The dependence of the relative deformation from the tensile load values showed an initial intensive growth of the gradient of its increase up to strain stress value of 0.7 *MPa*, probably because of intensive destruction of macromolecular segments in the studied samples. A similar information was emitted and from the other performed structural analyses. The results led us to include new elements into the idea of the experiment and in particular to eliminate the adverse action of destructive tensions above those causing bundle deformation 290 %. Results showed that above loading of 1.2 *MPa* the relative samples elongation falls below the above mentioned value of the bundle reletive elongation and is no need to limit it.
In the new version the thermal deformation experiment was carried out without limitation of the bundle extension at combination of the samples gravitational loading in the range from 0 *MPa* to 3.0 *MPa* with a good resolution of 0.12 *MPa* at a linear heating with the same heating rate (3.5 $^0C/min$) and again in the temperature range from 20 ^0C to 200 ^0C. The structural tests of the SHMM samples in this preliminary experiment showed the disadvantages of the wide temperature range. Therefore, were tested modifications of PET fibers at well defined temperatures of 80 ^0C, 85 ^0C, 90 ^0C and 95 ^0C in the temperature range just above the glass transition temperature of the objects defined in our other investigations of 74 ^0C. The samples were loaded gravitationally (with different orientation tensions with initial values of 40 *MPa*, 80 *MPa* and 120 *MPa*, varying during the deformation downloading) as well as with constant rate of loading 0.1 *m /min* up to various relative elongations of 20 %, 40 % and 60 %.

2. Experimental

2.1 Materials

PET undrawn multifilament yarns produced by melt spinning on the industrial spinning installation Furnet (France) have been selected as a precursor samples. The technological parameters and basic characteristics of the original filaments are shown in Table 1.
It can be seen from the Table 1 that within the group of the selected samples have both amorphous and partially crystalline filaments. The selected specimens are spun at different spinning speeds and thus with different preliminary orientation. So they are suitable for the achievement of the above-defined purpose of the present study.

Sample	V_L, *m/min*	d, *μm*	Δn	α, %
A	1100	44.0	0.006	0.8
B	1150	44.0	0.008	1.7
S1	2280	14.5	4.32	23.7
S2	2805	13.0	5.35	28.8
S3	4110	11.0	5.82	36.9

Table 1. Basic characteristics of the investigated PET fibers. 1. Sample; 2. V_L, m/min - spinning speed; 3. d,μm - diameter of the single fiber; 4. Δn - birefringence; 5. α, % - degree of the sample crystallinity.

2.2 Methods

2.2.1 Simultaneous heat-mechanically modification (SHMM)

Different versions of simultaneous thermal and mechanical treatments of the studied yarns were performed using devices constructed and produced in our laboratory.
The first version of SHMM includes linear samples heating from room temperature up to 200 ^{0}C accompanied by applied to the fiber bundle strain stress. The heating rate was 3.5 ^{0}C/min. The used gear consists of a vertically located cylindrical furnace, which moves around a rolled up PET bundle fixed by special holders and subjected to needed tensile stress. The temperature reaching of 200 ^{0}C was followed by a simultaneous termination of the tensile stress and the yarn remove from the oven at room temperature. Highly supercooled i.e. deep tempered and isothermally crystallized at temperatures close to the melting temperature thin films PET, used for forming of the investigated fibers are shown in Fig. 1a, b, c and d respectively.

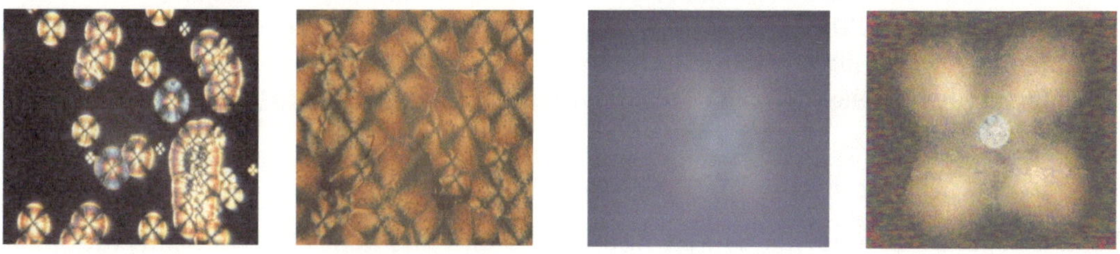

Fig. 1. a. Fig. 1. b. Fig. 1. c. Fig. 1.d.

Fig. 1. a, b - polarization microphotography; Fig.1. c, d - diffraction pictures.

In the second variant of SHMM the investigated filaments were subjected to tensile stresses with different values under certain constant temperatures. The simultaneous heat mechanical samples modification was carried out using an apparatus created in our laboratory. The device involves a movable cylindrical oven located on the horizontal rails and a setup for the sample deformation reading. The heat-mechanical treatment begins when the preheated oven was rapidly shifted around the studied PET bundle that was simultaneously stretched with the needed strain stress. The experiment involves annealing of an as-spun PET yarns at four different temperatures in a narrow temperature range from 80 ^{0}C to 95 ^{0}C closely above its glass transition temperature while they are subjected to a well-defined tensile stress.

In the next version of SHMM the studied PET filaments were subjected to extension at a constant speed and constant temperatures in the same temperature range from 80 ^0C to 95 ^0C. The structural characterizations of the studied fibers after the above described heat-mechanical treatments were realized using differential scanning calorimetry (DSC) and wide-angle X-ray scattering (WAXS) measurements.

2.2.2 Differential scanning calorimetry (DSC)

Part from the calorimetric studies was performed on a Mettler-Toledo heat-flux calorimeter DSC 820 with liquid nitrogen accessory. The furnace was purged with nitrogen at a flow rate of 80 *ml/min*. Temperature calibration was done using the onset melting temperatures of indium and zinc, and the energy calibration was based on the heat of fusion of indium. Fibers were cut in pieces of less than 1 *mm* and sealed in standard 40 μl aluminum pans.
Another part of the calorimetric analysis was carried out using a NETZSCH heat-flux calorimeter STA 449 F3 Jupiter (TG/DSC) in static air atmosphere. Temperature calibration was done using the onset melting temperatures of indium, tin, bismuth and zinc, and the energy calibration was based on the heat of fusion of the same metals. Fibers were cut in pieces of less than 1 *mm* and sealed in standard 85 μl platinum pans.

2.2.3 Wide-angle X-ray scattering (WAXS)

The fiber structure was studied by wide-angle X-ray scattering (WAXS), too using two different apparatus namely:

1. Diffractometer HZG 4 (Freiberger Präzisionsmechanik, Germany) and Ni-filtered Cu K_α radiation with wavelength λ = 1.5418 Å. Equatorial scattering was monitored in transmission mode. The fiber samples were prepared as a layer with 2 *mm* thickness and 10 *mm* width, and mounted on the sample holder of the diffractometer;
2. Diffractometer URD - 6 (under license of SIEMES) of the company "Freiberger Präzisionsmechanik" (Freiburg im Breisgau, Baden-Württemberg, Germany). Used is β-filtered with Ni-filter $Cu\ K_\alpha$ radiation with a wavelength λ = 1.5418 Å.

3. Results and discuss

3.1 Investigation of amorphous PET fibers simultaneous heat - mechanically modified at linear heating and constant strain stress values

The study of the relationships between the SHMM modes and subsequent structural development in the PET filaments includes different versions of experiments.
In the first one amorphous fibers marked as sample A (Table 1) were subjected to SHMM at conditions as follows: Heating with linear increasing of the temperature in a wide range from 20 ^0C to 200 ^0C with heating rate of 3.5 $^0C/min$ under constant strain stress from 0 *MPa* to 1.7 *MPa* (increasing step of 0.1 *MPa*). It should be noted the additional experimantal conditions for some of the samples. The extension of the yarns loaded with tensile stress from 0.7 *MPa* to 1.2 *MPa* was limited up to 290 %. Moreover after the reaching of the limited bundle length the sample continues to be heated up to 200 ^0C.
The length changes of the investigated yarns were registered during their combined heat mechanical treatment. As expected the filaments retain initial dimensions in the temperature range from room temperature up to 75 ^0C. In this temperature interval samples remain in glassy state and the structural changes are negligible.

The changes of the bundle dimensions depend on the applied strain stress level considerably and strongly at temperatures between 80 ^0C and 130 ^0C. The observed dependence can be explained with the sample transition from glassy to rubbery state. The deformation behaviour demonstrated by the samples at a level of applied tensile stress up to 0.7 *MPa* is expectable. Experiments showed a decrease of the final bundle length at small stress values. The filaments shrinkage can be logically explained with the process of frozen internal stresses relaxation in the samples at the temperature range of the transition from glassy to rubbery state.

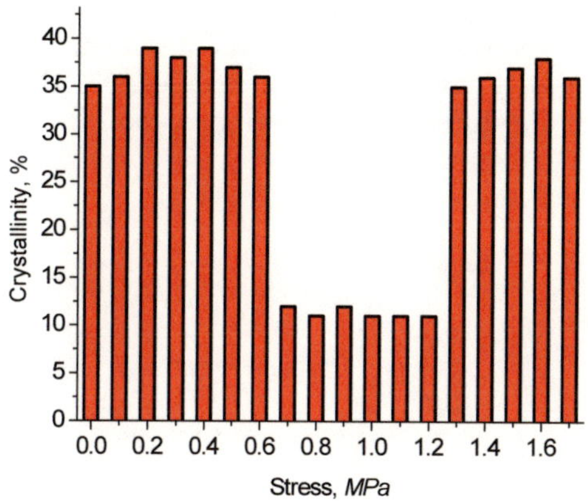

Fig. 2. Degree of crystallinity of PET fibers (sample A) depending on the SHMM conditions.

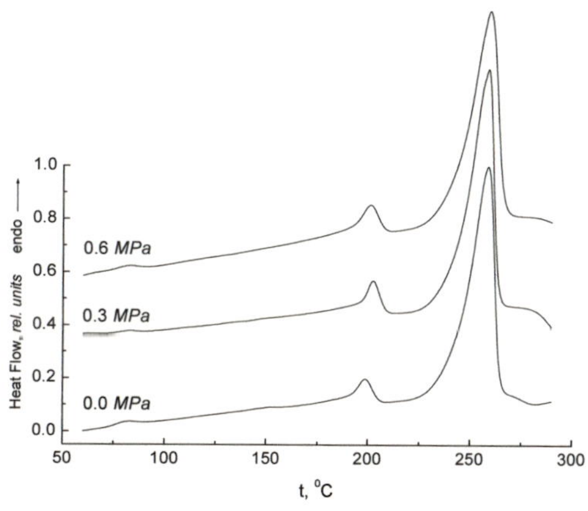

Fig. 3. Representative DSC curves of partially crystalline PET fibers (the curves are shifted vertically for clarity).

Unexpected and quite interesting was the deformation behavior of the bundles subjected to stresses in the range from 0.7 *MPa* to 1.2 *MPa*. As it was mentioned above the extension of the samples tested with tensile stress from 0.7 *MPa* to 1.2 *MPa* was limited up to 290 %. The observed deformation behavior strongly corresponds to the so-called fluid-like deformation. The bundle length was kept constant when the extension reached 290 %.

It is important to underline that such an intensive fluid-like deformation process was not observed for the samples subjected to strain stresses above 1.2 *MPa*. The received experimental data showed a decrease of the elongation with the stress values increasing.

Structural changes in the PET fibers as a consequence of the simultaneous thermal and mechanical treatments were studied using DSC, wide angle X-ray scattering (WAXS) and density measurements. The changes of the samples degree of crystallinity estimated on the basis of the DSC data depending of the strain stress values are presented in Figure 2. As it can be seen from Figure 2, some of the studied specimens are semi-crystalline while others are practically amorphous. The comparison with the SHMM conditions shows that the samples subjected to tensile stresses in the intervals from 0 *MPa* to 0.6 *MPa* and from 1.3 *MPa* to 1.7 *MPa* posses semi-crystalline structure. At the same time, the filaments with limited ability for extension treated in the stress interval from 0.7 *MPa* to 1.2 *MPa* are practically amorphous. Density measurements and WAXS diagrams proved the same crystallization properties of the studied PET specimens, too.

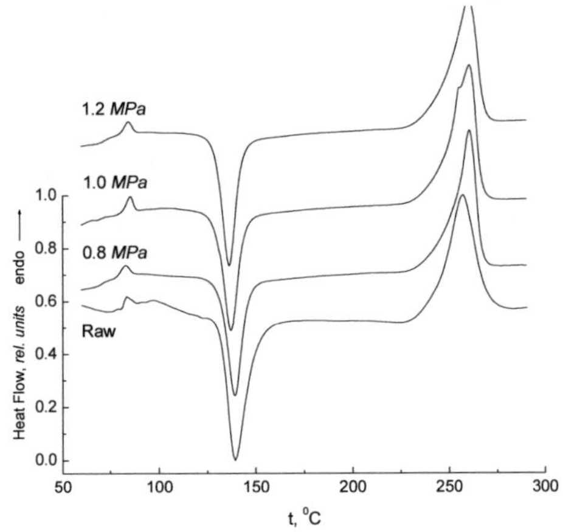

Fig. 4. Representative DSC curves of untreated and amorphous PET fibers (the curves are shifted vertically for clarity).

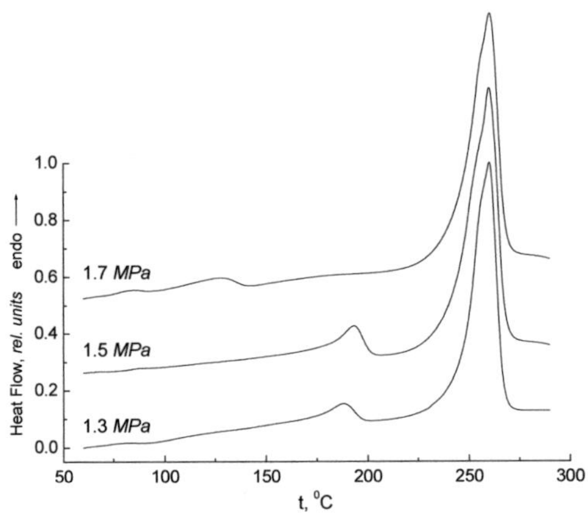

Fig. 5. Representative DSC curves of partially crystalline PET fibers (the curves are shifted vertically for clarity).

Representative DSC thermograms of partially crystalline and amorphous PET fibers subjected on heat mechanically treatments under the above decriebed conditions are present on Fig. 3, 5 and 4 respectively. As expected the DSC curve of the raw amorphous sample show pronounced cold crystallization and melting peaks. Unlike the untreated fibers, DSC thermograms in Fig. 3 and Fig. 5 show only preliminary melting and melting endotherms without cold crystallization peaks. Moreover the peak temperature of the premelting and melting endotherms in Fig. 3 smoothly shifts to higher temperatures with stress increasing. Multiple melting peaks in PET pellets (Kong & Hay, 2003) and filaments are observed and studied in earlier investigations. Similar to the raw PET filaments the DSC curves presented on Fig. 5, of the bundles subjected to SHMM at limited extension show glass - rubber transition, cold crystallization and melting peaks. Also it can be seen from Fig. 5 that the tensile stress increasing leads to fluently displacement of the cold crystallization peak to lower temperatures and to sliding to higher temperatures of the melting peak.

In conclusion it can be said that the heating with linear temperature rise, accompanied by application of external strain stresses strongly influences the nature of structural rearrangements in the investigated uncrystallized PET filaments. The observed fibers net deformation at tensile stress values less than 0.7 *MPa* and more than 1.2 *MPa* can be explained with a faster crystallization of the amorphous PET bundle from rubbery state, as a consequence of the influence of the applied tensile stress. The fluid-like deformation process predominates when the applied stresses are from 0.7 *MPa* to 1.2 *MPa*. It was found that after heating up to 200 ^0C amorphous PET filaments could preserve the amorphous state when the applied external strain stresses are in the same range.

At the same time questions having fundamental and practical aspects remain without clear answer and namely: What is the role of the restrictions and mechanical stress in obtaining of such qualitative different results? What would be the bundle deformation behaviour if there were no restrictions? What is the influence of the regime of heat treatment?

With purpose to clarify the role of the applied strain stress on the fibers structure development it was interesting to realize the above-described experiment without the mentioned limitations.

№	σ, *MPa*	№	σ, *MPa*
1	0.00	14	1.56
2	0.12	15	1.68
3	0.24	16	1.80
4	0.36	17	1.92
5	0.48	18	2.04
6	0.60	19	2.16
7	0.72	20	2.28
8	0.84	21	2.40
9	0.96	22	2.52
10	1.08	23	2.64
11	1.20	24	2.76
12	1.32	25	2.88
13	1.44	26	3.00

Table 2. Values of the applied strain stress during the SHMM.

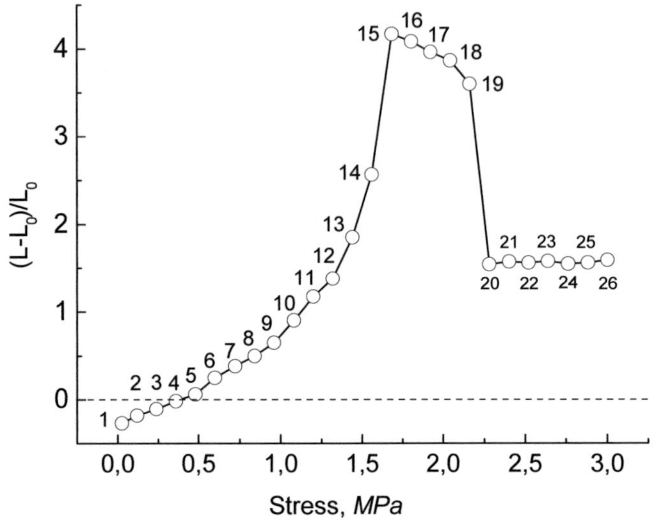

Fig. 6. Relative change of the bundle length (sample B) depending on the tensile stress values (here L_0 and L are the initial and final fibers length respectively).

In the next thermal deformation experiment amorphous PET fibers named sample B (Table 1), were linearly heated with rate of 3.5 $^0C/min$ from 20 ^0C to 200 ^0C. During the filaments heating they were subjected to constant tensile stress in a wider range from 0 *MPa* to 3.0 *MPa* (increasing step of 0.12 *MPa*, Table 2) without restrictions of the bundle deformation. The bundle length obtained after the heat mechanical treatment as a function of the applied strain stress is presented in Figure 6, where the dashed line marks the initial sample length.

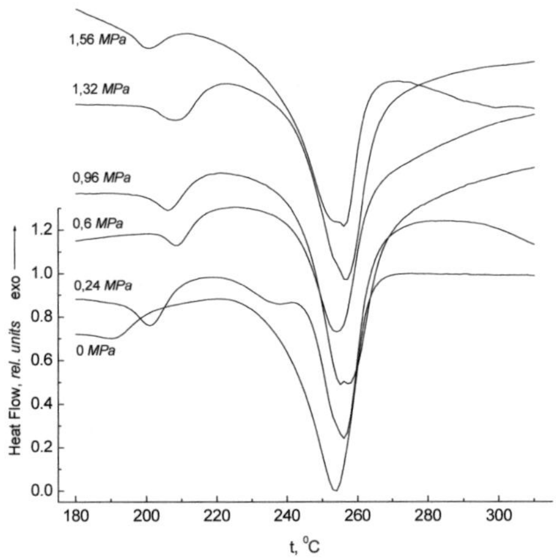

Fig. 7. Representative DSC curves of melting peaks of SHMM PET fibers from the first group (the curves are shifted vertically for clarity).

The deformation behaviour demonstrated by the samples at a level of applied stress up to 1.68 *MPa* is expectable. Experiments showed entropy shrinkage of the first four samples at small stress values up to the level of 0.36 *MPa*. The filaments shrinkage is a consequence of the frozen internal stresses relaxation at the temperature range of the sample glass transition. It could be supposed that the applied (external) stresses in our experiments up to value of 0.36 *MPa* do not compensate the emerging shrinkage forces.

Significant sample extension is stimulated by the stress increasing from 0.36 *MPa* up to 1.68 *MPa*. As it can be seen from the results presented in Figure 6 only increment of the final bundle length can be observed in this case. Obviously such of dependence can be detected when the applied strain stress is higher than the potential entropy shrinkage forces in an amorphous uniaxially oriented sample within the temperature range of glass transition. The received experimental data strongly corresponds to the so-called fluid-like deformation. At stress value of 1.68 *MPa* is reached more than fivefold bundle monotonic download. This is the maximum achievable prolongation by used method and conditions of SHMM.

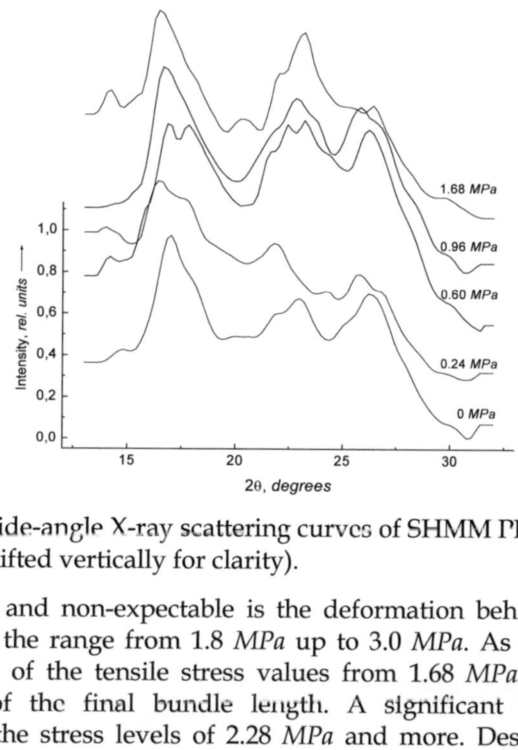

Fig. 8. Representative wide-angle X-ray scattering curves of SHMM PET fibers from the first group (the curves are shifted vertically for clarity).

Much more interesting and non-expectable is the deformation behaviour of the samples subjected to stresses in the range from 1.8 *MPa* up to 3.0 *MPa*. As it is illustrated on the Figure 6 the increasing of the tensile stress values from 1.68 *MPa* to 2.16 *MPa* leads to gradually decrement of the final bundle length. A significant reduction of the net deformation occurs at the stress levels of 2.28 *MPa* and more. Despite of the rise of the applied stress values the samples extension decreases considerably. Moreover the change of the tensile stress does not affect the deformation behaviour of the last seven yarns. Their ultimate length is more than twice less than the maximum achieved nder stress value of 1.68 *MPa*. Depending on the deformation behavior the investigated samples can be conditionally divided into three groups as follws. First one includes the bundles with numbers from one to fifteen. In the second one are the yarns from sixteen to twenty, and the third group includes the last six specimens which despite of the stress values increasing are extended less. Structural rearrangements occurred in the PET fibers as a result of the SHMM were studied using DSC and WAXS. It should be underlined that in contrast to the previous experiment the performed structural analysis show that all of the heat mechanically modified PET filaments are partially crystalline. Representative DSC curves of melting peaks of the above defined three groups of samples are present in Figures 7, 9 and 11. As it is visible from Figure 7 during the SHMM are formed three types of structures with three different types of perfection and stability. Depending on the melting temperature can be distinguished entities with a higher level of order forming an easy fusible mesophase, middle crystalline phase with lower perfection and main crystalline phase. Samples heating whether without load, forms easy fusible structure, which melts at about 190 ^0C. Just small increasing of the stress values leads to the structure improvement and stabilization and to the moving of the mesophase melting temperature to higher temperatures up to around 210 ^0C - 215 ^0C. The intermediate crystalline phase with lower perfection is observed as splitting of the main melting peak which visible migrate to the higher temperatures. With the tensile stress increasing the first melting peak as well as the main melting peaks are deformed with a tendency to split. The melting peaks also fluctuate around an average melting temperature significantly higher in comparison with the obtained without load. The observed shifting of the endo effects at higher temperatures possibly is a consequence of more organized structure formation due to the applied orientated pulling load. Only the sample from this group loaded with stress of 1.68 *MPa* show a slightly different melting behavior.

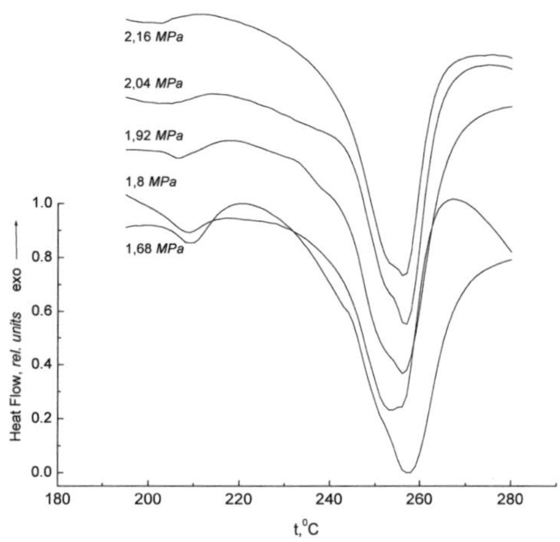

Fig. 9. Representative DSC curves of melting peaks of SHMM PET fibers from the second group (the curves are shifted vertically for clarity).

Representative wide-angle X-ray scattering curves of SHMM PET fibers from the above defined three groups of samples are present in Figures 8, 10 and 12. The diffraction curves are presented to illustrate the change in the fibers degree of crystallinity and orientation with the samples load increasing as well as their compliance with the DSC curves of the same objects for comparison of the structural information from the both methods wich are respectively geometric and energetically-structural sensitive. The first group of samples is characterized by a monotonic, although nonlinear elongation increase with the strain stress increasing up to the specimen with number 15 (Fig. 6). As is evident from Fig. 8, with the load increasing within this group the intensity distribution in the diffraction pattern shows noticeable changes with the stress increase, which is evidence for the significant structural reorganization without strict consistent trend observed in a specific type of amendment.

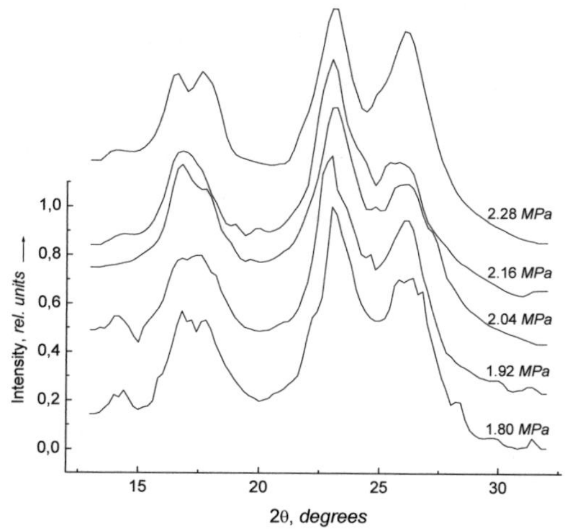

Fig. 10. Representative wide-angle X-ray scattering curves of SHMM PET fibers from the second group (the curves are shifted vertically for clarity).

Representative DSC curves of melting peaks from the second group of samples are given in Figure 9, where is observed a tendency on clearing of any thermo-effects with increasing of the strain stress values. That could be explained reasonably from viewpoint of the model for the homogenization of the structure at medium load values and respectively maximum deformation (Fig. 6). The samples deformation behavior corresponds to heat mechanicaly modified fibers stretched up to position in which the carrying are the most widespread fractions in the amorphous regions according to the normal Gaussian distribution. It is logically that the orientation processes at this group to be most effective with predominance of the orientation processes over destructive.

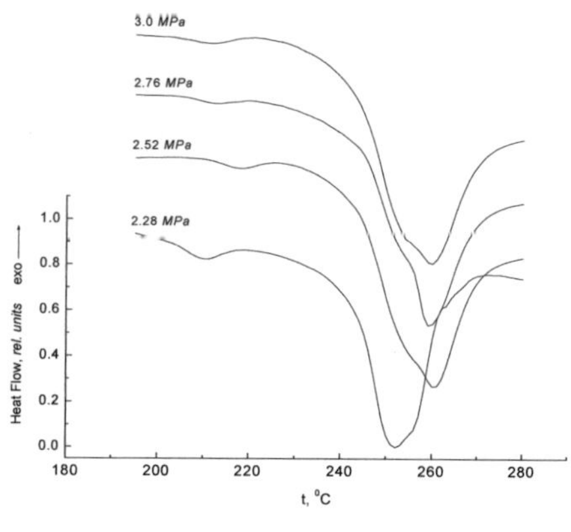

Fig. 11. Representative DSC curves of melting peaks of SHMM PET fibers from third group (the curves are shifted vertically for clarity).

In the second group of WAXS curves (Fig. 10) occurred more stable trend of increase in the intensity of the difraction radiation from improved crystalline and oriented regions in the samples. The DSC curves from the same group (Fig. 9) also showed a stable trend of deviation of the melting process to higher temperatures which confirms the suggestion for improvement of the crystalline phase. The same is valid and for the oriented amorphous regions. As from the diffraction and DSC curves, as well as from the stress - deformation dependence is confirmed the assumption for additional objects orientation allowing improvement of the crystalline phase and the supporting fraction in amorphous sections, which leads up to decrease in the total relative fibers deformation.

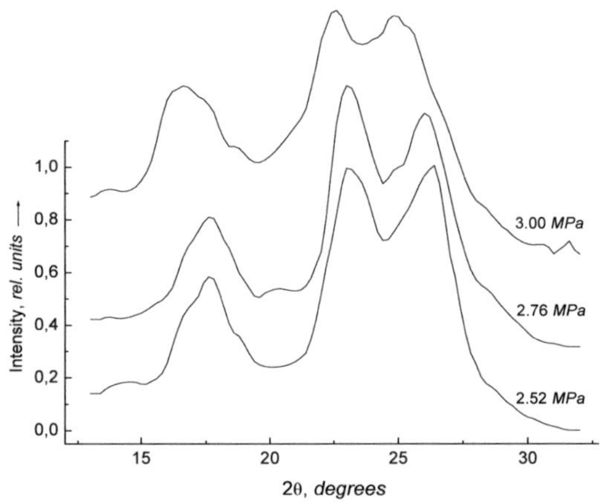

Fig. 12. Representative wide-angle X-ray scattering curves of SHMM PET fibers from third group (the curves are shifted vertically for clarity).

The DSC thermograms of the samples from the third group are present in Figure 11. It can be seen that with the tensile stress increasing endo peaks shift to lower temperatures. And it is valid for both before melting and the main melting processes. The most likely reason for the observed effect is that this third group of samples was withdrawn most suboptimal, with a predominance of destructive processes over the orientation. As a result, the obtained structure is mechanically and thermodynamically unstable, with the lowest density and perfection and therefore melts most easily at lower temperatures. As in Figure 9 as well as in Figure 11 are seen beginnings of split of the main melting peaks. The reasons for the splitting of the melting peaks may be different. In this case, at this type of heat mechanically fibers modification, the splitting occurs most probably due to structural reorganization during the melting process.

With the strain stress increasing at the samples with numbers from 20 up to 26 in which the relative fibers deformation almost does not change (Fig. 6) the intensity of the diffraction reflections significantly increase (Fig. 12). Probably for the account of low elastic deformation is realized a significant improvement of the crystalline phase on the surface of the lamellae or in the newly oriented regions. The exception occurs only in the sample subjected of the biggest tension stress of 3.0 *MPa*, where the intensity of the diffraction pattern falls strongly. Perhaps the increased destruction of separate fractions of macromolecular segments partially distorts the degree of the orderliness in the polymer system. At the same time the indestructable part of additional downloaded segments further improves their arrangement so that is realized more detailed infrastructure of distribution of the intensity of the powder diffraction pattern. The thermal curves (Fig. 11) show logically enhancement of the high temperature melting fractions with the tension stress increasing, as well their visible split.

On the basis on the carried out structural investigations of amorphous PET fibers simultaneous heat - mechanically modified at isothermal conditions and constant strain stress values it can be make the following conclusions: The mechanical strain force applied simultaneously with the linear heating of the studied PET yarns affects significantly the deformation behaviour and samples crystallization kinetics. Moreover in contrast to the results obtained in the first experiment, all of the so treated specimens are partially crystalline. The role of the tensile stress in the adjustment of the interacting processes of the fluid like deformation and stress-induced crystallization clearly reveals in the ultimate samples deformation. At stress values from 1.56 *MPa* to 2.16 *MPa* predominates the fluid like fibers extension, while the further stress increasing leads to the earlier crystallization start and thereby to decrease of the final fibers length.

3.2 Investigation of amorphous PET fibers simultaneous heat - mechanically modified at isothermal conditions and constant strain stress values

The glass transition temperature of the amorphous sample B (Table 1) was determined of 74 ^0C in our previous work. It was very interesting to follow the influence of SHMM on the structural development of studied samples in the transition temperature range between the glassy and rubbery fibers state.

With purpose to clarify the role of the temperature and tension stress values on the structure development in amorphous PET yarns, were carried out thermal deformation experiments at constant temperatures in a narrow temperature range from 80 ^0C to 95 ^0C. The experiment involves annealing of an as-spun PET bundles at temperatures of 80 ^0C, 85 ^0C, 90 ^0C and 95 ^0C accompanied by precisely defined tensile stresses from 0 *MPa* to 30 *MPa* with increment step of 3 *MPa*. The samples were loaded during two minutes with tensile stress after ten minutes annealing.

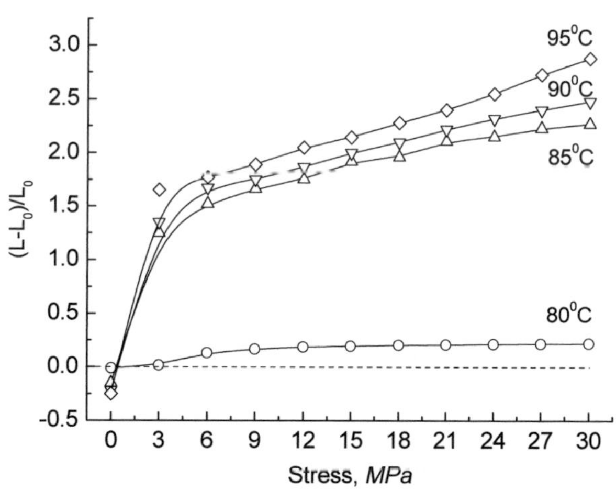

Fig. 13. Relative change of the bundle length annealed at different temperatures, depending on the tensile stress values (here L_0 and L are the initial and final fibers length respectively).

The yarn length obtained after the above described heat mechanical treatment as a function of the applied strain stress is presented in Figure 13 where the dashed line shows the initial sample length. As is apparent from Table 1 the sample B is almost completely amorphous, possess relatively low birefringence value and therefore weak pre-orientation of the amorphous phase. Specimen B is subjected to SHMM with a good resolution in the particularly effective temperature range above the glass transition. The curve obtained at temperature of 80 ^0C shows the initial level of the material deformation at selected conditions. The bundle deformation behavior (at 80 ^0C) is probably due to a slight stretch, uncoupling and disentangle of the free and slightly folded segments in the amorphous areas. Similar effects are observed up to stress value of 9 *MPa*, but mainly at stress levels between 3 *MPa* and 6 *MPa*. At loading above 9 *MPa* the curve comes to the a plateau and up to normal load of 30 *MPa* the structure is stable.
The temperature raising to 85 ^0C and 90 ^0C shows a common way of the obtained curves where can be seen an initially entropy contraction for the unladen samples and a significant increasing of the relative residual deformation at stress level of 3 *MPa*. After the initial sharply increase of the ultimate bundle length both curves indicate only small increments of the deformation and orientation with load increasing. Behavior of the curve at 95 ^0C displays much higher values of the sample deformation, which growing with the load values rising. The observed trend can be result from reinforced destruction and slip the segments instead of unfolding and orientation, i.e. from sub-optimal fibers straining. The differences in the bundles deformation behavior, depending on the load values and temperature show that the selected interval of the complex stress - temperature conditions is a suitable to realize the objectives of the thermal deformation experiment.

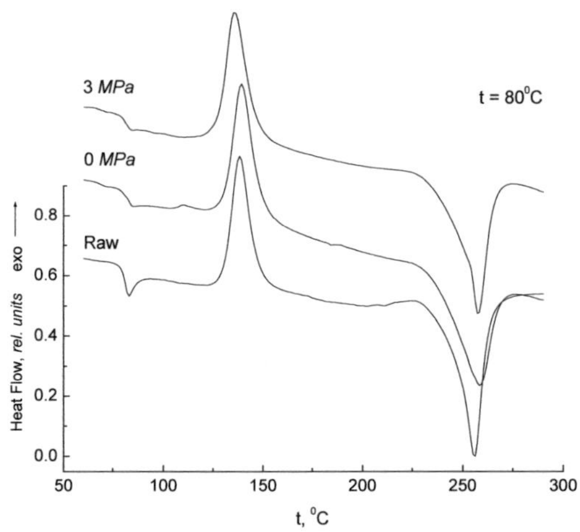

Fig. 14. DSC curves of untreated sample B and heat mechanically modified under different strain stresses at temperature of 80 ^0C (the curves are shifted vertically for clarity).

The structural analyses of the amorphous PET yarns heat-mechanically treated at the above-described conditions were carried out using differential scanning calorimetry and birefringence measurements. The clearly expressed cold crystallization (Fig. 14) of the samples modified at 80 ^0C is due to the presence of disordered and unstable structure in the unorientated amorphous fibers. In such structures, heated up to temperatures around 130 -

140 ^0C in the absence of external stress (0 *MPa*) is released the segment mobility in macromolecules, which is a precondition for the cold crystallization. For similar structures the applied stress of 3 *MPa* at temperature of 80 ^0C is not enough to cause noticeable fibers orientation and structure stabilization. The melting peaks of these samples are broader and imperfect because the melting polymer system is imperfect, contains less and more defective crystalline phase.

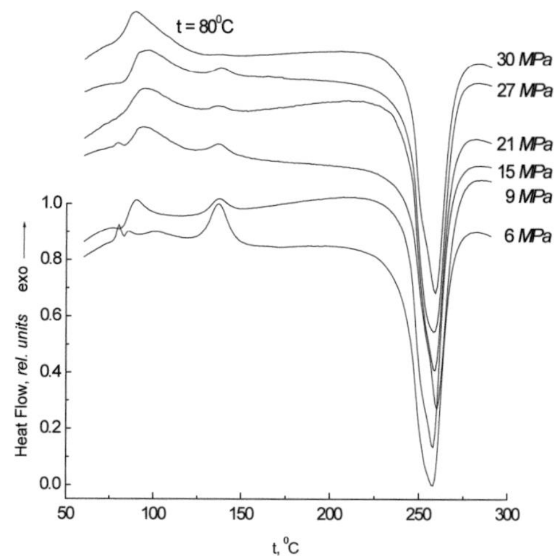

Fig. 15. DSC curves of sample B heat mechanically modified under different strain stresses at temperature of 80 ^0C (the curves are shifted vertically for clarity).

With increasing of the tensile stress values from 6 *MPa* and more (Fig. 15) the cold crystallization peak at about 140 ^0C gradually disappears but appears a new similar peak at much lower temperatures, immediately after the transition from glassy to rubbery state. Probably the higher stress further destabilize the structure and facilitates the glass transition proces. Furthermore it is observed an increase of the melting enthalpy and monotonous proportional dependence of the melting temperature increasing with the load growth and respectively the quality and quantity of the crystalline phase. And more essential, in accordance with the crystalline phase and structure improvement the low temperature cold crystallization moves to the lower temperatures. Consequently the more improved and arranged structure crystallizes at lower temperatures in the amorphous phase, while more difficult at higher temperatures melts the more perfect crystal phase.

Except the study by differential scanning calorimetry was measured and the birefringence as a function on the stress of the heat - mechanically treated at 80 ^0C fibers. It is interesting to note the appearance of "resonance" filaments birefringence with a pronounced maximum at tension stress value of 6 *MPa* which is an additional evidence for the role of the applied stress in the filaments structure rearrangement. With increasing of the SHMM temperature from 80 ^0C to 85 ^0C (Fig. 16) and under minimum load of 3 *MPa* losing the effect of the samples cold crystallization at temperatures around 130 - 140 ^0C. Instead appears the above described cold crystallization at lower temperatures around 110 ^0C. It is observed an increasing of the melting temperature and enthalpy as well as the perfection of the melting crystalline phase. At middle and higher loads a cold crystallization at 130 - 140 ^0C does not happen. It is observed the same effect of lowering of the temperature and intensity of the earlier cold crystallization. With the loads increasing is enhanced the tendency for splitting of the melting peaks as well as homogeneous proportional increases of the melting temperature. An interesting fact is that at load of 6 *MPa* has spilled, extended over a wide temperature range endo effect of enhanced segmental mobility and entropy and at stress of 12 *MPa* has rudiment of one.

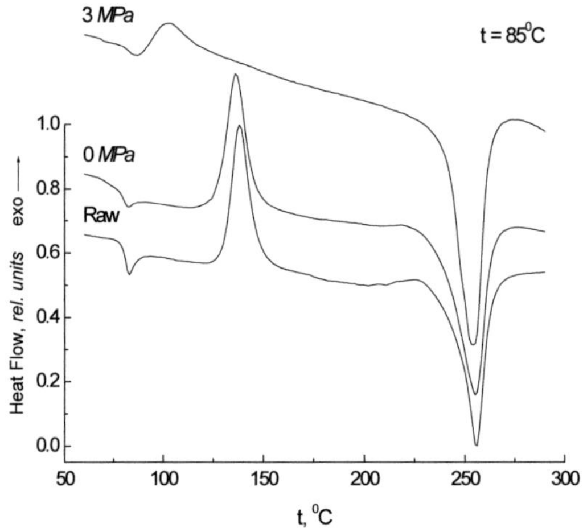

Fig. 16. DSC curves of untreated sample B and heat mechanically modified under different strain stresses at temperature of 85⁰C (the curves are shifted vertically for clarity).

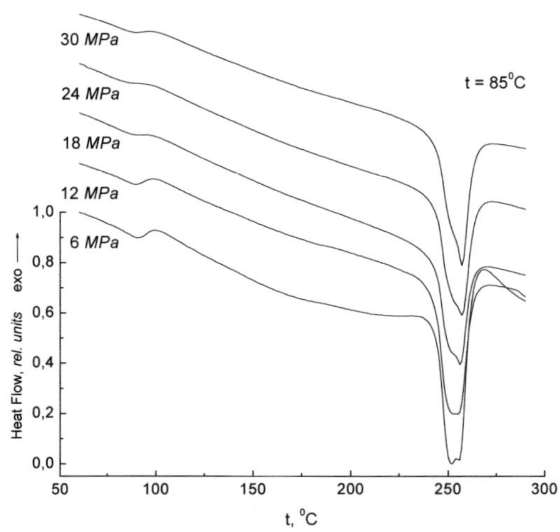

Fig. 17. DSC curves of sample B heat mechanically modified under different strain stresses at temperature of 85⁰C (the curves are shifted vertically for clarity).

With purpose to specify the occurred in the samples structure developments caused by the SHMM at temperature 85 ⁰C has been measured and the fibers birefringence. The obtained results pertaining to the birefringence values depending of the applied load show a wide basic peak at stresses of 12 *MPa*, 15 *MPa* and 18 *MPa* with two attendant smaller peaks at stresses of 3 *MPa* and 27 *MPa* correspondingly.

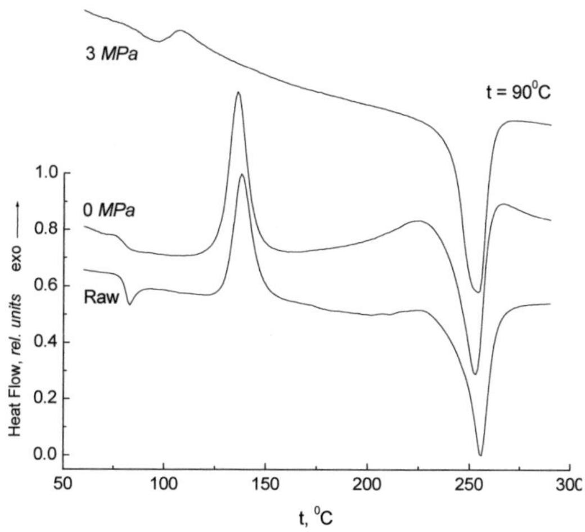

Fig. 18. DSC curves of untreated sample B and heat-mechanically modified under different strain stresses at temperature of 90 ^0C.

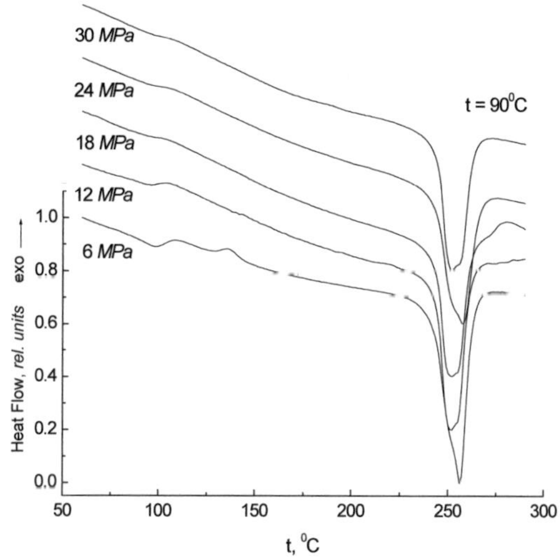

Fig. 19. DSC curves of sample B heat -mechanically modified under different strain stresses at temperature of 90 ^0C.

Very interesting is the DSC curve obtained without loading (strain stress value of 0 *MPa*) and temperature of 90 ^0C (Fig. 18). It is impressive much larger peak of cold crystallization in comparison with all other temperatures. This is probably the most suitable temperature for heat mechanical modification of the studied amorphous PET fibers, with the relevant background. At this temperature are released large resources for structural reorganization. From the same graph (Fig. 18) is seen that the effect of restoration of the segmental motion, known also as alpha-relaxation transition is practically suppressed. Perhaps during of the

SHMM directly achieves some effect of low-temperature arrangement in the free state of the object at load absence. The fact that only at this temperature takes place so intensive cold crystallization, with pronounced additional crystallization (or pre-crystallization), just before the samples melting is evidence for the optimal combination of the initial structure of the object with the selected temperature of SHMM. As at temperatures of 80 ^0C and 85 ^0C (Fig. 15 and 17) as well at 90 ^0C is observed a complex nature of the cold crystallization again under load of 6 *MPa*.

As is evident from Fig. 20 at temperature 95 ^0C the amorphous samples heat mechanically processed at lower loading values indicate a decreasing intensity of cold crystallization and

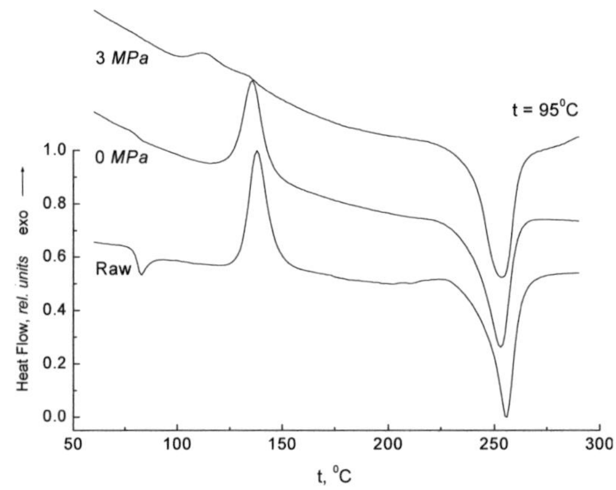

Fig. 20. DSC curves of untreated sample B and heat-mechanically modified under different strain stresses at temperature of 95 ^0C.

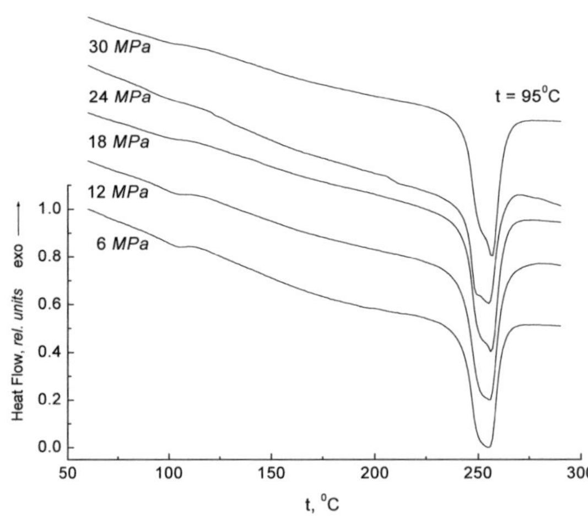

Fig. 21. DSC curves of sample B heat -mechanically modified under different strain stresses at temperature of 95 ^0C.

rising thermograms inclination. Possible reason for the observed tendencies are due to the overall increase of the entropy of the system with the samples heating, resulting in the amend of the thermal conductivity and heat capacity of the objects. In the field of the medium and high loads (Fig. 21) appears the same effect of the lowering of the temperature and intensity of the cold crystallization and homogeneous proportional increasing of the melting temperature. With the loads increasing are monitored and enhanced split of the melting peaksp too. It is reasonable that with the stress values rising to differentiate two phases in the studied samples with structural differences between them and accordingly with different melting temperatures.

№	σ, *MPa*	α, % (80 ^{0}C)	α, % (85^{0}C)	α, % (90^{0}C)	α, % (95^{0}C)
1	Raw	1.7	1.7	1.7	1.7
2	0	2.0	11.8	4.0	13.7
3	3	2.9	39.5	36.1	40.7
4	6	34.7	38.5	39.2	42.1
5	9	34.9	39.3	41.3	43.3
6	12	35.0	41.6	42.5	45.0
7	15	34.0	42.5	43.3	44.7
8	18	33.3	41.8	42.6	44.5
9	21	33.7	41.5	41.5	44.4
10	24	34.0	39.7	40.6	44.6
11	27	35.8	40.5	40.0	44.9
12	30	34.2	40.7	40.9	43.8

Table 3. Degree of crystallinity of sample B subjected to SHMM at isothermal conditions and constant strain stresses.

On the basis of the data obtained by DSC was calculated and the degree of crystallinity of the studied PET fibers. The obtained results are presented in Table 3. If compare the data given in Table 3 with the relative change of the bundle length (Fig.22) will be seen certain similarity in the change the samples degree of crystallinity and fibers deformation behavior. The bundle deformation behavior observed at temperature of 80 ^{0}C (Fig. 22) shows a relatively large yarns extension with the stress level increasing from 3 *MPa* to 6 *MPa*. As it is visible from the Table 3 the samples degree of crystallinity at the same temperature (80 ^{0}C) and load levels (3 - 6 *MPa*) grows with more than thirty percents. Further more, at loadings above 9 *MPa* both the final bundle length and filaments crystallinity comes to a plateau up to load value of 30 *MPa*.

Very interesting are the changes of the filaments degree of crystallinity obtained with the temperature rising to 85 ^{0}C and 90 ^{0}C where a sharp increase of the samples crystallinity occurs even at stress value of 3 *MPa*. Like the bundle deformation behavior (Fig. 22) after the initial sharply increase of the filamemts crystallinity, both dependencies practically follow a common course. At loadings above 6 *MPa* they are coming to a plateau up to stress values of 30 *MPa*. Similar is the course and of the fibers degree of crystallinity at temperature 95 ^{0}C. It can be concluded that the samples degree of crystallinity reasonably good follows the bundle deformation, which is additional proof for the role of the strain stress in the crystallization of the studied PET fibers.

3.3 Investigation of partially crystalline PET fibers simultaneous heat - mechanically modified at isothermal conditions and constant strain stress values

The next stage from the the realized studies was to investigate the role of the SHMM conditions on the structure developments of partially crystalline PET filaments with different preliminary orientation.

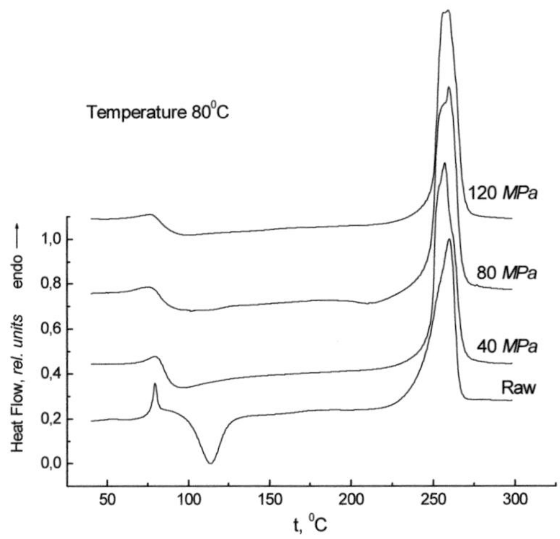

Fig. 22. DSC curves of untreated sample S1 and heat-mechanically modified under different strain stresses at temperature of 80^{0}C (the curves are shifted vertically for clarity).

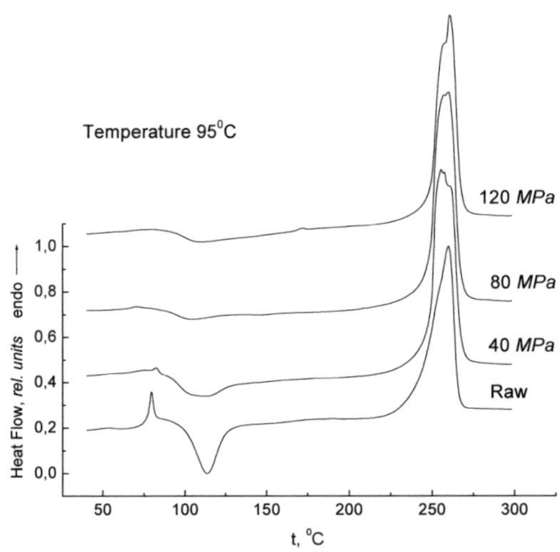

Fig. 23. DSC curves of untreated sample S1 and heat-mechanically modified under different strain stresses at temperature of 95^{0}C (the curves are shifted vertically for clarity).

With that purpose the samples named S1, S2 and S3 (Table 1) were heat - mechanically treated at constant temperatures in the same transient range from glassy to rubbery state. The thermal deformation experiment involves bundles annealing at temperatures of 80 ^{0}C, 85 ^{0}C, 90 ^{0}C and 95 ^{0}C accompanied by constant gravimetric tensile loadings of 40 *MPa*, 80

MPa and 120 *MPa* during two minutes. In comparison with the above described experiments the applied strain stress values are significantly larger. The experimental conditions allow to trace the combined unfluence of the tensile stress, temperature and preliminary fibers orientation on the going in the samples structural changes.

The structural analyses of the PET filaments subjected to SHMM at the above-described conditions were carried out using differential scanning calorimetry and birefringence measurements.

Representative panels of DSC curves of sample S1 obtained at temperatures 80 ^{0}C and 95 ^{0}C are present in Figures 22 and 23. The DSC curves of the heat mechanicaly treated samples show very important differences due to the structural reorganization occurred as a result of the modification. There is a radical change of the objects behavior in the temperature regions of the segmental motions defrosting (glass transition) and an increase in the macromolecules mobility (melting). In the areas of the hardening (change of the thermodynamic, entropy mobility of the polymers) and liquefying (increased viscous-liquid mobility) are observed visible structural changes with a steady trend with increasing the load and temperature. Probably due to the regulating effect of the SHMM on the structural organization the glass transition disappears together with the cold crystallization relatively monotonous. After samples SHMM there are no large differences in conformation state of the frozen and with very high entropy segments or larger ensembles of them. With the load and temperature increasing there is a slight shift towards the higher temperatures and increased splitting of the melting peaks. With the load values increasing before everything is lost the low temperature component of the melting peak. It is difficult to define the contributions of the unidirectional influence of the heating and loading in the phases forming during the high temperature multiple melting proces. Sometimes these effects are slight and not quite as a regular visible trend but such is the nature of the processes of structural transformation, too. Much more impressive is the objects behavior in rubbery state in the whole temperature range of increasing segmental mobility and relaxation effects between the processes of glass transition and melting in comparison with the above discussed objects modified under linear heating.

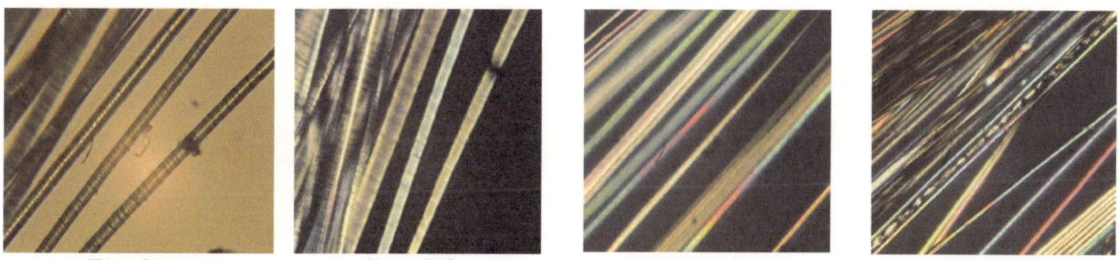

Fig. 24a. Fig. 24b. Fig. 24c. Fig. 24d.

Fig. 24. a) light microphotography of suboptimal drawn PET fibers with pronounced stripes due to enhanced destructive processes; b) polarization microphotography of similar objects; c) polarization microphotography of not homogeneous oriented sections; d) similar objects with filler defects.

Different types of defects and orientation inhomogeneity of PET fibers samples S1, S2 and S3 heat - mechanically treated at constant temperature of 95 ^{0}C and under constant gravimetric tensile loading of 120 MPa are present in Fig. 24.

In contrast to the rich on thermal effects relaxation area of the filaments rubbery state at the linear heated objects, here is observed fully stable samples behavior. The fibers modification under present conditions forms more homogeneous stable structure more compact (oriented) and poor of thermal effects during scanning proces.

3.4 Investigation of partially crystalline PET fibers simultaneous heat - mechanically modified at isothermal conditions and constant extension speed

After the examination of different variants of thermal treatment of PET fibers combined with different levels of constant gravimetric tensile loadings it was interesting to clarify the role of the mechanical treatment mode on the structure reorganization in the studied filaments. Therefore in the following variant of the thermal deformation experiment the studied partially crystalline PET yarns S1, S2 and S3 (Table 1) were subjected to extension at a constant speed under constant temperatures again in the same temperature range above the glass transition temperature from 80 ^0C to 95 ^0C. In the present investigation PET bundles were annealed during ten minutes at temperatures 80 ^0C, 85 ^0C, 90 ^0C and 95 ^0C then were withdrawn with a constant rate of 0.1 *m/min* up to relatively elongation (ε) of 20 %, 40 % and 60 %. Immediately after the treatment the sample was removed from the furnace at room temperature. The structural developments of the so heat mechanically modified filaments were investigated using differential scanning calorimetry.

Representative stacks of DSC curves of sample S1 subjected to SHMM at different levels of elongation and temperatures 80 ^0C and 95 ^0C are present in Figures 25 and 26.

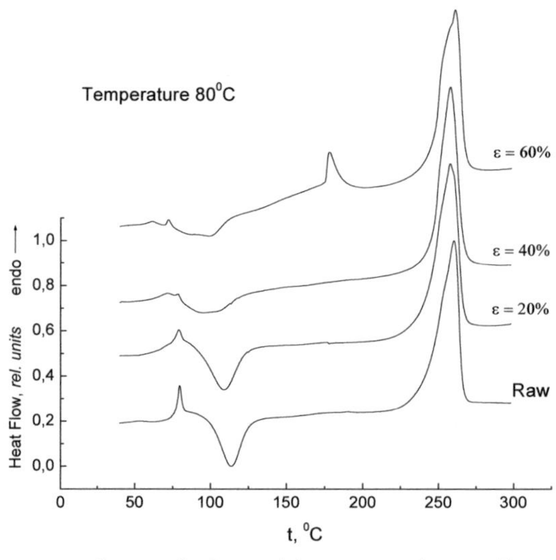

Fig. 25. DSC curves of untreated sample S1 and heat - mechanically modified at temperature of 80 ^0C and constant extension rate (the curves are shifted vertically for clarity).

Received by DSC results for the samples S1, S2 and S3 show that at the gravitational load of the fibers the attenuation and the practical disappearance of the effects of softening and cold crystallization follow monotonic trend with the temperature increasing, while for the drawn samples the temperature dependence is weaker, seems secondary. For these samples has a stronger dependence of the observed effects from the mechanical influence, respectively from the withdrawal degree. In each panel of DSC curves is observed highly reduction of the intensity of the softening and cold crystallization running with increasing of the fibers elongation. At the highest temperature (95 ⁰C), these processes are less pronounced even at the small degrees of withdrawal but here also is observed damping of the effects, proportional to the extension degree increasing.

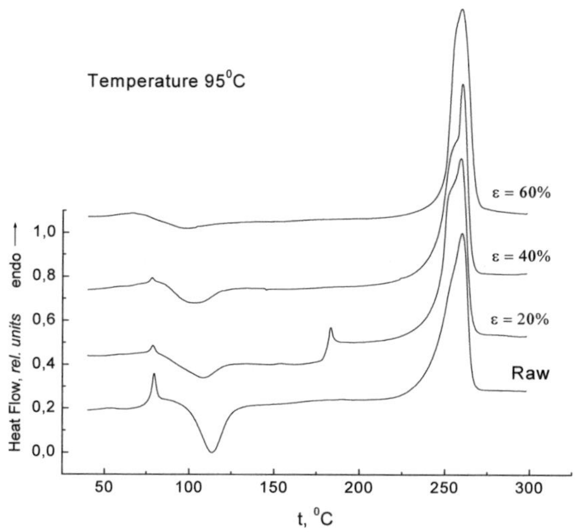

Fig. 26. DSC curves of untreated sample S1 and heat - mechanically modified at temperature of 95 ⁰C and constant extension rate (the curves are shifted vertically for clarity).

The appearance on effects of pre-melting in relaxation area around 190 ⁰C - 200 ⁰C (Fig. 25 and 26) in the fibers subjected to extension at a constant rate is very interesting. It is observed at all temperatures but is less pronounced in the intermediate temperatures of 85 ⁰C and 90 ⁰C. At the lowest temperature (80 ⁰C) the effect of pre-melting occurs at the largest deformation, respectively the largest structural reorganization. At the intermediate temperatures the effect is more slightly marked and is observed at the mean degree of the fibers elongation while at the highest temperature it occurs even at the smallest deformations. In addition all DSC curves with the observed pre-melting effects are steeper, with a larger gradient of the specific heat capacity increasing of the polymer structure with the temperature rise.

All this makes it possible to conclude that the forced modification of the studied PET fibers allows to form inhomogeneous and unstable stressed and overstressed structures which under certain conditions are thawed and melted during the DSC scanning. Similar entities could represent permanently stable in the time more stressed micro-localized areas, resulting in the observed during DSC measurement effects. For the gravitational loading similar effects are not monitored and this is reasonable. In the terms of this version of SHMM in the conditions of mutually coordinated influences and interactions of the mechanical affects the structural reorganization occurs more smoothly and slowly and turns off the formation and permanent freezing of overstressed regions. The DSC curves of both the filaments extended at a constant rate and filaments loaded at constant levels show beginnings of split of the melting peaks, which grow with the increase of the load and temperature.

It can be concluded that both the temperature and strain stress enable the structural reorganization and orientation of the polymer, but the forced withdrawal shows stronger orientation effects than the SHMM under constant loadings. Only concrete tests of the fibers strain-strength properties can arbitrate which samples are more optimally orientated and with less micro destructions.

4. Conclusion

Based on the analysis of the results obtained from the recent large-scale experiment it can draw the following conclusions:
There are studied PET fibers with different starting structure as amorphous as well partially crystalline samples with different pre-orientation;
Have been applied different regimes of thermal treatment as linear heating of the objects in a wide temperature range from 20 ^0C to 200 ^0C and annealing at constant temperatures in very important in terms of the structural changes temperature band from 80 ^0C to 95 ^0C, located close above the glass transition temperature;
The thermal treatments were accompanied from different levels of constant gravimetric tensile loadings and by withdrawal of the fibers to different levels at a constant extension rate;
The structural analyses of the investigated PET fibers subjected to SHMM show very important results concerning the role of the temperature and the applied strain stress and in particular:
the deformation behaviour of amorphous PET fibers subjected to linear heating from 20 ^0C to 200 ^0C show the role of the tensile loading in the management of the rival processes of fluid like deformation and strain induced crystallization;
significant increase of more than 30 % of the fibers degree of crystallinity in conditions of annealing in temperatures from 80 ^0C to 95 ^0C, with the strain stress increasing from 3 *MPa* to 6 *MPa*;
the received by DSC results for the partially crystalline samples S1, S2 and S3 show the role of the mechanical treatment mode on the running relaxation and phase transitions. At the gravitational fibers loading the attenuation and disappearance of the effects of the glass transition and cold crystallization follow monotonic trend with the temperature increasing, while for the drawn samples the temperature dependence is weaker, seems secondary. For these samples has a stronger dependence of the observed effects from the mechanical influence, respectively from the withdrawal degree.
The further steps are specific strain-strength tests that may arbitrate, which samples are optimally modified and such studies ahead.
The purpose of future research is by comparison between the gravitational load and different type of orientation drawing and thermal treatment to choose the optimal combination between them.

5. Acknowledgment

Part of the present work has been supported by Grant РД 07 - 340/15.03.2011 from Konstantin Preslavsky University, Shumen, as well as by Grants DO-02-61/08 and DDVU-02-106/2010 from University "Prof. Dr. Assen Zlatarov", Burgas.

6. List of the used symbols

α	degree of the sample crystallinity
d	diameter of the single fiber
ε	value of the relatively bundle elongation
λ	wavelength of the radiation in the WAXS apparatus
L_0	initial bundle length of the raw sample
L	final bundle length obtained after heat mechanically treatment
Δn	birefringence of single fiber
σ	applied to the objects tensile stress
V_L	speed of the filaments formation

7. References

Bai C., Spontak R. J., Koch C. C., Saw C. K., Balik C. M. (2000). Structural changes in poly (ethylene terephthalate) induced by mechanical milling. *Polymer*, Vol. 41, pp. 7147-7157, ISSN 0032-3861

Betchev Ch. (1995). On the Characterization by Shrinkage Forces of As-spun Poly (ethylene terephthalate) Filaments. *Polymer Testing*, Vol. 14, pp. 163–172, **ISSN** 0142-9418

Dupaix R. B., Boyce M. C. (2005). Finite strain behavior of poly(ethylene terephthalate) (PET) and poly(ethylene terephthalate)-glycol (PETG). *Polymer*, Vol. 46, Issue 13, pp. 4827 - 4837, ISSN 0032-3861

Guzatto R., Da Roza M. B., Denardin E. L. G., Samios D. (2009). Dynamical, morphological and mechanical properties of poly (ethylene terephthalate) deformed by plane strain compression. *Polymer Testing*, Vol. 28, pp. 24–29, ISSN 0142–9418

Karagiannidis P. G., Stergiou A. C., Karayannidis G. P. (2008). Study of crystallinity and thermomechanical analysis of annealed poly (ethylene terephthalate) films. *European Polymer Journal*, Vol 44, pp. 1475-1486, ISSN 0014–3057

Keum J. K., Song H. H. (2005). Thermal deformations of oriented noncrystalline poly(ethylene terephthalate) fibers in the presence of mesophases structure. *Polymer*, 46, pp. 939–945, ISSN 0032-3861

Kong Y., Hay N. J. (2003). Multiple melting behaviour of poly (ethylene terephthalate). *Polymer*, Vol. 44, pp. 623-633, ISSN 0032-3861

Llana P. G., Boyce M.C. (1999). Finite strain behavior of poly(ethylene terephthalate) above the glass transition temperature. *Polymer*, Vol. 40, pp. 6729–6751, ISSN 0032-3861

Ma H., Zeng J., Realff M. L., Kumar S., Schiraldi D. A. (2003). Processing, structure, and properties of fibers from polyester/carbon nanofiber composites. *Composites Science and Technology*, Vol. 63, pp. 1617-1628, ISSN 0266-3538

Raabe D., Chen N., Chen L. (2004). Crystallographic texture, amorphization, and recrystallization in rolled and heat treated polyethyleneterephthalate (PET). *Polymer*, Vol. 45, pp. 8265-8277, ISSN 0032-3861

Safapour S., Seyed-Esfahani M., Auriemma F., Ruiz de Ballesteros O., Vollaro P., Di Girolamo R., De Rosa C., Khosroshahi A. (2010). Reactive blending as a tool for obtaining poly (ethylene terephthalate)-based engineering materials with tailored properties. *Polymer*, Vol. 51, pp. 4340-4350, ISSN 0032-3861

Shabana H. M. (2004). Refractive index-structure correlation in chemically treated polyethylene terephthalate fibers. *Polymer Testing*, Vol. 23, pp. 291-297, ISSN 0142-9418

Sharma V., Desai P., Abhiraman A. (1997). Crystallinity Vis-a-Vis Two - Phase Models of Oriented Polymers: Inferences from an Experimental Study of Poly (ethylene terephthalate). *Journal of Applied Polymer Science*, Vol. 65, pp. 2603-2612, ISSN 1097-4628

Sulong A. B., Park J., Azhari C. H., Jusoff K. (2011). Process optimization of melt spinning and mechanical strength enhancement of functionalized multi-walled carbon nanotubes reinforcing polyethylene fibers. *Composites: Part B*, Vol. 42, pp. 11-17

Wijayathunga N. V., Lawrence C. A., Blackburn R. S., Bandara M. P. U., Lewis E. L. V., El-Dessouky H. M., Cheung V. (2007). Influence of laser irradiation on the optical and structural properties of poly (ethylene terephthalate) fibres. *Optics & Laser Technology*, Vol. 39, pp. 1301–1309, ISSN 0030–3992

Wu J., Schultz J. M., Samon J. M., Pangelinan A. B., Chuah H. H. (2001). In situ study of structure development during continuous hot-drawing of poly (trimethylene terephthalate) fibers by simultaneous synchrotron small- and wide angle X-ray scattering. *Polymer*, Vol. 42, pp. 7161-7170, ISSN 0032-3861

Zhang Z., Wu Sh., Ren M., Xiao Ch. (2004). Model of cold crystallization of uniaxially oriented poly(ethylene terephthalate) fibers. *Polymer*, Vol. 45, pp. 4361-4365, ISSN 0032-3861

Ziabicki A., Jarecki L. (2007). Crystallization-controlled limitations of melt spinning. *Journal of Applied Polymer Science*, Vol. 105, pp. 215-223, ISSN 1097-4628

6

Conception of an Absorption Refrigerating System Operating at Low Enthalpy Sources

Nahla Bouaziz, Ridha BenIffa, Ezzedine Nehdi and Lakdar Kairouani
Engineering National School of Tunis
Tunisia

1. Introduction

Since the work of Duhem in 1899, relating to binary mixtures, that the absorption chillers have been growing significantly.
In 1930, Borzig had developed a machine using the couple water-ammonia. This system is interesting in that it works by supplying heat energy regardless of its origin (waste heat, geothermal water, solar energy...).

2. Operating principle

The operating principle of such a machine is briefly describe below (Figure 1). The absorption system differs from the vapor compression machine by providing a third heat source which is the generator (Qg). The absorption machine uses a binary mixture; one fluid is more volatile than the other and constitutes the refrigerant. Couples most commonly used are: Water-Ammonia (H2O/NH3) Ammonia is the refrigerant. Lithium Bromide-Water (LiBr/H2O), water is the refrigerant. The elements of an absorption machine are shown in Figure 1. They are: Boiler or generator: the refrigerant rich solution is heated to the temperature Tg, which is higher than the temperature of vaporization of refrigerant to the pressure considered. Condenser: similar to that of a vapor compression machine.
Evaporator: similar to that of a vapor compression machine.

- Absorber: steam from the evaporator is absorbed by the existing solution which will be enriched in refrigerant. The absorber is also connected to the generator. The weak solution coming from the boiler enters to the absorber.
- Inter-exchange solution: all current machines include a heat exchanger (sometimes called internal transmitter) between the rich solution leaving the absorber to Tab and the weak solution leaving the boiler at Tg. This exchanger is used to preheat the rich solution before entering at the generator.

A pump is used to lead the rich solution to the generator, and an expander is used to return the weak solution to the absorber. In general, the coefficient of performance (COP) of such a machine is around 0.7. To improve the COP or adapt the machine to any source of energy, some purpose can be considered such a multiple effects machines, combined machines (absorption-compression, integration of ejectors...) [1-7]. The COP is defined as:

$$\mathrm{COP} = \frac{\dot{Q}_F}{\dot{Q}_g + \dot{W}_P} \tag{1}$$

$\dot{W}_P$ is low so we can write:

$$\mathrm{COP} = \frac{\dot{Q}_F}{\dot{Q}_g} \tag{2}$$

Where $\dot{Q}_F$ is the amount of cold produced and $\dot{Q}_g$, the heat energy supplied to the generator.

The mass balance at the generator provides for 1kg of refrigerant vapor, (*f*) kg of solution as:

$$f = \frac{x_v - x_p}{x_r - x_p} \tag{3}$$

x_v is the mass fraction of steam (x_v =1 for $LiBr/H_2O$ and is close to 1 for the couple NH_3/H_2O). Où x_p and x_r are respectively the titles of the weak and rich solutions, determined from the diagrams of Merkel and Oldham. (*f*) is called entrainment ratio, he must have reasonable values in order to reduce the energy consumption of the pump. In what follows, we present the performance of absorption chillers using couples NH_3/H_2O or $LiBr/H_2O$.

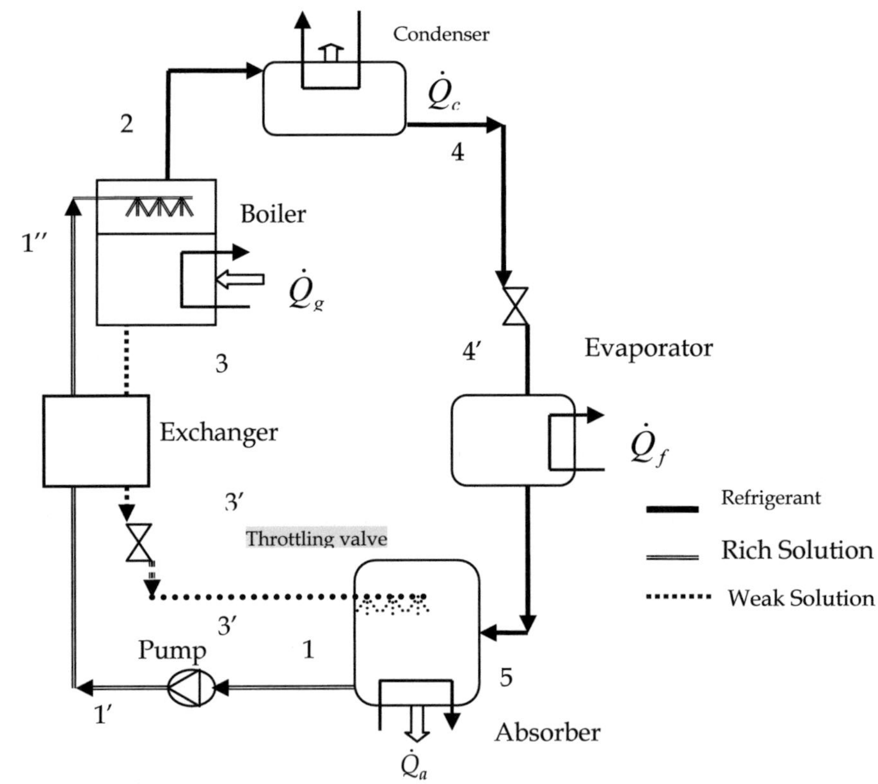

Fig. 1. Operating principle of an absorption machine

2.1 Oldham diagram

The refrigeration cycle is shown in the Oldham diagram (Log P, $\frac{1}{T}$) on which we can trace the iso-titles of the solution. By choosing the pair of pressure evaporation and condensation (P_e, P_c), it follows the pair of corresponding temperatures (T_e, T_c). From the saturation line (x=100%), we draw a vertical line to determine the rich solution (x_r). The intersection of the line of the rich solution and isobaric P_c indicates the threshold temperature (T_s). The threshold temperature (T_s) is the minimum temperature of the generator, below which the installation does not work. The generator temperature determines the line of the weak solution and hence its title (xp) (Figure 2).

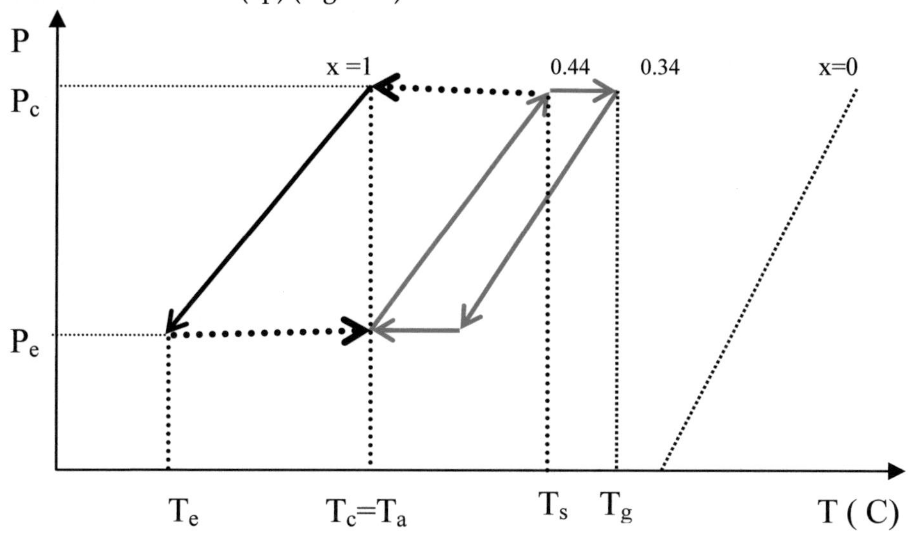

Fig. 2. Oldham Diagram

NH_3/H_2O installations must be equipped with a rectification column to remove water entrained with the refrigerant to prevent it from solidifying in the pipes of the evaporator. The cooling capacity is:

$$\dot{Q}_e = \dot{m}_{ff}\Delta h_{év} \quad (4)$$

$$\dot{Q}_e = COP\dot{Q}_g \quad (5)$$

$\dot{Q}_g$ is the heating power:

$$\dot{Q}_g = \dot{m}_{ff}[-f\,h_{abs} + (f-1)h_g + h_v] \quad (6)$$

(*f*) is the driving factor, it is the mass of rich solution is likely to emit one kg of refrigerant vapor.

h_v is the heat of vaporization of refrigerant in the solution.

h_{abs} is the enthalpy of the rich solution leaving the absorber.

h_g is the enthalpy of the weak solution leaving the generator.

These enthalpies are taken from the diagram of Merkel or can be obtained from empirical correlations [8-9]

3. Combined and multi-stage installations

We have presented a single absorption plant; it is conceivable to study the combined or hybrid systems.
For the combined system, the absorption cycle, serve to ensure the condensation of the refrigerant for the vapor compression cycle. The latter can operate between temperatures of condensation and evaporation desired.
For the hybrid system, a compressor acts as a liaison between two stages of absorption.

3.1 Combined installations

The system presented as an example, Figure 3, uses for the installation of R134a vapor compression and the couple water-ammonia absorption for installation.
Condensing temperature is 30 C and evaporating temperature of R134a is -10 C. The COP of the plant absorption is only 0.64 [1].
This system can be profitable if you have a free source of energy or recovery such as solar, thermal discharges of gas power plants or geothermal energy.

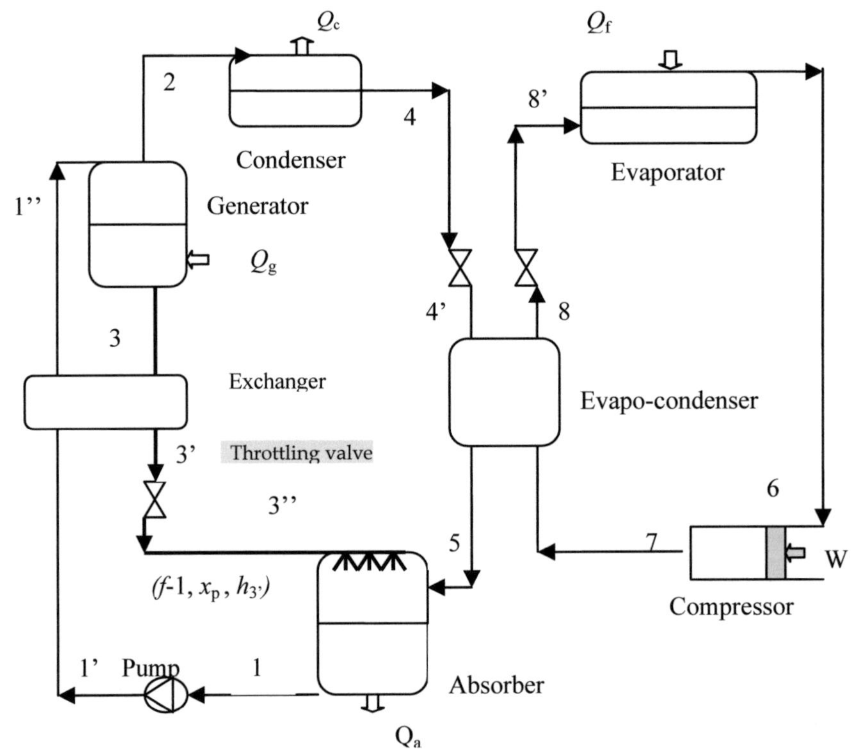

Fig. 3. Combined Installation

This absorption/compression refrigeration system is proposed to improve the overall cycle efficiency. The COP excluding the pump work and the generator energy required is as high as 5.4–6.2, which is higher than that of the single vapor compression cycle and absorption cycle, under the same operating conditions (evaporation temperature at 263 K and condensation temperature at 308 K).

This system presents an opportunity to reduce the continuously increasing electrical energy consumption.

Further investigations are needed to optimize the combined system design and operating parameters and to assess the efficiency and the feasibility of the system. A pilot installation can be built near geothermal, solar or waste energy sources [1].

3.2 Double stage system

In the absorption system at double stage, shown in Figure 4, the displacement of the refrigerant from low pressure to high pressure by means of two thermo-compressors 1 and 2 combined in series. To analyze the cycle of transformations, we consider the following assumptions:

- Temperatures of output rich solutions from absorbers Ab_1 and Ab_2 are equal and identical to the condensation temperature T_c.
- Temperatures of output weak solutions from generators Ge_1 and Ge_2 are equal.

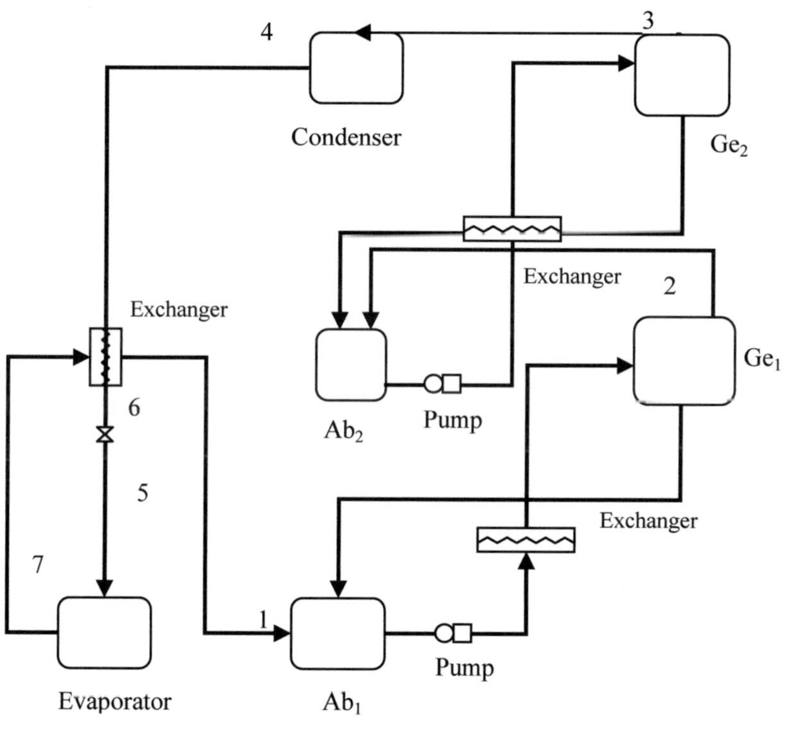

Fig. 4. Absorption machine with double-stage

There is no wall friction in the circuit.
In this installation, the first thermo-compressor transports the refrigerant from low pressure P_F to an intermediate pressure Pi corresponding to a saturation temperature of the refrigerant, T_i. Mass titles are respectively x_{r1} and x_{p1} for rich and poor solutions.
The second thermo-compressor transports the refrigerant of intermediate pressure Pi to the condenser pressure Pc. Mass titles are respectively xr2 and xp2 for rich and poor solutions.
The addition of one or more intermediate stages has a direct influence on lowering the generator temperature. But if the number of stages increases, the coefficient of performance decreases. Multi-stage systems have been studied by several authors; the results show that the COP is about 0.37 but a generator temperature is less than that of a single stage. The generator temperature can reach 65 C when T_c is 40 C and the COP of the plant is 0.26 which is relatively higher than that of a single stage that does not exceed 0.25 for an evaporation temperature of -10 C. These operating conditions can be profitable for valorization of energy sources at low enthalpy.

3.3 Other systems with multi-stages

We develop other configurations double-stage, we detail the calculation of energy and mass balance for some of them.
Several authors have considered different absorption machine configurations. Some systems are composed with simple stage machine [10-15] and others are formed by a succession of stages with various component associations and sometimes inserting other new components [16-20]. In the following section, we present three different configurations of the multi-stage refrigeration system and we develop a novel absorption hybrid configuration. We will explain and quantify it's adaptability to low-enthalpy sources. All configurations object of this work are followed by the representation of the corresponding cycle on the Oldham diagram.

3.3.1 System AGEcAG

The cascade system is composed of two elementary cycles, each one is considered as a single stage with the main difference, that the second stage is operating at a higher evaporative and condenser temperatures (see Figure 5). Figure 6 shows the Oldham diagram of the cycle.
In the following, we develop the energy and mass balance for the system AGEcAG. The entrainment factor f_i is the necessary rich solution flow able to move 1 kg of NH_3 from generator.

$$f_i = \frac{x_{vi} - x_{pi}}{x_{ri} - x_{pi}} \tag{7}$$

The proportion of the hot and cold flows at the mixer (evapo-condenser) is noted y and defined as follows:

$$y = \frac{h_{v1} - h_{sor\_mé\_liq}}{\left(h_{sor\_mé\_va} - h_{sor\_cd}\right)} \tag{8}$$

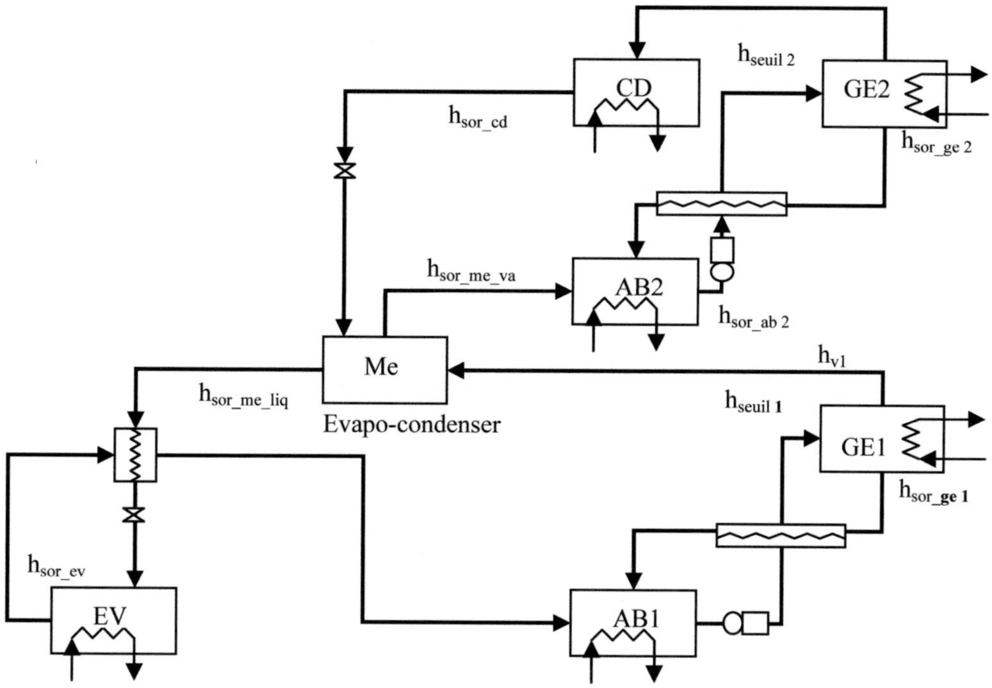

Fig. 5. System AGEcAG

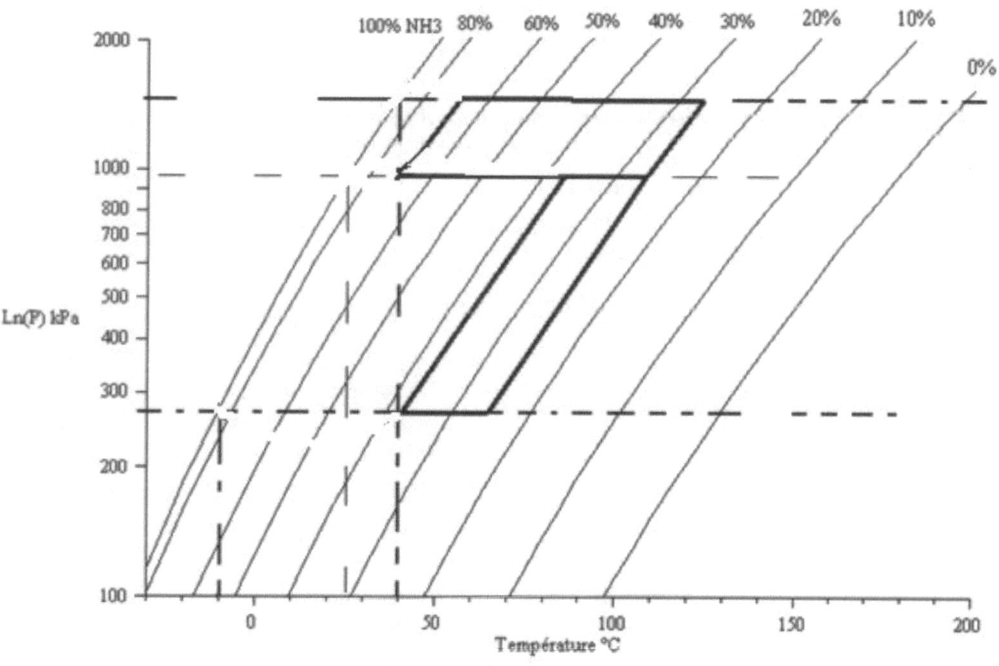

Fig. 6. Oldham Diagram of system AGEcAG

Energy balance for each installation component is presented. By neglecting the rectifier, we get for, Condenser, Evaporator, Generator and Mixer respectively:

$$\dot{Q}_{CD} = \dot{m}_{NH3} y \cdot (h_{v2} - h_{sor\_cd}) \tag{9}$$

$$\dot{Q}_{EV} = \dot{m}_{NH3} \, (h_{sor\_ev} - h_{sor\_mé\_liq}) \tag{10}$$

$$\dot{Q}_{GE1} = \dot{m}_{NH3} (h_{v1} + (f_1 - 1) \cdot h_{sor\_ge1} - f_1 \cdot h_{seuil1}) \tag{11}$$

$$\dot{Q}_{GE2} = \dot{m}_{NH3} y \cdot \left(h_{v2} + (f_2 - 1) \cdot h_{sor\_ge2} - f_2 \cdot h_{seuil2}\right) \tag{12}$$

$$\dot{Q}_{AB1} = \dot{m}_{NH3} (h_{sor\_ev} + (f_1 - 1) \cdot h_{ent\_ab1} - f_1 \cdot h_{sor\_ab1}) \tag{13}$$

$$\dot{Q}_{AB2} = \dot{m}_{NH3} y \cdot \left(h_{sor\_mé\_liq} + (f_2 - 1) \cdot h_{ent\_ab2} - f_2 \cdot h_{sor\_ab2}\right) \tag{14}$$

$$\dot{Q}_{Mé} = \dot{m}_{NH3} (h_{v1} - h_{sor\_mé\_liq}) \tag{15}$$

After developing the energy and mass balance for the cascade system of AGEcAG,
The COP's system is defined as follows:

$$COP = \frac{\dot{Q}_{EV}}{\dot{Q}_{GE1} + \dot{Q}_{GE2}} \tag{16}$$

Where $\dot{Q}_{GE1}$,$\dot{Q}_{GE2}$ and $\dot{Q}_{EV}$ are the generators and evaporator power, respectively.
We note that for the considered absorption refrigerating system, the second stage is used to lower the operating temperature of the first stage and not to increase the COP. It is evident that the amount of the system required energy is higher than that required for a single stage, because of the two generator components.

3.3.2 System AGAG

This machine is composed by two absorbers, a condenser working at the same temperature TAB, two generators operating at the same temperatures (T_{GE1}=T_{GE2}) and an evaporator. Besides the absorber $AB_1$and the evaporator EV, the absorber AB_2 and the generator GE1, the condenser CD and the generator GE_2 operate respectively at the pressures P_{EV}, P_{moy} and P_{CD}. The connection between the two stages is provided between the generator G_{E1} and the absorber AB_2 (see Figure 7). Figure 8 shows the Oldham diagram of the cycle.
We develop bellow, the energy balance and mass for the cascade system AGAG
To calculate the entrainment factors, we use equation (6) and after determining x_{ri}, x_{pi} for (i = 1 or 2) that are the titles of the rich solutions and the poor solution for the first stage (i = 1) and the second stage (i = 2).

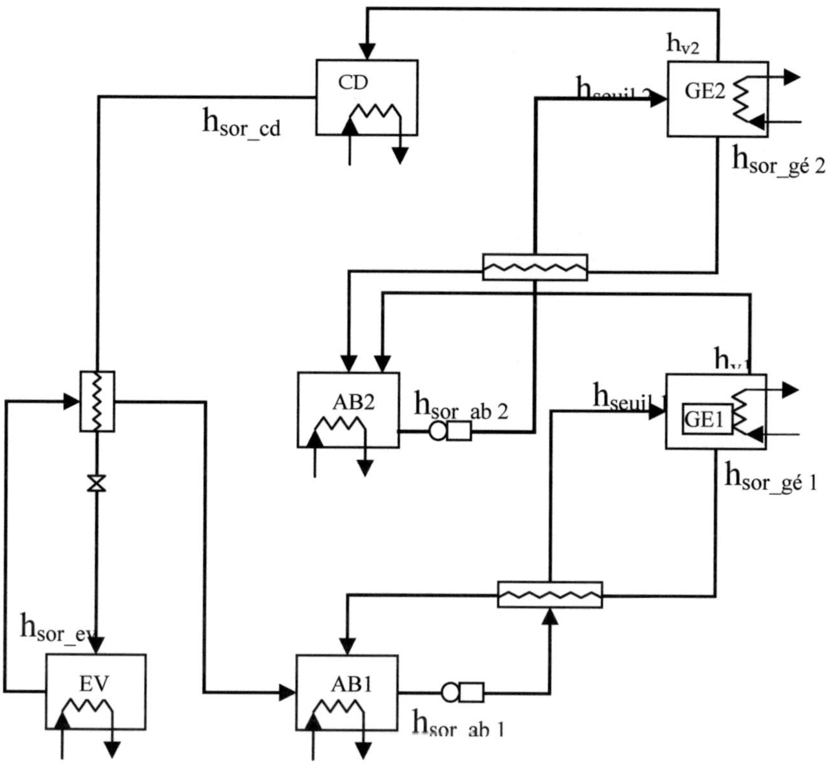

Fig. 7. System AGAG (connection generator - absorber)

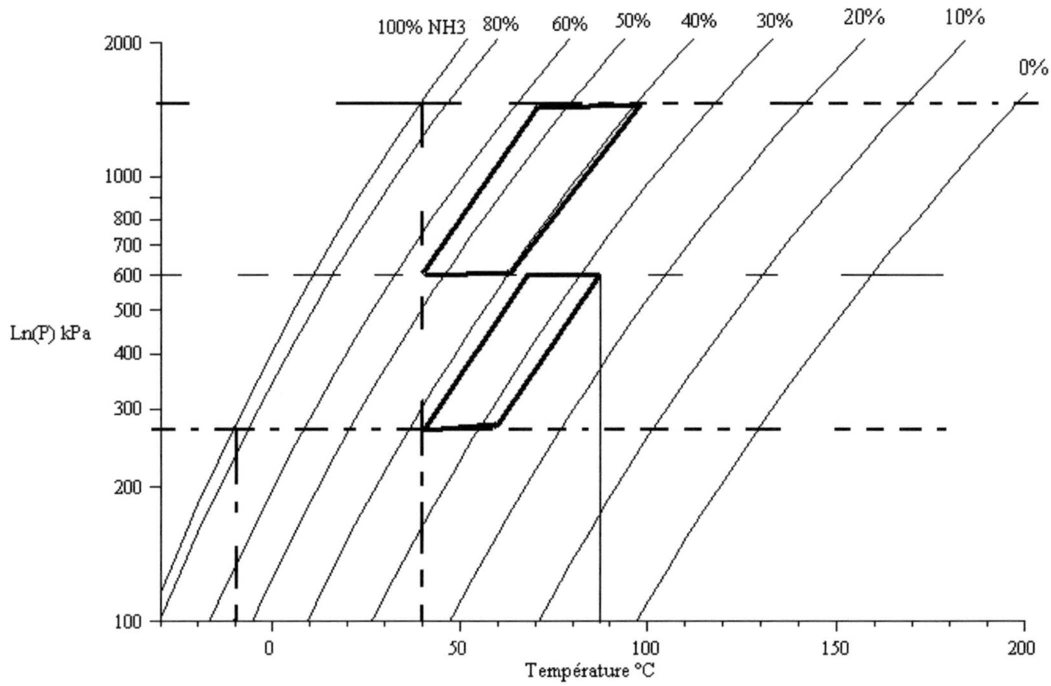

Fig. 8. Oldham diagram of system AGAG

$\dot{m}_{NH31}$ and $\dot{m}_{NH32}$ are respectively the mass flow of the refrigerant at the 1st and the 2nd stage. The mass balance for the two stages, gives:

$$\dot{m}_{NH3} = \dot{m}_{NH3i} \tag{17}$$

Two equations can be deduced from (17):

$$\dot{m}_{SRi} = f_i \cdot \dot{m}_{NH3i} \tag{18}$$

$$\dot{m}_{Spi} = (f_i - 1) \cdot \dot{m}_{NH3i} \tag{19}$$

For i=1 or 2

In order to establish the energy balance, we consider the same assumptions and we neglect the work of the pumps and the thermal power of the two rectification columns.

Heat released from the condenser is:

$$\dot{Q}_{CD} = \dot{m}_{NH3}(h_{v2} - h_{sor\_cd}) \tag{20}$$

Heat added to the evaporator is:

$$\dot{Q}_{EV} = \dot{m}_{NH3}(h_{sor\_ev} - h_{sor\_cd}) \tag{21}$$

Heat added to the generators is:

$$\dot{Q}_{GE1} = \dot{m}_{NH3}(h_{v1} + (f_1 - 1) \cdot h_{sor\_ge1} - f_1 \cdot h_{seuil1}) \tag{22}$$

$$\dot{Q}_{GE2} = \dot{m}_{NH3}(h_{v2} + (f_2 - 1) \cdot h_{sor\_ge2} - f_2 \cdot h_{seuil2}) \tag{23}$$

Heat released from the absorbers is:

$$\dot{Q}_{AB1} = \dot{m}_{NH3}(h_{sor\_ev} + (f_1 - 1) \cdot h_{ent\_ab1} - f_1 \cdot h_{sor\_ab1}) \tag{24}$$

$$\dot{Q}_{AB2} = \dot{m}_{NH3}(h_{v1} + (f_2 - 1) \cdot h_{ent\_ab2} - f_2 \cdot h_{sor\_ab2}) \tag{25}$$

We deduce the coefficient of performance as:

$$COP = \frac{\dot{Q}_{EV}}{\dot{Q}_{GE1} + \dot{Q}_{GE2}} \tag{26}$$

Using the expression of $\dot{Q}_{EV}, \dot{Q}_{GE1}$ and $\dot{Q}_{GE2}$, the explicit formula of the coefficient of performance becomes:

$$COP = \frac{(h_{sor\_ev} - h_{sor\_cd})}{\left(h_{v1} + (f_1 - 1)h_{sor\_ge1} - f_1 h_{seuil1}\right) + \left(h_{v2} + (f_2 - 1)h_{sor\_ge2} - f_2 h_{seuil2}\right)} \tag{27}$$

In this system, it is remarkable that there is a single evaporator and two generators so there is higher energy consumption. It is twice the consumption of a single stage, but the added value of this system is to lower the generator temperature. So we can conclude that for this preliminary study, the COP of this system is lower than that of a single stage.

3.3.3 System AAG

The system works as follows; rich solution is pumped from the absorber AB_2 (at the temperature T_{AB} and intermediate pressure Pmoy) and enters to the generator. Ammonia vapor goes to the condenser CD and the poor solution discharges through the absorber AB_1. The connection between the double-stages is insured between the absorbers AB_1 and AB_2 (see Figure 9). Figure 10 shows the Oldham diagram of the cycle. In such case, the COP is defined as.

$$COP = \frac{\dot{Q}_{EV1} + \dot{Q}_{EV2}}{\dot{Q}_{GE}} \tag{28}$$

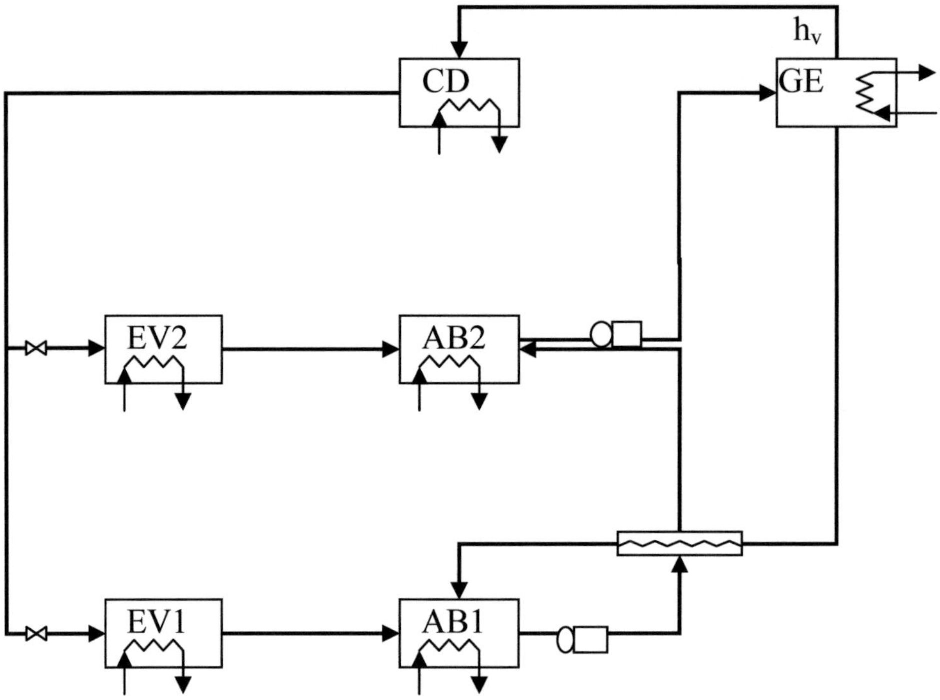

Fig. 9. System AAG (connection absorber-absorber)

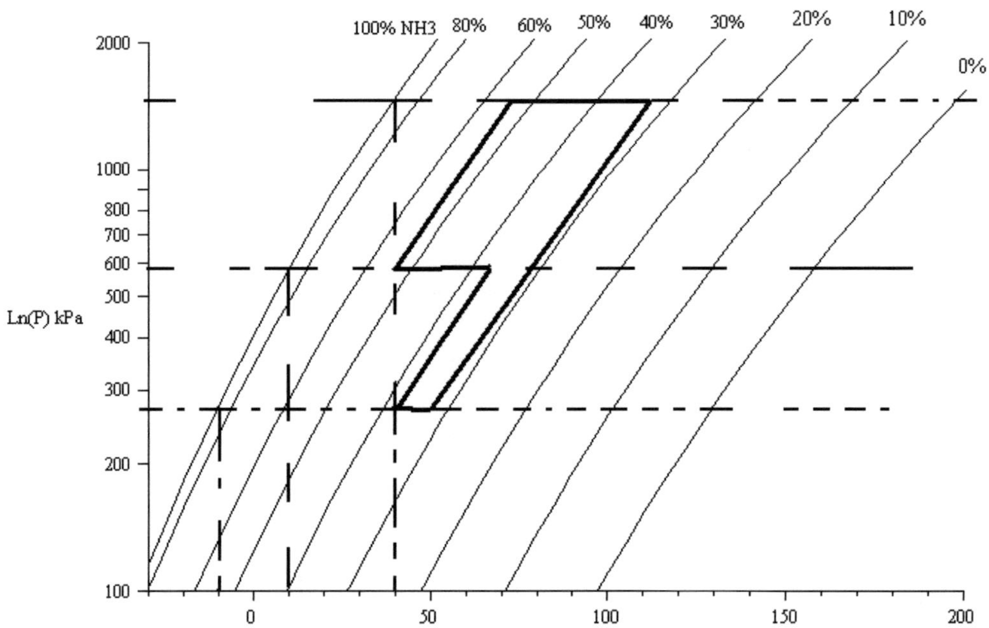

Fig. 10. Oldham diagram of system AAG

3.3.4 New system

The preceding sections, present different possible combinations with connection between the various components of the double-stage absorption refrigeration system; we propose a new designed system. The configuration consists to introduce a compressor between the first and the second stage. The considered new system is composed by two generators, two absorbers; a condenser, an evaporator and a compressor (see Figure 11).

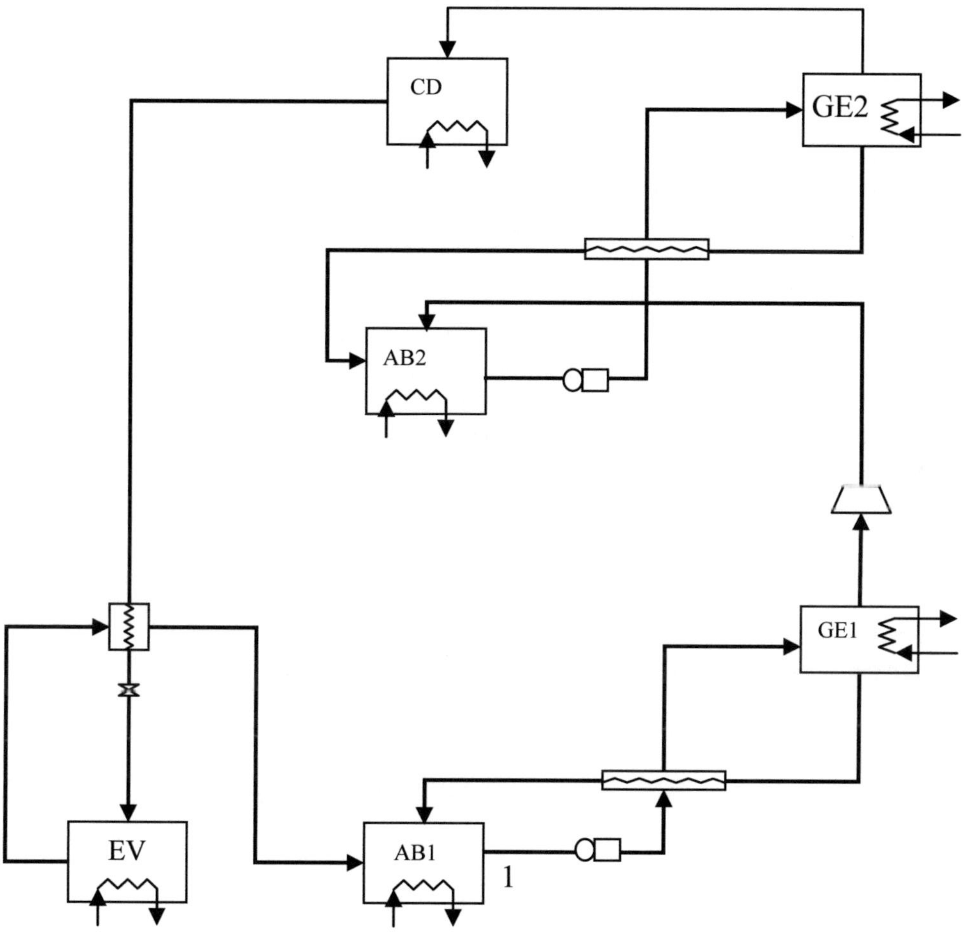

Fig. 11. New system

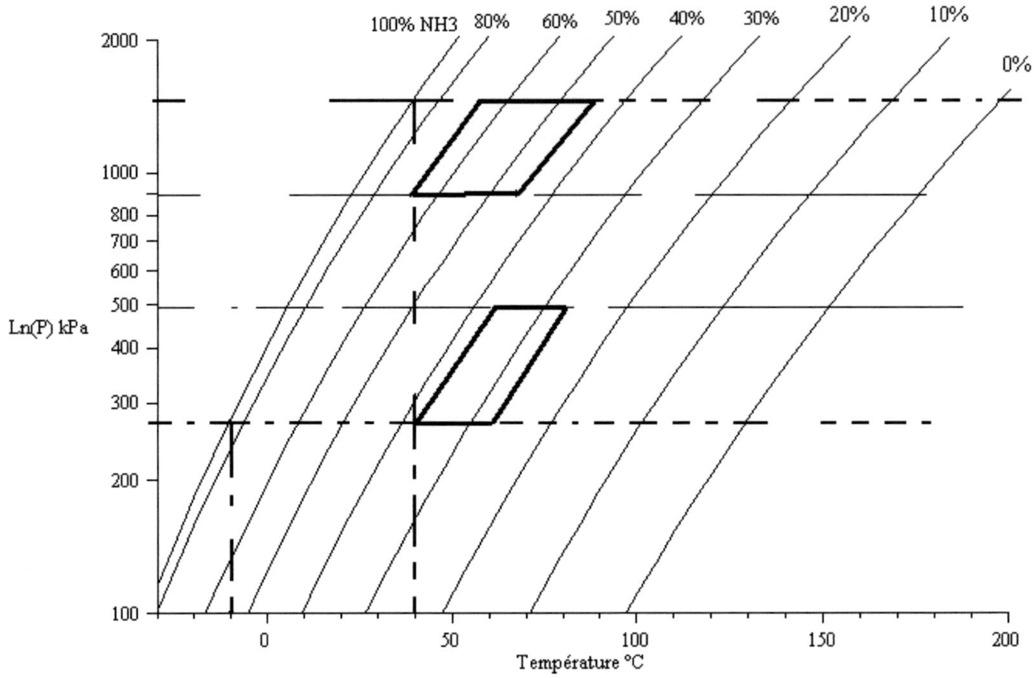

Fig. 12. Oldham diagram of new system

To determine the entrainment factors, mass flow rates and heat of the various components of such machine, we use the same equations as for the cascade AGAG, except the compressor's modeling, which must be studied separately. In fact, to determine the compressor power, we consider the ammonia at the generator exit (GE_1) as an ideal gas. For an isentropic process Laplace relation gives:

$$T_{ent\_comp} \times P_{ent\_comp}^{(1-k)/k} = T_{sor\_comp} \times P_{sor\_comp}^{(1-k)/k} \tag{29}$$

Where: T_{ent_comp}, P_{ent_comp} and, T_{sor_comp}, P_{sor_comp} are the compressor temperature and pressure inlet and outlet respectively.
Under assumption of isentropic processes (ideal case), the consumed power is given by:

$$\dot{Q}_{is} = \dot{m}_{HN3} cp_{NH3} \times (T_{sor\_comp} - T_{ent\_comp}) \tag{30}$$

But we must take into account the isentropic η_{is} where the real power:

$$\dot{Q}_{réel} = \frac{\dot{Q}_{is}}{\eta_{is}} \tag{31}$$

$$\dot{Q}_{réel} = \dot{m}_{HN3} (h_{sor\_comp} - h_{ent\_comp}) \tag{32}$$

So we can deduce from (20) and (21), the value of the steam enthalpy at the compressor outlet:

$$h_{sor\_comp} = h_{ent\_comp} + \frac{\dot{Q}_{is}}{\dot{m}_{HN3}\eta_{is}} \tag{33}$$

With

$$\eta_{is} = 0.874 - 0.0135 \cdot \tau \tag{34}$$

$$\tau = \frac{P_{sortie}}{P_{entrée}} \tag{35}$$

In this case, we note that the COP's formulation is different from other systems, since it depends on the mechanical work that is no longer negligible. Therefore, in addition to the two generators power, the compressor power ($\dot{Q}_{comp}$) is considered. The COP's expression becomes:

$$COP = \frac{\dot{Q}_{EV}}{\dot{Q}_{GE1} + \dot{Q}_{GE2} + \dot{Q}_{comp}} \tag{36}$$

3.4 Results and discussions

Several studies have been devoted to determine the COP and limitations of absorption system operating conditions [21-23]. In order to evaluate the refrigeration absorption system performance, relative to different previously presented configuration, we have developed a numerical program. The calculating procedures of the fluid thermodynamic properties and the performance coefficient were obtained using MAPLE computer tools.

3.4.1 Single-stage machine

By setting the three temperature levels T_{EV}, T_{AB}, T_{GE} and different operating conditions we determine the thermodynamic properties of the studied refrigerating system allowing the evaluation of its performance coefficient.

We note that the COP depends mainly on the evaporating temperature (necessary for the production of desired cold), the condensation temperature (function of cooling temperature of the absorber and condenser components) and finally generator temperature.

For a fixed generator temperature T_{GE} with a condensation temperature data, we analyze numerically the COP's variation of the single stage machine versus the evaporator temperature (see Figure 13).According to figure 13 and 14, we note that the coefficient of performance of a single-stage absorption system increases with the evaporator temperature rising and increases with the condenser temperature decrease.

It is noted from Figure 13, that the COP's system is higher for low values of T_{CD} and high values of T_{EV}. It is apparent that the range of the single stage machine operating conditions is adaptable to different generator temperatures. We note that for a generator temperature of 100 C and a condensing temperature higher than 40 C, the machine can operate at an evaporator temperature above -5 C. Under these conditions, the corresponding COP is approximately 0.45. We can conclude that for a temperature of 100 C at the generator, 40 C or higher for condensation, the single-stage machine is rather favorable to the air conditioning (T_{EV}> 0) than refrigeration. The COP may reach 0.55 for a condensing temperature of 45 C and an evaporator temperature of 15 C. Besides, by increasing the generator temperature of 10 C, the same system can work at a temperature of -15 C (evaporation) with the same constraint in the condensing temperature (40 C) in order to reach a COP of 0.3. To produce cold, the absorption system loses almost one-third of the COP's machine.

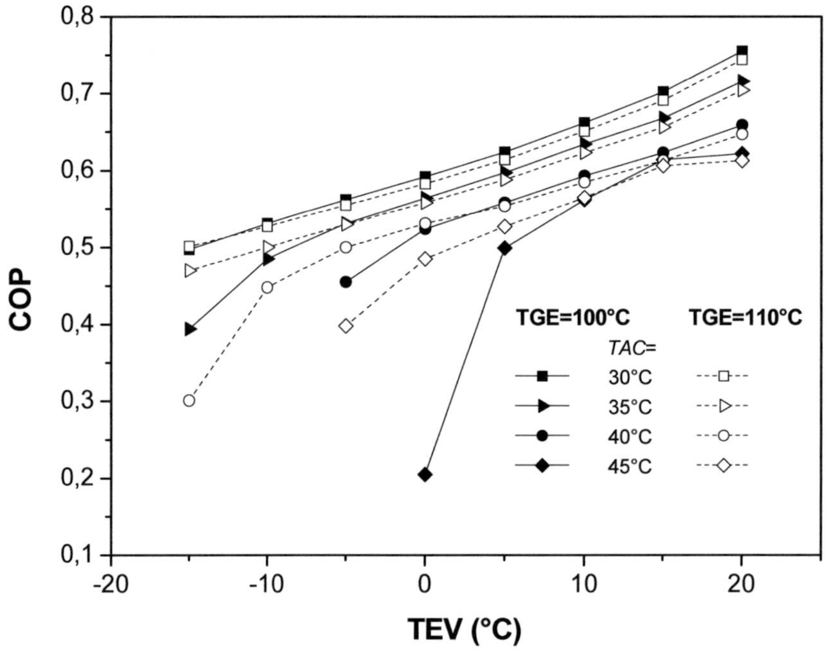

Fig. 13. COP evolution versus T_{EV} for different T_{AC} temperature and T_{GE} =100 C and 110 C

In the following, we fix the evaporator temperature for each family of curve. The numerical results illustrate the evolution of the performance coefficient for different generator temperatures. Each family has different curves for different condensation temperatures chosen between 30 C and 40 C (see Figure 14).

We note that the coefficient of performance increases with the generator and the evaporator temperature increase. While the optimal functioning depends on the condensation temperature, in fact, if it increases, the COP decreases. The cold production begins at a generator temperature greater than 110 C. On the other hand, Figures 13 and 14 show that the single stage absorption system has limited operating evaporation, condensation and generator temperatures.

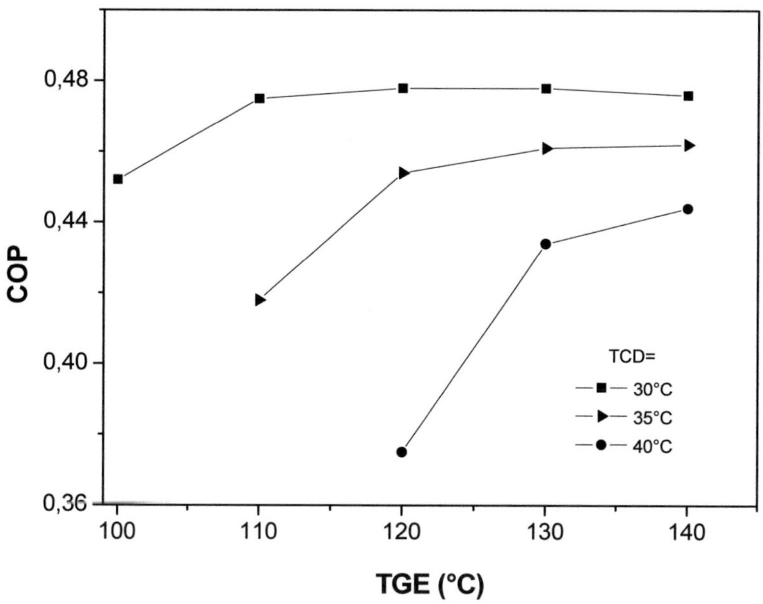

Fig. 14. COP evolution versus T_{GE} for T_{EV}=-20 C

3.4.2 Multi-stage machine

For cascading cycles, we note that there are many configurations, the difference between them is the connection between the various components of the refrigeration installation, Figures 15 and 16 show the evolution of the COP versus the possible generator and evaporation temperatures.

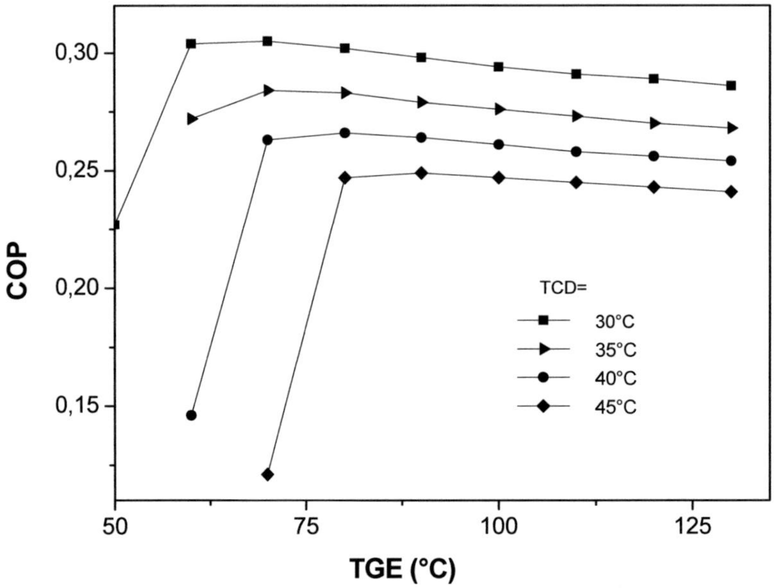

Fig. 15. COP evolution versus T_{GE} for P_1=500 kPa, P_2=900 kPa and T_{EV}=-10 C.

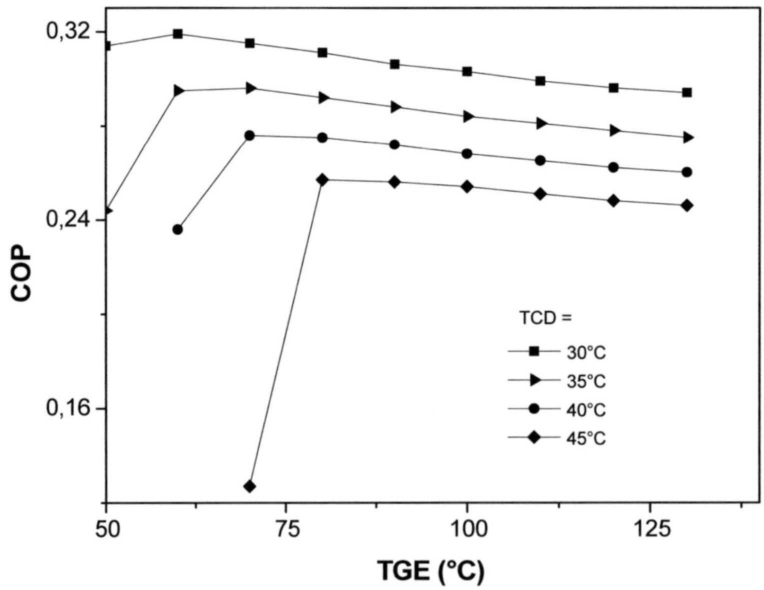

Fig. 16. COP evolution versus T_{GE} for P_1=500 kPa, P_2=900 kPa and T_{EV}=-5 C

The analysis of the COP's evolution versus different temperatures shows that the coefficient of performance increases when the condensation temperature decreases and the temperature evaporation increases. Besides, we note that in the first part of the curve, the COP increases with an increase of the generator temperature until the value of 85 C.
Figures 15 and 16 show that the COP is maximum for the generator temperature range varying between 60 and 85 C.
Figures 17 and 18 represent the COP's evolution versus the intermediate pressure, P_1, P_2. They show that the pressure averages don't have a great influence on the increase of the absorption refrigerating system performance; the advantage of this installation is that it can increase the difference of title between rich solution and weak solution. It is remarkable that this machine can operate at low temperatures. The efficiency of the hybrid absorption system proposed can reach 8.2, while the efficiency, proposed in literature which cannot exceed 5.
In the following section, we present respectively the COP evolution versus the generator temperature and the intermediate pressure (figures 19,20 and 21).

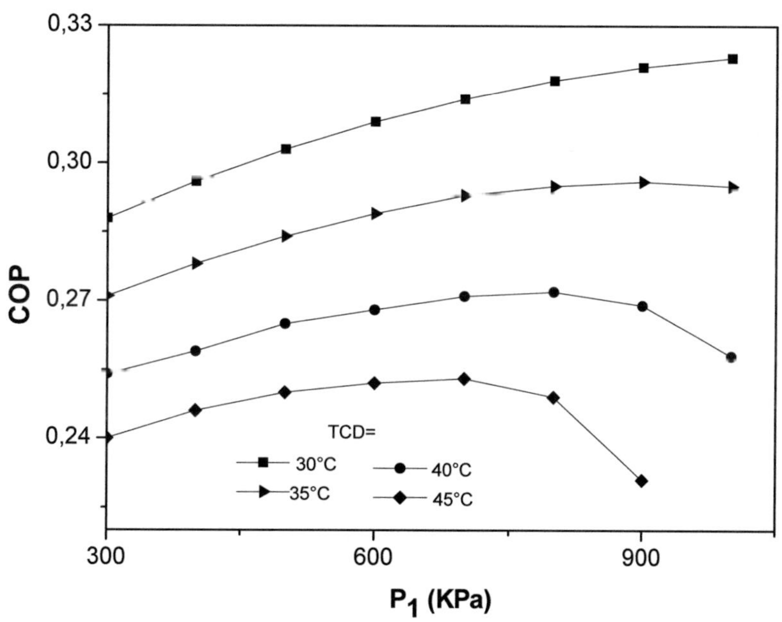

Fig. 17. COP evolution versus P_1 for T_{EV}=-10 C, P_2=1000 kPa and T_{GE}=90 C

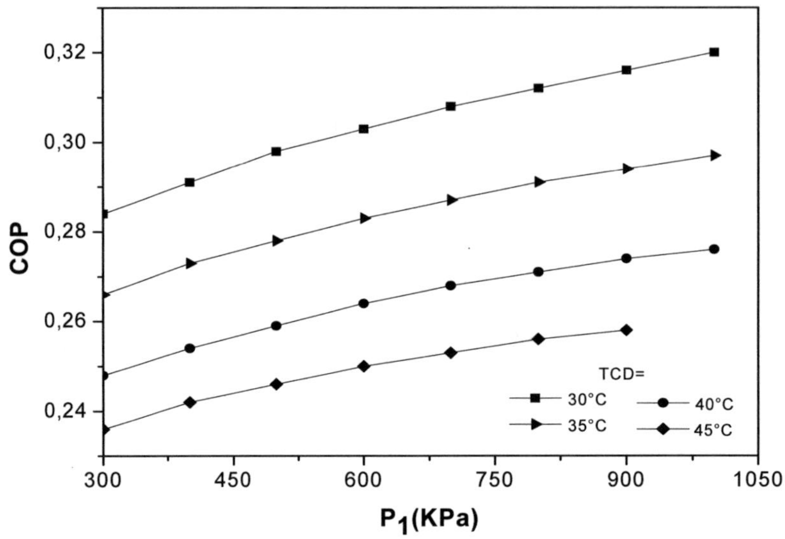

Fig. 18. COP evolution versus P_1 for T_{EV}=-10 C, P_2=1000 kPa and T_{GE}=110 C

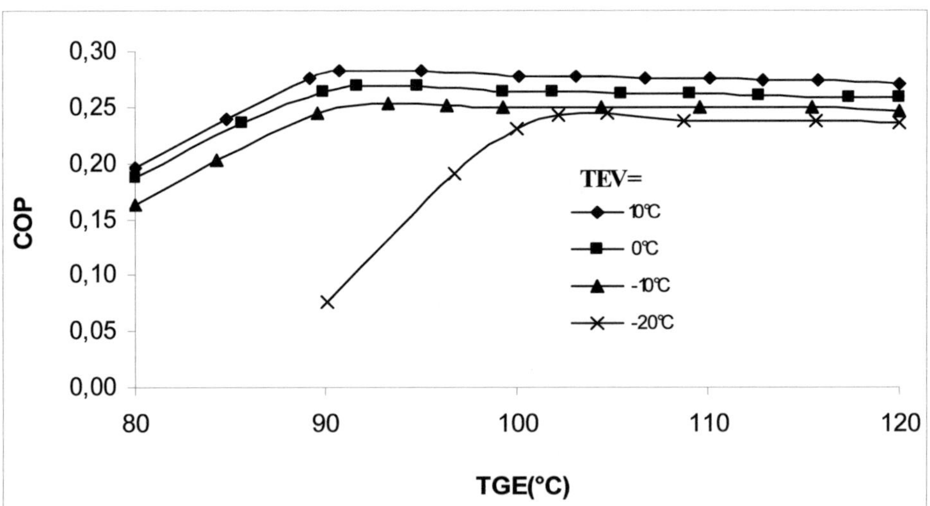

Fig. 19. COP evolution versus T_{GE} with T_{CD}=45 C and P_{moy}= 700 kPa

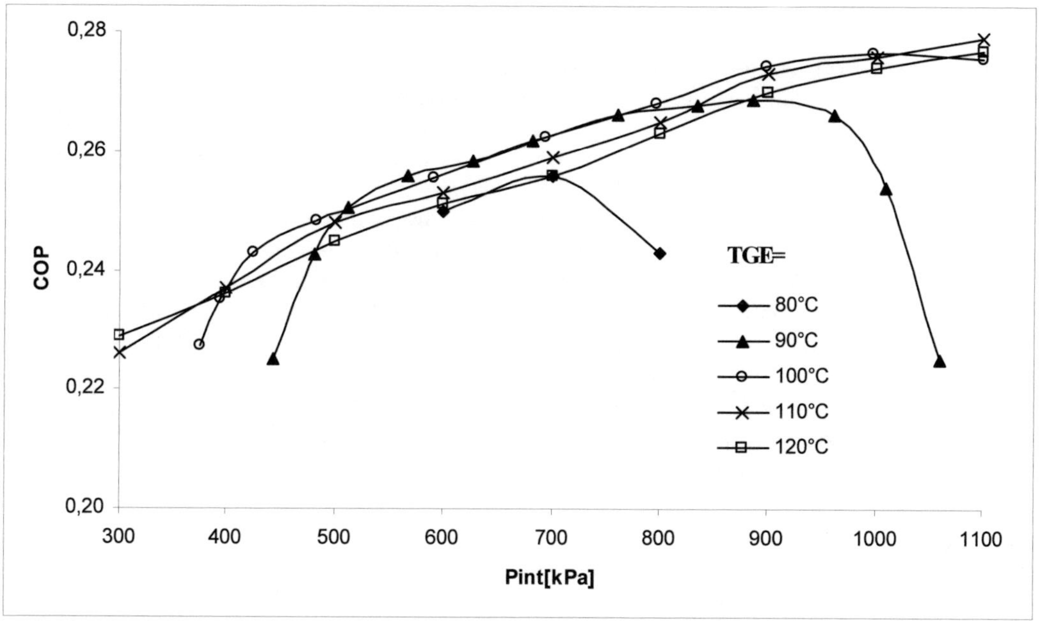

Fig. 20. COP evolution versus P_{int} with T_{CD}=40 C and T_{EV}= -10 C

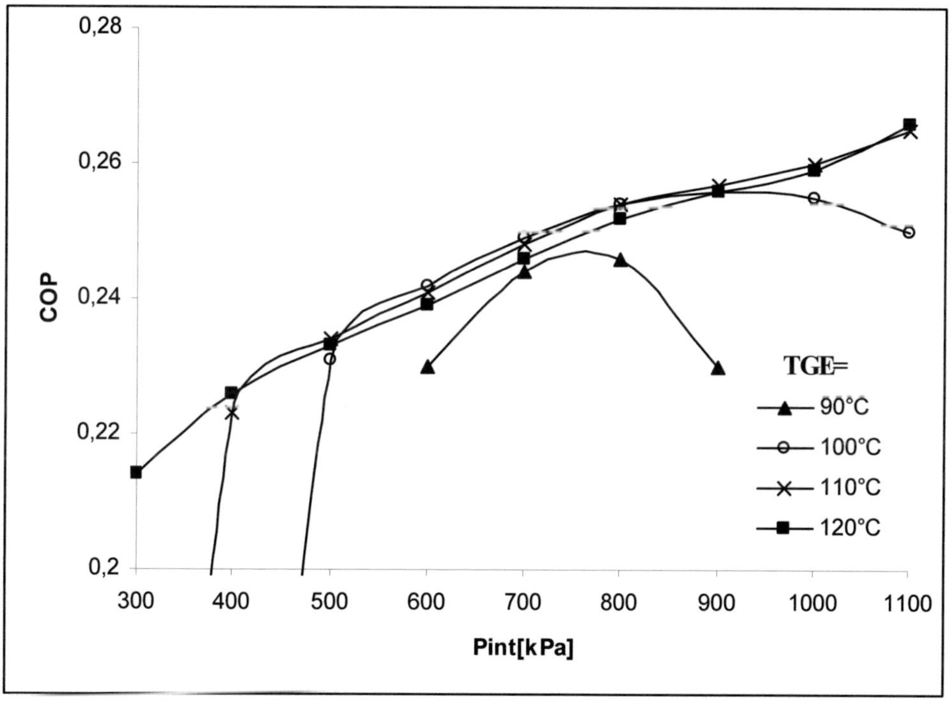

Fig. 21. COP evolution versus P_{int} with T_{CD}=45 C and T_{EV}= -10 C

From figures 19, 20 and 21, we conclude that the COP is always less than 0.28.

4. Conclusion

In this investigation, single and double-stage absorption cycles using water-ammonia are analyzed. We presented different configurations and we proposed a novel hybrid absorption refrigeration cycle. The proposed absorption/compression refrigerating system object of this work is studied in details. In order to evaluate the performance of the invoked machine, a procedure based on the MAPLE software is set up to compute accurately the thermodynamic properties of different states. The comparative study of the performance relative to different absorption cycles is carried out and the numerical results highlight that the single-stage machine has a COP higher than that of the double-stage absorption system; however, it requires a temperature generator relatively high. On the other hand, the average pressure has no influence on the increase of the system performance while it reveals very important on reducing the generator temperature.
Finally, we can conclude through this study that sources at moderate temperatures (solar, geothermal or other) can be used to power refrigeration systems absorption. The COP is acceptable and it is approximately 0.28.

5. Nomenclature

COP	= Coefficient of Performance
f	= Circulation ratio
h	= Specific enthalpy (Jkg^{-1})
$\dot{m}$	= Mass flow rate (kgs^{-1})
P	= Pressure (bar, Pa)
$\dot{Q}$	= Heat-transfer rate (W)
T	= Temperature (K, C)
x	= mass fraction
Greek symbols	
Δ	= Variation
ξ, η	= Efficiency
Subscripts	
1	= First stage
2	= Second stage
AB	= Absorber
CD	= Condenser
EV	= Evaporator
GE	= Generator
p	= Poor
r	= Rich
Sor	= Outlet
v	= Vapour

6. References

[1] L. Kairouani, E. Nahdi. Cooling performance and energy saving of a compression – absorption refrigeration system assisted by geothermal energy. App. Th. Eng. 26(2-3), 2006, 288-294.

[2] L. Kairouani, E.Nahdi. Thermodynamic Investigation of Two-Stage Absorption Refrigeration System Connected by a Compressor. A.J.A.S. 2(6): 1036-1041, 2005.
[3] H. Daliang, C. Guangming, T. Limin, H. Yijian. A novel ejector-absorption combined refrigeration cycle. Int J. Refrig., In Press, Available online 16 July 2010.
[4] J. Fernández-Seara, J. Sieres, M. Vázquez. Compression–absorption cascade refrigeration system . App. Th. Eng., 26(5-6), 2006, 502-512.
[5] A. Sözen, M. Özalp Performance improvement of absorption refrigeration system using triple-pressure-level. App. Th. Eng. 23(1-3), 2003, 1577-1593
[6] A.K. Pratihar, S.C. Kaushik, R.S. Agarwal. Simulation of an ammonia–water compression–absorption refrigeration system for water chilling application. Int J. Refrig., 33, 2010, p. 1386-1394.
[7] M. Jelinek, A . Levy, I. Borde. Performance of a triple-pressure level absorption/compression cycle. App. Th. Eng. In Press, Available online 1 February 2011
[8] A. Zohar, M. Jelinek, A. Levy, I. Borde, Numerical investigation of a diffusion absorption refrigeration cycle. Int. J. Refrig. 28, 2005, p.515–525.
[9] R. Manuel, P. Conde. Thermophysical properties of NH_3 + H_2O solutions for the industriel design of absorption refrigeration equipment. Conde Engineering. 2004.
[10] D. Boer, M .Valles, A. Coronas, Performance of double effect absorption compression cycles for air-conditioning using methanol–TEGDME and TFE–TEGDME systems as working pairs. Int J. Refrig. 21, 1998 p. 542–555.
[11] R-M. Tozer, Fundamental thermodynamics of ideal absorption cycles. Int J. Refrig. 20, 1997, p.120-135
[12] K. Joudi, H. Lafta Simulation of a simple absorption refrigeration system. Energ. Conv. and Manag. 42 (13), 2001, p. 1575-1605.
[13] M.I. Karamangil, S. Coskun, O. Kaynakli, N. Yamankaradeniz A simulation study of performance evaluation of single-stage absorption refrigeration system using conventional working fluids and alternatives. Ren. and Sus. Energy Reviews. 14 (7), 2010, p. 1969-1978.
[14] L.J. He, L.M. Tang, G.M. Chen. Performance prediction of refrigerant-DMF solutions in a single-stage solar-powered absorption refrigeration system at low generating temperatures. Solar Energy, 83(11), 2009, p. 2029-2038.
[15] A. Keçeciler, H.I. Acar, A. Dogan. Thermodynamic analysis of the absorption refrigeration system with geothermal energy: an experimental study. Energ. Conv. and Manag. 41 (1), 200, p. 37-48.
[16] S.Arh, Gaspersic B, Development and comparison of different advanced absorption cycles. Rev. Int. Froid. 13, 1990, p.41-50.
[17] A. Levy, M. Jelinek, I. Borde, F. Ziegler. Performance of an advanced absorption cycle with R125 and different absorbents. Energy, 29(12-15), 2004, p. 2501-2515.
[18] M.V. Rane, K. Amrane, R. Radermacher. Performance enhancement of a two-stage vapour compression heat pump with solution circuits by eliminating the rectifier. Int J. Refrig. 16(4), 1997, p. 247-257.
[19] R. Best, W. Rivera, M. J. Cardoso, R. J. Romero, F. A. Holland. Modelling of single-stage and advanced absorption heat transformers operating with the water/carrol mixture. App. Th. Eng. 17 (11), 1997, 1111-1122.

[20] Yong Tae Kang, Hiki Hong, Kyoung Suk Park. Performance analysis of advanced hybrid GAX cycles: HGAX.International. Int J. Refrig. 27 (4), 2004, p. 442-448.
[21] Laouir A, legoff P, Hornt J.M, Cycle de frigopmpes à absorption en cascades matérielles–détermination du nombre d'étages optimal pour le mélange ammoniac–eau. Int J. Refrig. 25; 2002, p.136-148.
[22] Sahina B, Kodal A, Thermoeconomic optimization of two stage combined refrigeration system: a finite-time approach. Int J. Refrig. 25; 2002, p.872-877.
[23] Fernàndez-Seara J, Vazquez M, Study and control of the optimal generation temperature in NH3-H2O absorption refrigeration systems. App. Th. Eng. 21; 2001, p.343-357.

7

Non-Equilibrium Thermodynamics, Landscape Ecology and Vegetation Science

Vittorio Ingegnoli
Dpt. of Biology, Natural Sciences Faculty, University of Milan
Italy

1. Introduction

As underlined by Ingegnoli (2002), scientists have to avoid two representations of nature which tend to a world of alienation: (1) the deterministic one, with no possibility of novelty and creation, (2) the stochastic one, which leads to an absurd world with no causality principle and without any ability to forecast. Possibly, the major incentive toward a new conception of nature comes from scientists like W. Ashby (1962), Von Bertalanffy (1968), Weiss (1969), Lorenz (1978, 1980), Popper (1982, 1996) and Prigogine (1977, 1996), who observed how nature creates its most fine, sensitive and complex structures through non-reversible processes which are time oriented (time arrow). No doubt that thermodynamics becomes the most important physical discipline when complex adaptive systems exchanging energy, matter and information are involved with life processes.

Mainly starting from the System Theory and the study of complex systems, this group of scientists asserts that: (a) an organic whole is more complex than the sum of its parts (emergent properties principle) and (b) the description of the behaviour of a dynamic system presents more solutions than the classical ones. Therefore, they reach the conclusion that "life is only possible in a Universe far away from equilibrium" and that "indeterminacy is compatible with reality". The self-organising properties of non-equilibrium dissipative structures and the basic feature of indeterminacy show the real nature of our universe.

Following these scientific paradigms we can focalise a new course of Landscape Ecology[1], related to a new definition of landscape. The need of a widening foundation of this discipline brought to the school of Biological Integrated Landscape Ecology (Ingegnoli, 2002), recently named Landscape Bionomics (Ingegnoli, 2010, 2011). All these premises allow to understand the extant scientific situation in vegetation science, in which phytosociology presents serious limitations, especially in landscape evaluation.

A theoretical revision of life organisation characters and basic transformation processes of ecological systems open this chapter, leading to consider more advanced transformation and metastability processes in vegetation (from community dynamics to biological territorial capacity of vegetated units). This more theoretical and critical section is followed by an innovative section, proposing new criteria to overcome deterministic concepts (e.g. potential vegetation) in the study of vegetation and landscape. The first statements by Braun-Blanquet

---

[1] The discipline of Landscape Ecology has been defined as "a study of the structure, functions and change in a heterogeneous land area composed of interacting ecosystems" (Forman & Godron, 1986).

(1928) maintain their significance as basic concepts in studying vegetation, but are in need to be integrated in new scientific theories (Naveh, 1984, 1990; Pignatti, 1994; Pignatti, Box & Fujiwara 2002; Ingegnoli, 1997, 2002; Ingegnoli & Giglio, 2005; Ingegnoli & Pignatti, 2007).

We will see that, following scientific paradigms like thermodynamics, it is possible to relate the landscape equilibrium to the concept of metastability, that is the state of a system oscillating around a central position (steady or stationary state), but susceptible to being diverted to another equilibrium state. Therefore different types of landscapes (or their parts) may be correlated with diverse levels of metastability. This statement has a very important dynamic significance, because it allows knowledge of the transformation modalities of a landscape and consequently (as we will see further) allows the diagnosis of its healthy state.

Trying to evaluate the metastability of a landscape, one has to refer to the concept of biodiversity (i.e. landscape diversity) and to the concept of latent capacity of homeostasis of an ecocoenotope (or tessera). Referring to a vegetation ecocoenotope, it has been possible to define a magnitude, named *biological territorial capacity* or BTC (Ingegnoli 1991, 2002; Ingegnoli and Giglio 1999, 2005, Ingegnoli and Pignatti, 2007), which represents the flux of energy that an ecocoenotope must dissipate to maintain its proper level of order and metastability. Therefore, the linkage of vegetation science with landscape ecology and with thermodynamics becomes more effective. An example of application of the discipline on the territory of Mori (Trento, Italy) is shown at the end of this chapter.

2. Main characters of biological systems

Between life and its environment we can discover strict relationships, exchange of matter and information and *a priori* knowledge. Energy can be transformed in matter or information, depending on different codifications of the Chronotope[2].

In the frame of the Theory of Relativity (Einstein) not only energy and mass are transmutable, but even space and time. Therefore the Chronotope shows 4 dimensions. Energy can be organized as matter or information, depending on different codifications of the chronotope. When energy is transformed in matter it assumes 3 spatial dimensions (x, y, z) plus one temporal dimension (t); while, if energy is transformed in information it assumes 2 spatial dimensions (e.g. plane wave) and 2 temporal dimensions (t_1, t_2). We have to underline these concepts, because the development of neg-entropy is needed in the evolution of natural systems, like landscapes and vegetation ones.

As expressed by P. Manzelli (1994, 1999), professor at the University of Florence, when the visible light frequencies cross a transparent medium, the associated plane wave remains dimensioned as information (2 spatial and 2 temporal dimensions); on the contrary, when the wave encounters the retina, the photochemical reaction is done through the conversion into a particle of the plane wave, which assumes a form available to interact with the three-dimensional structure of the matter.

It is important to underline these facts, because every transformation between energy and matter needs a catalisys through an information system, to increase the neg-entropy and to proceed toward ordered forms. We know that the exchanges energy-matter-information, which allowed the emergence of life on Earth, are of the maximum importance and changed completely the evolution of the entire Planet. A mutual interaction and an information

[2] Chronotope (literally: space-time), term used both in science (Einstein's Relativity) and literature (Baktin on Novels).

exchange are present between life and his environment: a sort of "*a priori*" knowledge. As Karl Popper (1994) underlined: "From the beginning, life must have been equipped with a general knowledge, the one which we usually name 'knowledge of the natural lows'". Note that the current definition of adaptation is Darwinian, but it must be changed, because it is not seen as a form of *a priori* knowledge.

In facts, the definition of life contains both biological systems *and* their environment: therefore every living system follows life processes and exhibits systemic attributes.

Life is a complex self-organising system, operating with continuous exchange of matter and energy with the outside; the system is able to perceive, process and transfer information, to reach a target, reproduce itself, have an history and participate in the process of evolution. In an evolutionary view, structure and function become complementary aspects of the same evolving whole. Consequently life can not exist without its environment: both are the necessary components of the system, because life depends on exchange of matter and energy between a concrete entity, like an organism, and its environment (Ingegnoli and Pignatti 1996; Pignatti and Trezza, 2000; Ingegnoli, 2002). That is the reason why the concept of life is not limited to a single organism or to a group of species, and therefore life organisation can be described in hierarchic levels.

The world around life is made also by life itself; so the integration reaches again new levels. This is another reason why biological levels can not be limited to cell, organism, population, communities and their life support systems: life also includes ecological systems such as ecocoenotopes (Ingegnoli 2002), landscapes, ecoregions, and the entire ecosphere.

A short exposition of the main modern scientific paradigms (from hierarchic structure to non-equilibrium thermodynamics) and the new importance of history is necessary to better understand these characters of living systems and to update ecology.

2.1 Hierarchic and dynamic systems

The central concept of the hierarchical System Theory (Pattee,1973; Allen & Starr, 1982; O'Neill et al. 1986) is that the organisation of a system results from differences in process rates, which change with the scale. Levels within the hierarchy are isolated from each other because they operate at distinctly different rates. Boundaries, which are not only the physical ones, separate the set of processes from components in the rest of the system. As an example, for the investigation of a woodland, the first approximation will be to study in what kind of vegetational landscape it is growing, what are the climatic constraints, etc.; then this woodland has to be investigated on even a more detailed scale, e.g. single trees, if the interest shifts to the components of the plant association and the reason of their existence Note that one of the most important consequences of the hierarchical structure of systems is the concept of *constraint,* deriving from the complex interaction of several factors: it is more correct than the concept of limiting factor, i.e., a single negative action producing a linear reaction. Constraints affect the behaviour of an ecological system though the behaviour of its components and with environmental bonds imposed by superior levels of organisation. Remember that there is a linkage between constraint and information.

The System Theory states that an evolving system is first of all defined as *dynamic*. In consequence, the output (y) depends on the *history* of the system, not linearly on the input (*a*). A third element has to be introduced: the *state*, which includes information on the past, present and potential evolution of the whole. The value x (t), assumed by the state at the instant *t*, must be sufficient to determine the value of output in the same instant: knowing the values of $x\ (t_1)$ and $a\ (t_1,t_2)$, the state (then the output) in the instant t_2 can be calculated. The couple state-time (x, t) has great significance because the set X , T is the set of events, the *history* of the system. The space containing the points corresponding to the states of the system is called the 'space of the phases'. Once an instant t, an initial state $x\ (t_0)$, an input function $a\ (.)$ are fixed, the transition function $f\ [t,\ t_0,\ x\ (t_0),\ a\ (.)]$ is univocally determined, and named "movement" of the system:

$$x\ (t) = f\ [t,\ t_0,\ x\ (t_0),\ a(.)] \tag{1}$$

A function of output transformation $u\ [t,\ x(t)]$ brings to:

$$y(t) = u\ [t,\ x(t)] \tag{2}$$

Thus, a dynamic system can be described using 6 sets of variables, correlated by 2 functions.

2.2 Dissipative systems

Systems which experience dynamic changes consume energy, therefore the photosynthesis (or chemio-synthesis in primeval systems) becomes necessary.
Photosynthetic processes have the main responsibility of energy transfer in biological systems. This is possible because living systems are open systems, otherwise, the free energy F would not be available. In open systems, variations of entropy can be the consequence of different processes: d_eS , is the entropy exchanged with the environment, and d_iS , is the entropy variation due to irreversible processes within the system. The second term is clearly positive, but the first term does not have a definite sign. So the inequality of Clausius-Carnot becomes:

$$dS = d_eS + d_iS \quad (being\ d_iS > 0) \tag{3}$$

In a period in which the system is stationary $(dS = 0)$, thus

$$d_eS + d_iS = 0 \quad and \quad d_eS < 0 \quad (\ being\ d_eS = -\ d_iS) \tag{4}$$

In evolutionary processes, when the system reaches a state of lower entropy (new stationary state) $S\ (t_1) < S\ (t_0)$, it is able to maintain it in balance by "pumping out" the disorder. But this is possible only in non-equilibrium conditions of *dissipative* systems: a dissipation of energy into heat is necessary to maintain the system far from equilibrium and to create order, as can be observed in thermodynamics, but also in the mediterranean vegetation (Pignatti, 1979; Naveh & Lieberman, 1984). The amount of entropy "pumped out" is indicated as negentropy.
An energy dissipation, which allows work to be done, has to be coupled, for instance, with the transformation of the system from state A_0 to state A_1. The process able to perform this transformation is an example of *operator* (*Op*), a *rule* of action on a given function. If we express it in the form $A_1 = (Op)\ A_0$, the complete transformation process is

$$A_1 = [(Op)\ A_0] \cup (e_w \rightarrow e_d\ ) \tag{5}$$

where: e_w = available energy, e_d = dissipated energy.
If the state of the system becomes an auto-function for a certain operator (i.e. a function able to remain as before when applied to an Op) the system does not undergo further changes.

This state is called a *fixed point* of the system, and it may represent a stationary state or an *attractor.*

2.3 Self-organisation and chaos

Complex interacting systems in which cycling, structuring and auto-regulation are realised from the inside, may be called *self-organising* systems. In living systems the capacity to maintain a dynamic equilibrium as a whole is called *homeostasis*. It is ensured by a large number of closely interrelating cybernetic feedback mechanisms, hierarchically ordered. These biological and ecological processes of auto-regulation can be active also at the landscape level.

Auto-regulation needs information, deriving from biological and technological processes, which can be carried out both in energetic and/or in material way: that is, energy structures itself with the help of information. Positive and negative feedbacks coupling are fundamental, too. Their dynamics can be synthetically expressed by:

$$x_t = f\,(x_0\,,\,t,\,\lambda), \qquad (6)$$

where x_t is the state of the system at time t, x_0 is the state of the system at time 0, λ is a specific parameter for the examined system indicating the acquisition of energy and matter from outside.

Depending on the parameter λ and its values (Pignatti & Trezza, 2000), X may tend toward a temporary stationary state (*metastable* state) or a chaotic one. Note that the *uncertainty* given by *chaos* does not depend on complexity: in fact, even a simple deterministic system can be chaotic.

A system is chaotic when it amplifies *initial* conditions, thus magnifying small differences, for instance between two trajectories. It is impossible to shorten the description of a chaotic system because of its unpredictable behaviour due to branching possibilities of evolution, thus to a manifold of attractors.

Highly chaotic webs are so disordered that the control of complex behaviours is impossible, while highly ordered webs are so rigid that they can not express a complex behaviour. But if "frozen" components begin to melt, it is possible to have more complex dynamic behaviours leading to a complex co-ordination of activities within the system. Thus, the maximum complexity is reached in a "liquid" transition between solid and gaseous states, where the best capacity of evolution is expressed. For instance, it is possible to see a similar situation in DNA and its capacity to maintain a ordered structure but also to change by mutations. As shown by Prigogine (1996), if we consider the Bernoulli equation:

$$x_{n+1} = 2\,x_n\ (Mod\ 1) \qquad (7)$$

where: *Mod 1* = numbers between 0 and 1, it is easy to see that very short differences of the initial conditions can brought to very different trajectories, as shown in Fig. 1.

The threshold between order and chaos seems to be an essential requisite of complex adaptive self-organising systems (order at the edge of chaos). As these systems are dissipative, an order through fluctuations is effective in working between the above mentioned conditions.

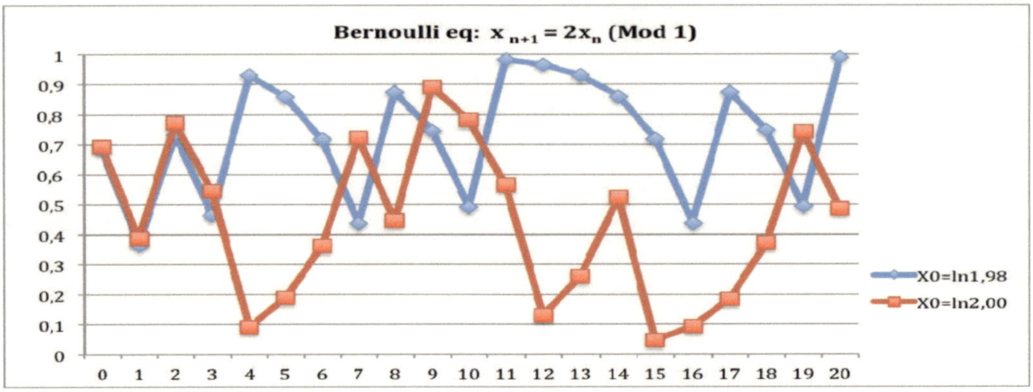

Fig. 1. An example of deterministic chaos. Starting from two very similar initial conditions (x_0 = ln 1.98, x_0 = ln 2.00) the Bernoulli equation (7) shows very different trajectories, after time 3. Note that these lines may represent the projection of 2 possible movements of a dynamic system within the field of the states of the system itself.

3. Non-equilibrium thermodynamic and metastability in ecological systems

A self-organised living system is able to capture intense energy fluxes and to utilise its neg-entropic input to produce new structures. Prigogine showed (1972) that even simple physical systems present processes of order.

Figure 2 shows the concentration of the intermediate product X in a chemical reaction: going further on the stable thermodynamic branch, the intermediate product enters a field of instability with the appearance of subsequent bifurcations.

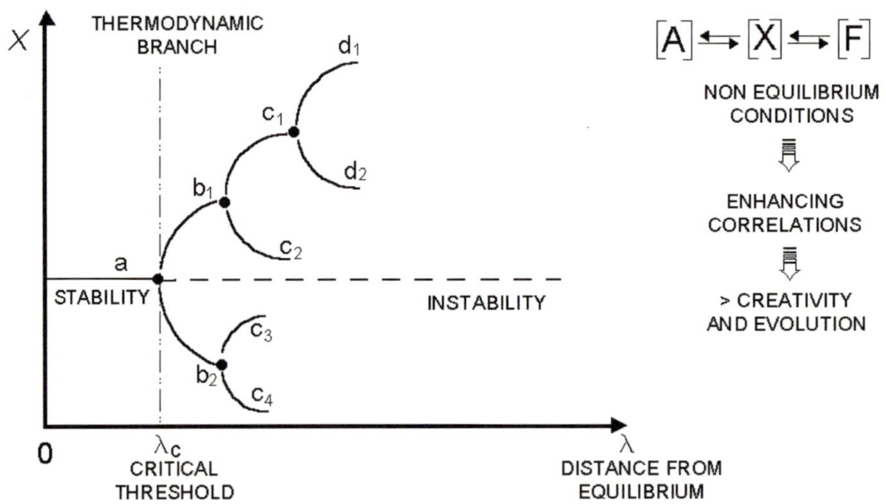

Fig. 2. Consecutive bifurcations in a non-equilibrium system. Going further on the stable thermodynamic branch, the intermediate product enters a field of instability with the appearance of subsequent bifurcations. Note that the point d_2 can be reached through the path a-b_1-c_1-d_2 but also a-b_1-c_2-d_2. So, an historical behaviour is shown in this process (from Ingegnoli, 2002).

Therefore, the result cannot be deterministic: when a system arrives at a branching point, disturbances, like fluctuations or strange attractors, become important, allowing the system to choose one of the two branches of new relative stability. So, the evolution of this kind of system has an *historic* criterion in itself.

The fluctuation-dissipation sequence can be viewed as a feedback process. A macro-fluctuation, due to a change of disturbances, produces instabilities leading to an increased dissipation of energy and the system becomes more difficult to maintain. When a threshold is reached, characterised by the prevailing of new structures over the former ones, a new organisational state results. That is why the Prigogine statement is "order through fluctuations". Ecological conditions are important for a system at a branching point, enabling it to choose one of the two branches of new relative stability (*metastability*).

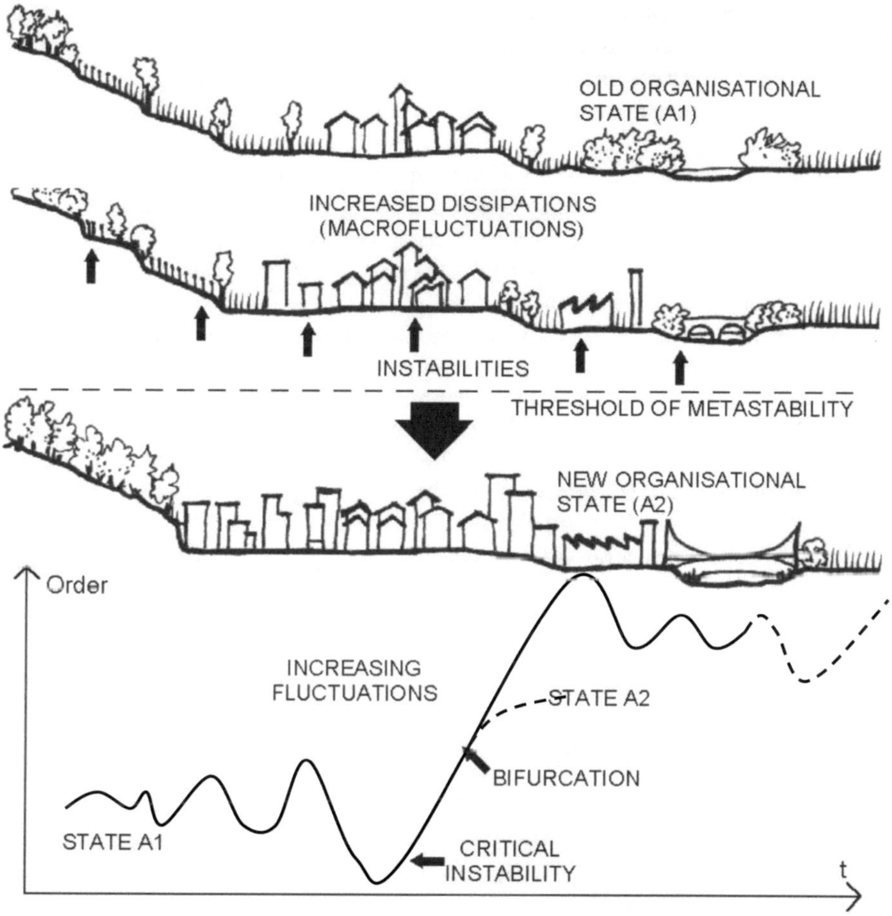

Fig. 3. Landscape transformation. From a state *A1* of lower order through increasing dissipation, a system reaches a critical threshold and, after a branching point, it arrives at the state *A2* of higher order. The old organisational state is a rural landscape; an increased flux of energy produces macro fluctuations of the local organisation and then some instabilities. These instabilities cause an increased dissipation of energy, the system becomes difficult to maintain: when a threshold is reached (e.g. a prevailing of urban structures over the former rural ones) a new organisational state results (from Ingegnoli, 2002).

Under these conditions, mutual relations of large range occur among the components. The matter acquires new properties, a new sensitivity of matter to itself, to information and its environment takes place, associated with dissipative and not reversible processes. The system, in the far from equilibrium condition, is able to self-organise through intrinsic probabilities, exploring its structure and realising one among the possible structures, but not a random one. This process takes place from cell proteins formation to the vegetation and the landscape transformation.

Let us show an example of landscape transformation (Fig. 3). From a state *A1* of lower order through increasing dissipation, a system reaches a critical threshold and, after a branching point, it arrives at the state *A2* of higher order. In this case, the old organisational state is an agricultural landscape. An increased flux of energy (e.g. agricultural improvement and social-economic richness) produces macro fluctuations of the local organisation and then some instabilities (i.e. land abandonment, use of the fluvial valley, building of the first industries, and so on). These instabilities lead to an increased dissipation of energy, the system becomes more difficult to maintain: when a threshold is reached, characterised by the dominance of urbanised structures over the previous rural ones, a new organisational state results, that needs a different kind of management.

When a system is oscillating around a steady attractor, but may even move toward another attractor, it presents the condition of *metastability* (Godron 1984; Naveh and Lieberman 1984; Forman and Godron 1986). Note that the concept of metastability is not a compromise between a form of stability and one of instability. Higher or lower metastability depends on the distance from the position of maximum stability and on the height of the thresholds of local (far from equilibrium) stability.

Ecological systems with low metastability have a low resistance, but a high resilience to disturbances. By contrast, high metastability systems have high resistance to disturbances. For example, a prairie patch has a higher resilience than a forest one. Note that the concept of metastability allows the traditional concept of ecological equilibrium to be updated: "equilibrium" does not stay around 0, but it identifies various stationary or equilibrium states far from 0. A system reaches a new organisation after instabilities and the passage to a new metastable level.

Remembering the hierarchic theory of systems, we know that some limitations on the dynamic of an ecological system come from inferior levels of scale and are due to the biological potential of its components. Other limits are imposed by superior levels as environmental constraints (Cfr. 2.1). Therefore, a wide range of conditions emerges for every kind of ecological system, for instance a vegetation complex in a landscape, and can be expressed as the *constraints field* or optimum set of existence.

Note that, in many cases, the majority of disturbances can be *incorporated* into ecological systems. The mentioned constraint field of an ecological system is based on a resistance strategy to a current regime of perturbations. Therefore, we can speak of 'disturbance incorporation' when the system organisation exerts control over some environmental aspects that are impossible to be controlled at a lower level of organisation. This process may limit possible alterations to its stationary state; meanwhile it may utilise perturbations as structuring forces.

3.1 The importance of history

Remembering the importance of the concept of time after the theories of Albert Einstein, this should be extended to all the modern science. As formerly mentioned, the state of a system is fundamental to understand the movement of the system itself; consequently, in the "order through fluctuation" process the evolution of a system presents an historic criterion in itself. Therefore, history assumes a new crucial importance even in ecological studies. Note that history (*historia* in Latin) derives from the Greek 'ιστωρ' which means "cognition and research" but today history is intended mainly in humanistic sense and -if not- in deterministic sense.

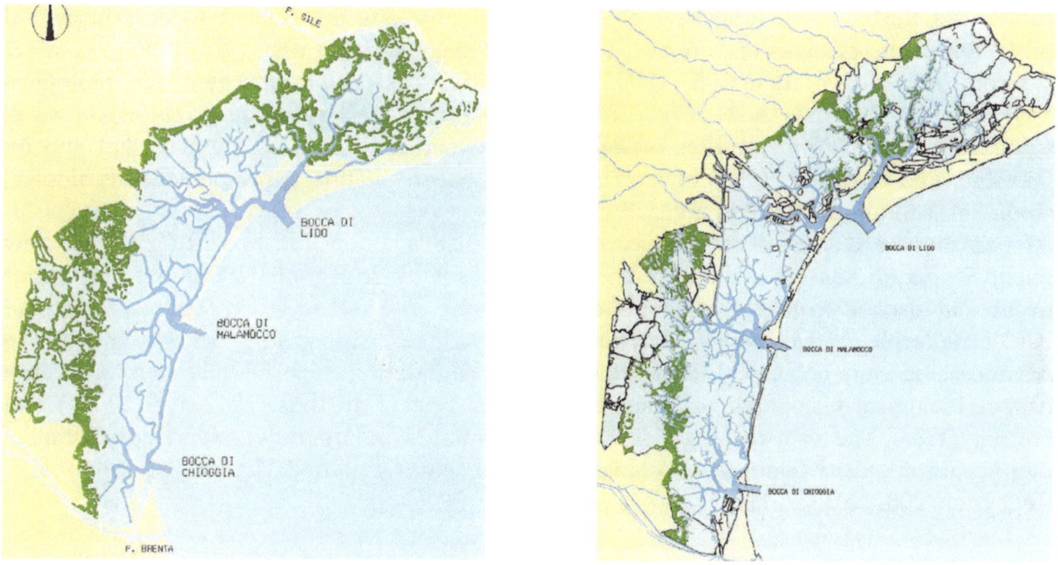

Fig. 4. Synthetic maps of the Venice lagoon, showing the distribution and the extension of the salt marsh prairies (green), called "barene". Note the sharp difference between 1930 (left) and 1998 (plots from CVN-Technital, 2002). Note the presence of a large harbour with an industrial area (west to Venice). In the last century (1900-2000) the barene formations decreased dramatically, from 13.2% to 4.6%.

In humanistic sense, history is the understanding on the human past. Without the presence of some cultural artefact, no natural system can be studied properly in historical way. A landscape is seen only as a "cultural product", thus a forest, for instance, can not be studied as an historical subject. In deterministic sense, history is the description of naturalistic frames from which being able to deduce temporal changes according to some typologies following some laws. A landscape, in this way, is studied considering its territory as a subject containing all its own determination parameters, in a way that will not be questioned.

Hence, the humanistic sense of history is obviously too limited. In deterministic sense history forces natural changes into mechanical succession schemes. For instance, some Author presumes to evaluate the ecological state of a landscape measuring the distance of the present vegetation from the potential one: a nonsense, as we will see later on.

These limited definitions of history may bring to severe methodological errors which depend on obsolete scientific paradigms. We have to remember that the real world is transforming itself following the time arrow, in a non-finalistic evolution and in a creative way. That is why history has becoming indispensable. Without it, it is simply impossible to understand properly the right sense of the events.

Related to time irreversibility the natural processes may be variant or invariant, anyway they form real systems the behaviour of which does not accept a full determinism. So, history is the research on the evolution occurred in natural systems, that is on the happening of the phenomena in a previous time (Zanzi, 1998) (Fig.4).

4. Landscape bionomics

In the last thirty years, following an increasing consciousness related to environmental problems, some scientists of different Countries (Naveh & Lieberman, 1984, 1990; Forman & Godron, 1986, 1995; Ingegnoli, 1980, 1991; Noss, 1983, 1997) identified the biological hierarchic level of the "system of ecosystems" -that is the landscape level- as the most suitable and sensible for studies on relations between man and his environment and on "positive and negative effects of men actions on nature". Thus, a new level of ecological studies was founded, named Landscape Ecology.

At present, the discipline of landscape ecology needs a revision according to the new scientific paradigms we enhanced before. That is why Ingegnoli (2002) tried to better focalize landscape ecological elements and processes, in order to widen the foundation of landscape ecology, as expressed through his Biological Integrated School. Indeed, to advance landscape ecological theory, a widening foundation must be able to relocate in a deeper biological vision the different approaches, first of all those by Naveh (1984) and Forman (1986). The term "ecology" is today both inflated and degraded. So, the discipline of Biological Integrated Landscape Ecology has been recently named "Landscape Bionomics" (Ingegnoli, 2002, 2010, 2011).

4.1 The new school of biological integrated landscape ecology, or landscape bionomics

First of all, it is necessary to reach a manifold but unique definition of landscape and also to recognise what is important about landscapes. In this framework, it is useful to understand that:

a. the landscape, as a level of hierarchical organisation of the life on Earth, *is a proper biological system;*

b. thus, the landscape is a complex, adaptive, dynamic, self-organising, hierarchical system;

c. its complex structural model can be based on the concept of tissue, thus being named *ecotissue* (Ingegnoli, 1993, 2002) (related concept: ecocoenotope);

d. we have to consider landscape bionomics (ecology) as a discipline like medicine, biologically based and transdisciplinary. Remember that we have to study the landscape pathologies, but also their influence on human health, which may be dangerous even in absence of pollution.[3]

e. Even *culture does not implicate the subjection of nature to the dominance of man*; we may demonstrate that in many cases cultural changes of landscapes express natural needs.

Being the landscape a biological level, it is the physiology (ecology)/pathology ratio which permits a clinical diagnosis of the landscape, after a good analysis and anamnesis. No doubt that landscape bionomics has its own predictive theory, nevertheless, it is necessary to

[3] The environmental stress brings to lower 24h mean cortisol excretion and to partial inhibition of feedback mechanisms.

develop this discipline not as a simple predictive science, but also as a prescriptive one - again just like medicine.

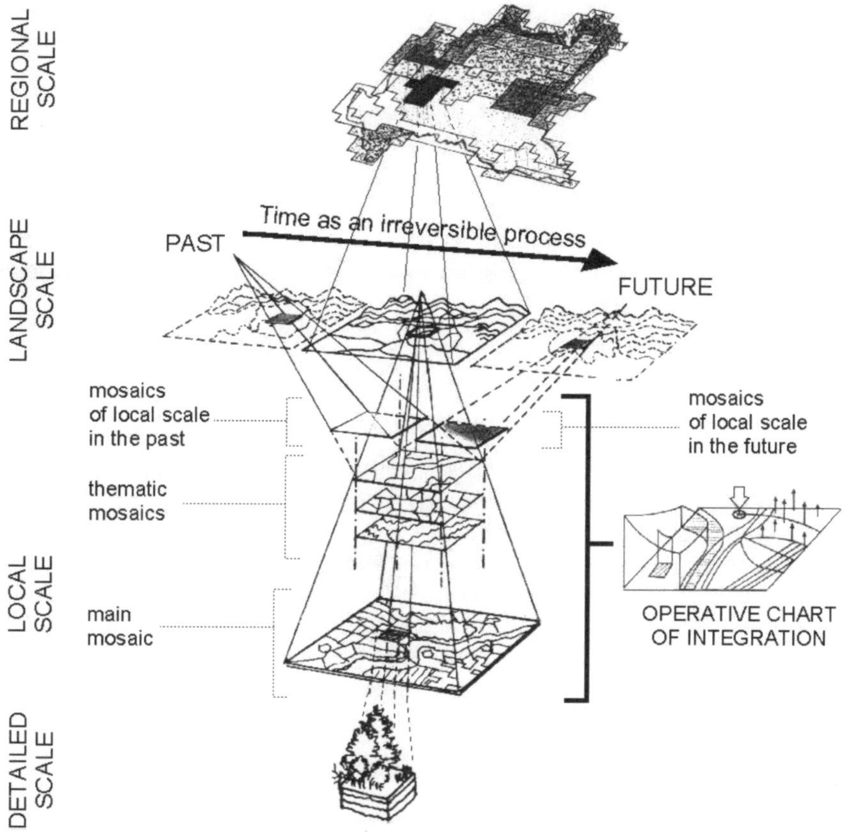

Fig. 5. The landscape ecotissue: the basic mosaic is generally the vegetation one. The complex structure of a landscape has to integrate diverse components: temporal, spatial, thematic. An operative chart of integration could be necessary to elaborate plans. Note that the integrations are intrinsic, that means they have to follow integration functions derived from the intrinsic characters of that level of life organisation (from Ingegnoli, 2002).

- Subsequent, it is necessary to define the ecocoenotope and the ecotissue, as follow:
- the *ecocoenotope* is an ecological system, composed by the *community* (biotic view), the *ecosystem* (functional view) and the *microchore* (spatial contiguity characters), while
- the *ecotissue* concept (or ecological tissue) represents a complex *multidimensional* structure built up by a main mosaic (generally formed by the vegetation *coenosis*) and a hierarchic set of mosaics and information of different temporal and spatial scales, correlated and integrated, constituting the *landscape structural model* (Fig.5).

In add, the mentioned school proposes:

- new complex integrated functions (e.g. biological and territorial capacity of vegetation; human habitat capacity evaluation, etc.),
- new methods and new applications (e.g. new evaluation of human habitat, new survey of vegetation, etc.).

4.2 BTC: The Biological Territorial Capacity of vegetation

Vegetation, as the most important component of the landscape, has to be related with the concept of metastability. The use of metastability concept enables (i) to study vegetation through new perspectives and (ii) to evaluate landscape transformation in a proper way. The evaluation of metastability in vegetation, implies the concept of landscape biodiversity (i.e. main types of vegetation communities) and the concept of latent capacity of homeostasis of an ecocoenotope (i.e. vegetation *tessera*[4]).

The *biological territorial capacity* or BTC (Ingegnoli 1991, 1993, 1999; Ingegnoli and Giglio 1999, Ingegnoli 2002; Ingegnoli & Pignatti, 2007), is referred to to vegetation tesserae, and it is a synthetic function defined on the basis of: (i) the concept of resistance stability ; (ii) the principal types of vegetation communities of the ecosphere ; (iii) their metabolic data (biomass, gross primary production, respiration, B, R/GP, R/B). Two coefficients can be elaborated:

$$a_i = (R/GP)_i / (R/GP)_{max} \tag{8}$$

$$b_i = (dS/S)_{min} / (dS/S)_I \tag{9}$$

where: R is the respiration, GP is the gross production, dS/S is equal to R/B and is the maintenance/structure ratio (or a thermodynamic order function, Odum 1971, 1983) and i are the principal ecosystems of the ecosphere.

The factor a_i measures the degree of the relative metabolic capacity of principal vegetation communities; b_i measures the degree of the relative antithermic (i.e. order) maintenance of the same main vegetation communities. The degree of homeostatic capacity of an ecocoenotope is proportional to its respiration (Odum 1971, 1983). So the a_i and b_i coefficients, even related in the simplest way, give a measure which is a function of this capacity:

$$BTC_i = (a_i + b_i)\, R_i\, w \tag{10}$$

where w is a variable necessary to consider the emergent property principle and to compensate the environmental constraints. Putting $\Omega = (a_i + b_i)\, R_i$, the value of w results: $w = 0.89 - 0.0054\, \Omega$, consequently:

$$BTC_i = 0.89\, \Omega - 0.0054\, \Omega^2 \; (Mcal/m^2/year) \tag{11}$$

Reference values of BTC have been calculated on the 30 main types of zonal vegetation of the ecosphere, as shown in Ingegnoli (2002): note that both natural and anthropogenic vegetation have been considered. Moreover, the BTC function becomes an ecological index which allows the recognition of regional thresholds of landscape replacement (i.e. metastability thresholds) during time, and especially the transformation modalities controlling landscape changes, even under human influence. This index is available even to measure the functional biodiversity of a landscape.

Remember that the concept of biodiversity, as defined by U.S. Office of Technology Assessment (1986), depends on two aspects: (1) the diversity of the components of ecological

---

[4] The name "*tessera*" (latin: component of a mosaic configuration) can be correlated with the delimitations of the principal types of ecosystems (i.e. biogeocenosis or, better, *ecocoenotopes*) constituting a sort of geographic map, some times apparently similar to the "land use" maps of the human territory, but with an ecological sense.

systems and (2) the diversity of their relations in the organisation of these systems (other 2 aspects: (2.1) *local* and (2.2) *context*). Biodiversity is also an attribute of an entire ecological system.

Therefore, to reach a better understanding of the ecological state of a forest, we have to check:

(1) species diversity (e.g. α, γ and β; Whittaker,1975) and landscape elements diversity (ψ, τ; Ingegnoli & Giglio, 2005);

(2.1) ecosystem-community diversity (e.g. tesserae) and

(2.2) landscape diversity (e.g. landscape unit), measuring the levels of their ecological organisation.

A better use of the BTC index derives from its very good correlation with the measure of human habitat (HH), which can be defined as areas where human populations live or manage permanently, limiting or strongly influencing the self-regulation capability of natural systems. As shown in Fig. 6, the polynomial line derived from about 50 case study of landscape units (LU) in the North of Italy (mainly in Lombardy, Trentino-Alto Adige, but even in Austria and Germany) presents a high R^2, so that the equation:

$$BTC = 0.0007\,x^2 - 0{,}152\,x + 0{,}86 \tag{12}$$

(where BTC is referred to the examined landscape unit and $x = HH$) may be used in the evaluation of the ecological state of the landscape. HH is expressed in % of the surface extension of the landscape unit.

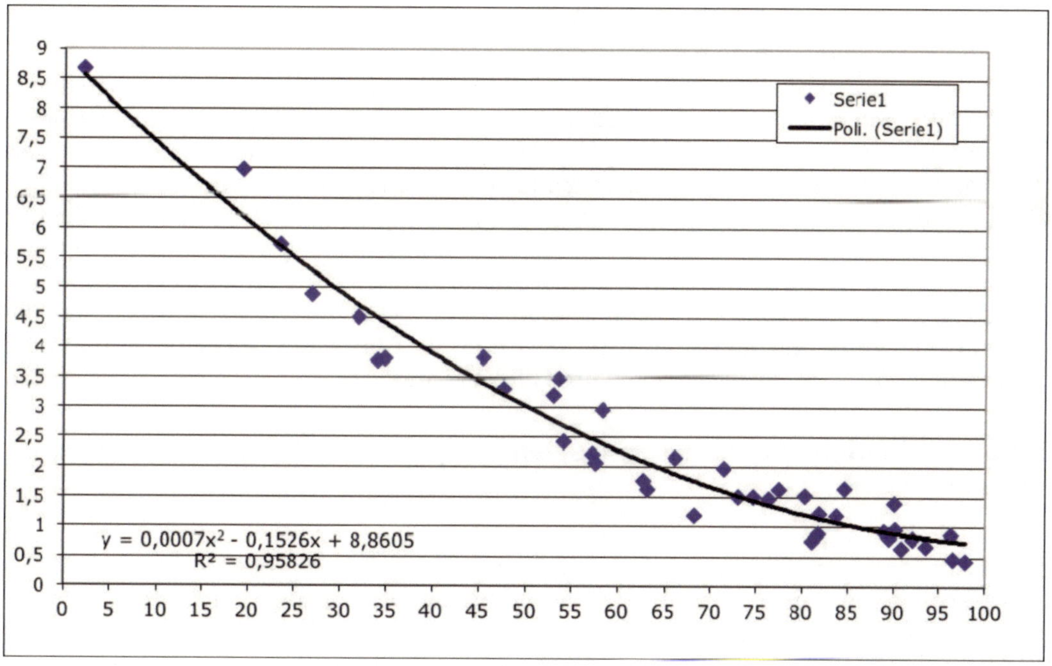

Fig. 6. Correlation between the BTC index (Mcal/m2/yr; Y axis) and the human habitat in about 50 case study of landscape units in central Europe (X axis : HH as %LU). Note the importance to utilise the equation (12) in the clinical diagnosis of the ecological state of the landscape.

4.3 Main transformation modalities in the landscape

In a landscape or in its subsystems (i.e. Landscape Units) the main transformation processes depend on the hierarchical structuring of an ecological system and its non-equilibrium thermodynamics, metastability, coevolution, evolutionary changes and ecological reproduction. Let us review the main steps, essential to revise later some basic concepts of vegetation science:

i. *Hierarchical structuring*. The behaviour of an ecological system is limited by: (a) the potential behaviour of its components on the lower level of scale, (b) the environmental constraints on the upper level of scale. This set of conditions represents the existence field in which the system of ecosystems must reside.
ii. *Non-equilibrium thermodynamic*. Thermodynamic bonds may determine an attractor, in its proper existence field, that represents a condition of minimum external energy dissipation. Possible macro-fluctuations produce instabilities, which move the system toward a new organisational state. These new states permit an increase of dissipation and move the system toward new thresholds to reach a new attractor. This could be represented as a cybernetic process of "order through fluctuation"(Cfr. Fig. 3 and 5).
iii. *Metastability*. An ecological system can remain within a limited set of conditions, but it may show alterations if these conditions change. The system may cross a critical threshold, approaching even radical changes. E.g. different types of landscapes or their parts may be correlated with diverse levels of metastability.
iv. *Coevolution*. The history of the interactions among the elements of a landscape in a given area shows a particular dominion that is characterised by the coherence of their reciprocal adaptation. This process leads to the stabilisation of different homeostatic and homeorhetic capacities of a landscape, which may be expressed with a particular degree of metastability of the entire system.
v. *Evolutionary changes*. The structuring of every biological system may be pursued, that is the information may be transmitted, only if the final state of the considered system is less unstable (i.e. more metastable) than its initial state. The modalities by which these processes are realised may be different and not limited to a single scale.
vi. *Reproductive processes*. Each level of life organisation presents tipical reproductive processes: (a) system available to maintain information, (b) mutation phase, (c) protection of new elements, (d) selection phase, (e) crucial disturbance eliminating the old structure (Oldeman, 1990; Ingegnoli, 2002; Bengtsson et al. 2003). Following previous points and ranked processes, each level of life has to renew: note that both assembly rules and dispersal filters need also a context.

5. Non-equilibrium thermodynamics, landscape bionomics and vegetation science

Ecological succession in general ecology, is the most important process related to transformation: through serial stages, an ecosystem changes in a predictable way toward a final stage, called climax. After an outside perturbation (or partial substitution of inner components), succession returns the ecosystem to the climax. For instance, an abandoned field near a forested patch is re-colonised from the forest edge and, in a given time, after the re-growth of shrubs and then of trees, the succession restores the "climax". Succession is a concept of primary importance in ecological theory: it has become the basis for dynamical explanations of many ecological phenomena, such as in phytosociological sygmeta. But this kind of succession is incompatible with the scientific principles underlined before, especially with non-equilibrium thermodynamics.

5.1 Limits with the reductionist concept of succession and the method of phytosociology

Remember the non-equilibrium thermodynamic with branching points after the instability threshold (Fig. 2), or the concepts of landscape metastability: in the first case, the history becomes the leading criterion of transformation; in the second, it is evident that, even when a succession to a climax may be considered valid at a single ecocoenotope scale, certainly it is not valid at a landscape scale.

Succession does not work as linear and mechanistic. According to Pignatti (1996), in the vegetational phytocoenosis of *Cytisus villosus* which follows after a fire of a *Viburno-Quercetum ilicis* patch, for instance in central Italy, or in the re-colonisation of *Picea abies* on abandoned alpine pastures in Central Europe (two cases in which normally succession is present) if more than one key factor becomes dominant, the ecological system and its transformation become unpredictable.

It should be always very important to remember that self-organising processes have to be considered at least on three scales: the one of interest, the upper (constraint) one and the lower one (significance). If some components of an autocatalytic set are excluded, the system will appear as linear. It is what happens to the classical theory of succession, because e.g. the landscape is never considered as a basic parameter. Therefore, in landscape bionomic the importance of ecological succession as linear and divided into primary and secondary phases is drastically reduced.

At present, especially in Europe, the vegetation is defined as a set of current vegetable individuals, growing in a determined site and in their natural disposition that it is assumed to be ordinated on the basis of self-organisation processes (Westhoff 1970): its study is principally founded on phytosociology (Braun-Blanquet, 1928). The logic of phytosociology derives from the correspondence between the existence of given environmental conditions of a site and the presence of plant species of a given statistical combination (Pignatti 1980, 1994). The relation between species and ecological factors, assumed as univocal, permits the definition of a n-dimensional ecological space: starting from a set of auto-ecological spaces, the synecological one is defined as the intersection set. For example, projecting on a plane the spaces of five species A,B,C,D,E, the frequencies of which are 0.6, 0.5, 0.4, 0.3, 0.2, the overlapping area may represent an association of these species: the probability of this set to be a casual one is only 0.0072 (Fig. 7). This limits the random character of the ecological relation obtained from the presence of species.

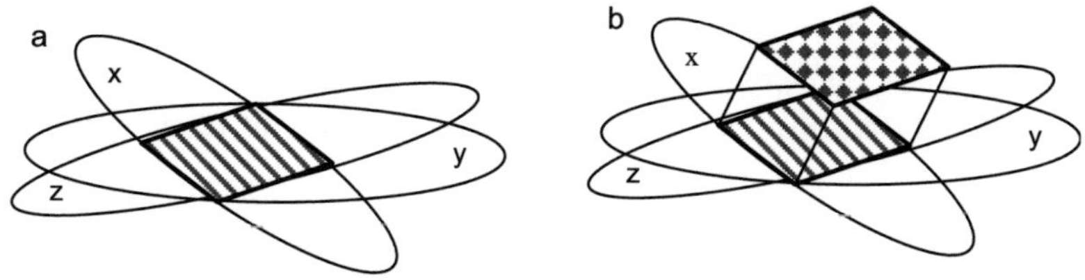

Fig. 7. Ecological space and the study of vegetation. **(a)** In the phytosociological model. **(b)** In the landscape ecological model. (from Ingegnoli, 2002).

Ecological information is often neglected, because of the supposed univocal ecological indication[5] of each species. Thus, the phytosociological model *presumes* the complete knowledge of an association only through its floristic description. This knowledge is then developed into syntaxonomy. The association dynamics is based on the concept of ecological succession and climax, assumed to be linear, with a deterministic sense, that is, on the concept of "potential natural vegetation" (Tüxen, 1956). Even the landscape is studied with the sygmetum method (Tüxen, 1978; Géhu 1988; Rivas-Martinez 1987).
At the ecosystem level, and for a formal description of the associations of vegetation, the method of phytosociology seems to give quite good results. Supporters of the use of this approach even in the study of the landscape are frequent in Europe: but not all scientists are in agreement. In fact, the described logic presents many limitations, especially from the point of view of landscape ecology (Naveh 1984; Ingegnoli 1997, 2002). The principal criticisms include at least these following points:

1. Phytosociology is based on too many deterministic aspects, first of all the importance given to the linear concept of ecological succession (seral steps), not compatible with the reality, being in contrast with the new scientific paradigms.
2. Until now, even in the representation of the ecological space, it has not been taken into consideration that an association must have an information content that is greater than the sum of the information acquired from the component species (Fig. 7). This is what allows an association to become an attractor within its context (i.e. ecotissue), in which it evolves and has to sustain a role (Ingegnoli 2002, 2005).
3. After about 100 years of investigations no true novelty changed the method of phytosociology, thus the results became more and more incoherent with the modern developments of science (Pignatti, Box & Fujiwara, 2002). Indeed this investigation remains in most cases a description of facts.
4. The method is scale dependent. What happens with relevés of 10 cm^2? What with 1 ha ?
5. Moreover, the observations of Ellenberg (1960) on relative *Standortkonstanz* of species (relative dependence on site factors) are often not considered. Note that an ecological interpretation of genome redundant size reinforces this concept (Bennett and Smith 1991).
6. It is impossible to show properly the order existing in a vegetational community only with a floristic description[6] (e.g. phytosociologic table). Rather, if the shorter algorithmic description of a system coincides with the description of the entire system, the system has to be classified as chaotic, dependent from the initial condition (Pignatti et al. 1998).
7. The aims of phytosociology are more linked with a description and typing of a *supposed* natural set of plants than with a study of vegetation in its complete reality. Without an integration, the use of phytosociology in landscape ecology could be in many cases too limited or impossible (Ingegnoli 1997, 2002).
8. Studying landscapes, we must consider as a proper entity also the vegetational new coenosis, created in anthropised landscapes even by sets of alien species which have

[5] E.g. Ellenberg bioindicator values of vascolar plants: light radiation, temperature, climate continentality, humidity, soil reaction, nutrients, salinity.

[6] Remember that the organisation of a vegetation coenosys concerns also the structuring of space-time (4 dimensions) and the relations with animals, human management, and so on.

replaced or are replacing autochthonous ones, especially with respect to former natural associations. This fact is confirmed also by cartography: many phytocoenosis can not be described in syntassonomy. Information related to natural species are not sufficient from a landscape ecological point of view.

9. We have also to consider the possibilities of random variations.

5.2 From landscape bionomics a new definition of vegetation

To understand the transformation of a landscape it is useful to study its vegetation, which characterises the main "landscape apparatuses" (Ingegnoli 2002), or "context role sub-systems" (CRS-S).

A landscape CRS-S concerns functional systems of tesserae and ecotopes which form specific configurations in the complex mosaic (i.e. ecotissue) of a landscape. A tessera is the smallest homogeneus unit visible at the spatial scale of a landscape: it corresponds to the former definition of ecotope (Naveh, 1984; Haber, 1990; Zonneveld, 1995) as the sum of physiotope and biotope. An ecotope is now the smallest landscape unitary multidimensional element that presents all the structural and functional characters of its landscape (formed by at least two tesserae).

These CRS-S are distinguished by a specific landscape function (and/or its range of sub-functions), not only by many local characters: e.g. productive, connective or stabilising functions. A first important landscape function results by the human habitat (HH) versus natural habitat (NH). The NH are the natural ecotopes, with dominance of natural components and biological processes, capable of normal self-regulation. Remember that the management role of human populations, if not directed against nature, may be considered in an ecotissue as semi-natural. Following the ecotissue model (Ingegnoli, 2002), the sum HH+NH > 1.

In this vision, the *definition* of vegetation has to be: the *whole of the plants* of a landscape element, considered in their aggregation capacities and in their relations with environmental and time-space factors. Thus, a cultivated tessera is to be considered as vegetation not only for its weeds (e.g. *Secalinetea, Chenopodietea*), but even for the cultivation itself (e.g. *Triticum aestivum, Hordeum vulgare*), without which the weeds does not succeed and the tessera does not become the habitat for many natural species (e.g. *Coturnix coturnix, Alauda arvensis*), besides to be a crucial ecological component for human population.

The frequent use of the concept of "potential natural vegetation" is not yet satisfactory for landscape ecological studies, because the word "potential" is intended to represent undisturbed conditions in a not defined time. The proposal of Ellenberg (1974), to distinguish among *zonal* vegetation, which expresses the responses of potential vegetation to climatic conditions; *extrazonal* vegetation, responding to local topoclimatic conditions; and *azonal* vegetation, responding to soil moisture conditions, was another good step, but it is again not sufficient for landscape bionomics theory, therefore even for vegetation science.

Remember that Ellenberg (1978) already perceived the ecosystem and man's dual part in the structure of a landscape, and Walter (1973) proposed to determine plant formations and types not only in their floristic aspect but also in stability, structure, human influence, diversity, productivity, etc. Note that the reasons for this criticism derive from the self-organisation processes especially when the role of disturbances is seen as structuring and when the transgressions in a linear succession are based on the interaction among landscape elements even in the same *zonal* area.

Trying to evaluate the actual vegetation on the basis of its ecological distance from the potential vegetation is not correct, because this implies the possibility that potential landscapes, reduced to very few, sometimes only one or two types of vegetation really exist. This is in contrast with all the main processes and dynamics of the landscape and it is a sort of "virtual ecology"! For instance, as pointed out by Pignatti even on the best primeval forest in Europe (i.e. the Perucica[7]), the large zonal ecosystems (e.g. tropical forests, taiga, savannah, Australian deserts, etc), nearly undisturbed, are never formed by a single association or very few ones!

In facts, it clashes with the non-equilibrium thermodynamic principle and the relative bifurcations of the state functions of a system in an instability field. Therefore, the concept of potential vegetation has to be strongly revised. It has to be defined not only for natural cases, but in relation to the main range of landscape disturbances (including man's) too, and with defined temporal conditions. It must never be considered as the optimum for a certain landscape (or part of it), but only as a general indication (never to be widely reached) in relation to the climate, the soil and the anthropisation of a certain limited period of time. It could be better named *the fittest vegetation for*.

This new concept refutes the general notion of 'potentiality' as the possibility of the coming into existence, in the absence of man and for large territories, of a deterministic, a priori fixed vegetation type and interpreted as the best condition for a place, independent of all other environmental and human factors in space and time. Moreover, no potential homogeneity can be a model for the develpoment of a landscape. On the contrary, the concept of *the fittest vegetation for* indicates the most suitable or suited vegetation for: the specific climate and geomorphic conditions, in a limited period of time and in a certain defined place; i.e. the main range of incorporable disturbances (including man's) under natural or not natural conditions. This could be a great change of perspective.

Note that it signifies also to eliminate, or at least declassify, the concept of primary succession and a revision of the concept of vegetation dynamics.

6. New method for vegetation evaluation in landscape bionomics

A new method of vegetation evaluation has been studied and proposed by Ingegnoli (2002, 2005), then discussed and completed with Elena Giglio and Sandro Pignatti (2005, 2007): it derives directly from the theoretical considerations reported here. This method can be named "Landscape Bionomics Survey of Vegetation" or LaBISV. A frame protocol is presented in Tab. 1: it is able to integrate three different criteria (a biotic one, an environmental one and a configuration one) with different temporal and spatial scales.

6.1 Frame protocol and parametric standard form

The below presented frame protocol uses a *parametric standard form* (a proper one for each type of vegetation) for the analysis and evaluation of a vegetated tessera. It is very helpful in the definition of the so called "*normal state*" for each specific type of tessera. Remember that landscape bionomics follows a clinical-diagnostic method and its main goal concern the evaluation of the healthy state of a landscape unit, in which the vegetation coenosis play a central role.

---

[7] The Perucica Primeval Forest is located in the Sutjesca National Park, in Bosnia-Herzegovina, and together with the Bialowieza forest in Poland is one of the few oldest forest landscapes in Europe.

Phase	Activities	Main operations	Notes
I	Identification of the landscape unit, (LU)	Following the Biological Integrated School of Landscape Ecology, recognition of boundaries and of composing ecotopes.	Ingegnoli, 2002, 2005 Ingegnoli & Giglio 2005
II	Choose of the vegetation tesserae (Ts)	Identification of the vegetation type, of its ecological (structural/functional) subdivisions and of the perimeter of the different tesserae.	Depending on ecological interest, Ts containing permanent plot too.
III	Collection of geographical data	Site and local data, e.g. climate, substrate, morphology, etc.	
IV	Collection of historical and human data	Old maps and books data, main human land uses, main historical changes.	
V	Survey of Ts characters	Vegetation height (canopy) and cover, structure, edge ratio, management, etc.	Ts as patch or corridor
VI	Survey of Plant Biomass parameters	Dead plant biomass, litter depth, biomass volume.	Above ground biomass
VII	Survey of Ecocoenotope parameters	Dominant sp, species richness, allochthonous sp, biological forms, stratification, threatened plants, renewal capacity, dynamic state, etc.	A phytosociological frame is needed
VIII	Survey of Ts/LU parameters	Contiguity, source/sink, functional role, disturbance incorporation, geophys. instability, fauna interest, transformation, etc.	
IX	Evaluation of vegetation parameters	Ordination of parameters in four classes in a standard form, then evaluation per column.	Scores depending on vegetation type
X	Evaluation of vegetation qualities (Q)	Evaluation (%) per group of parameters and/or the entire Ts	
XI	BTC estimation	Estimation through equations linked with the development models and BTC theory	
XII	Diagnostic activities	Comparison with other Ts and with the LU. Underline of the altered parameters. Integration with other ecological indicators ...	
Note: more information, especially on the interpretation of the parameters and score, may be founded in Ingegnoli & Giglio (2005). From: Ingegnoli V (2006) in ICP Forests Monitoring, Göttingen. pp. 241-259,			

Table 1. Landscape Bionomics Survey of Vegetation (LaBISV): frame protocol in synthesis

This form (Table 2) has been designed to check the organisation level and to estimate the metastability of a tessera considering both general ecological and landscape bionomical characters: T = landscape element characters (e.g. tessera, corridor); F = plant biomass above ground; E = ecocoenotope parameters (i.e. integration of community, ecosystem and microchore); U = relation among the elements and their landscape parameters.
The parameters for each T,F,E,U groups range from 3 to 12, thereby reaching the number of 26-33. The evaluation classes are four, the weights per class depending on an evaluation model (Fig. 8). Remembering the well known relationships among gross productivity, net productivity and respiration in vegetation ecosystems (Odum 1971, Duvigneaud 1977), the development of a vegetation community may be synthesised in: (1) the growing phases from young-adult to maturity, expressed by an exponential process; (2) the growing phase from maturity toward old age, expressed by a logarithmic process.

Example of the LABISV methodology synthesized in the present standard form					
BOREAL FOREST	**1**	**5**	**14**	**25**	**score**
T.TESSERA CHARACTERS (TS)					
T1 - Vegetation height (m)	< 9	9.1-18	18.1-29	**> 29.1**	Canopy
T2 - Cover of the canopy (%)	< 30	> 90	**31-60**	61-90	Ts surface
T3 - Structural differentiation	low	medium	**good**	high	Age, space groups, etc.
T4- Interior/edge (%)	none	< 30	**31-89**	> 90	(% Ts)
T5 - Management	simple coppice	coppice	wood	**natural forest**	Or similar
T6 - Permanence (years)	< 80	81-160	**161-240**	> 240	Old trees
F. VEGETATIONAL BIOMASS (ABOVE GROUND)					
F1- Dead plant biomass	near 0	> 10	**1-5**	5-10	% of living biomass
F2- Litter depth	near 0	< 1.5	**1.6-3.5**	> 3.5	cm
F3 - Biomass volume (m^3/ha)	< 200	201-500	**501-950**	> 950	pB = **696** m^3/ha
E. ECOCOENOTOPE PARAMETERS					
E1- Dominant species (n°)	> 3	3	2	**1**	As pB volume
E2- Species richness	< 15	16-30	31-40	**> 40**	n° sp./Tessera
E3- Key species presence (%)	< 5	6-40	41-80	**> 80**	Phytosociological
E4- Allochtonous species (%)	> 10	10-4	< 4	**0**	From other ecoregions
E5- Infesting plants %	near all	> 25	< 25	**0**	Coverage on Ts
E6- Threatened plants	evident	suspect	risk	**0**	Even acid rain damage
E7- Biological forms (n°)	< 3	4-5	**6-7**	> 7	Cfr. Box 1987, mod.

E8- Vertical stratification	2	3	**4**	> 4	traditional
E9- renew capacity	none	intense	**sporadic**	normal	Dominant species
E10- Dynamic state	Degrada-tion	recreation	**Regenera-tion**	Fluctua-tion	Cfr. Ingegnoli 2002
U. LANDSCAPE UNIT (LU) PARAMETERS					
U1- Similar veg. contiguity	0	< 25	**26-75**	> 76	% of perimeter
U2- Source or sink	sink	neutral	**Partial**	source	Species & resources
U3- Functional role in LU	reduced	minor	**evident**	important	Context & typology
U4- Disturbances incorporation	insufficient	scarce	normal	**high**	Local disturbances
U5- Geophisical instabilities	evident	partial	**risk**	none	On the phisiotope
U6- Permeant fauna interest	low	medium	**good**	attraction	Key species
U7- Tranformation modalities of the Ts	strong distuba-nces	gradual changes	temporal instabilities	**fluctuation**	Today + tendency
U8- Landscape pathology interference	serious	near chronicle	**easy to incorporate**	none	From landscape
U9- Permanance of analogous vegetation (years)	< 100	100-300	300-1200	**> 1200**	Historical presence
RESULTS OF THE SURVEY					
Total score Y (= h+j+k+w)	h = 0	j = 0	K = 17	w = 11	Y = 513
Quality of the Ts	Q = Y / 700				**Q = 73,3** [%]
Estimation of the BTC	BTC (b) = 0,01339 (y-28) + 0,12 (pB / 70)				**BTC = 7,69** [Mcal/m^2/yr]

Table 2. Example of the LaBISV methodology of survey synthesized in the present standard form. Forest permanent CONECOFOR plot TRE1 (Lavazè Pass[8]) *Piceion abietis*, 1.800 m. Survey: August 2004 by Ingegnoli and Giglio. Also the equation of estimation of the BTC derives from the model of Ingegnoli (2002).

Table 2 could be used also for Temperate deciduous forests, changing: (a) the parameters F3 (biomass volume) that become respectively < 150, 150-350, 350-600, > 600, and (b) the scores of the columns, which become 1,5, 12,22.

[8] The Pass of Lavazé is located between the Fiemme Valley and the Egentall, in the Region of Trentino-Alto Adige (Sud Tirol). The CONECOFOR is a programme of forest research ruled by CFS (State Forest Corp) of forest ecosystems monitoring.

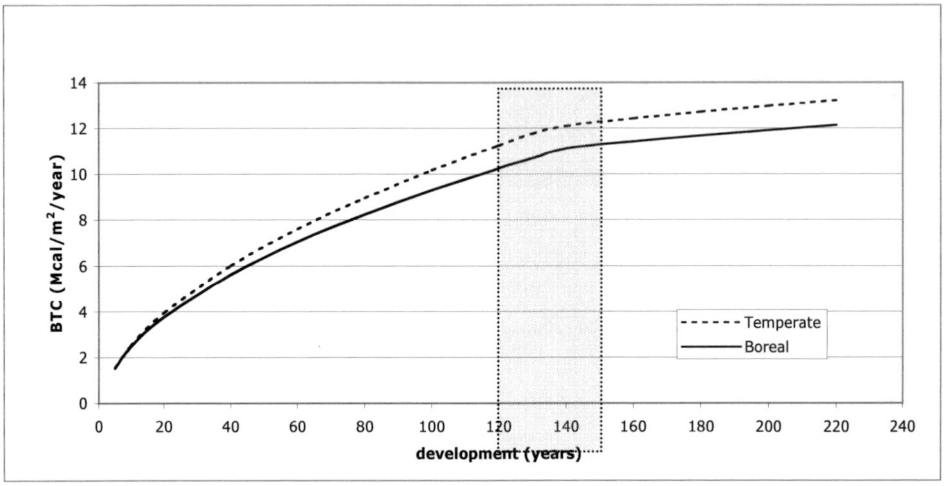

Fig. 8. Model for Temperate deciduous forest and Boreal coniferous forest. The vertical strip indicates the beginning of the maturity phase, from 120 to 150 years.

6.2 Main applications of the LaBISV

Following a similar method, a series of schedules (Ingegnoli & Giglio, 2005) have been designed, one for each main vegetation type (Table 3), other types are now in study.

Main vegetation types	Model BTC (max) $Mcal/m^2/yr$	Model development (years)	BTC Estimation equations ($Mcal/m^2/year$)
1. Boreal forest	11.0	120-150	0,01339 (y-28) + 0,12 (pB/70)
2. Temperate forest	12.0	120-150	0,01667 (y-28) + 0,13 (pB/65)
3. Sclerophyll forest	13.0	120-150	0,01705 (y-28) + 0,13 (pB/60)
4. Mediterranean pine forest	10,5	100-130	0,01510 (y-28) + 0,12 (pB/65)
5. Tall shrubland	4.0	30-40	0,00344 (y-30) + 0,10 (pB/17)
6. Low shrubland	2.6	25-35	0,00247 (y-30) + 0,03 (pB/0,2)
7. Prairie and pastures	1.4	20-24	0,001335 (y-29) + 0,02 (pB/0,14)
8. Reed	2.8	36-48	0,0023 (y-29) + 0,04 (pB/0,3)
9. Salt marshes	1,2	15-20	0,00260 (y-28) + 0,10 (pB/1,4)
10. Corridors with trees	9.5	90-130	0,0072 (y-33) + 0,10 (pB/75)
11. Wooded agrarian	4,5	30-40	0,00575 (Y-29) + 0,15 (Fm /35)
12. Agricultural field	2.0	10-20	0,00192 (y-26) + 0,09 B3
13. Urban garden	8.0	70-110	0,00526 (y-30) + 0,10 (pB/45)

Table 3. Synthesis of the main vegetation types considered by the model for vegetation survey proposed by Ingegnoli (1999, 2002, 2005).

This method and its schedules, to which the notes for each ecological parameter (Giglio & Ingegnoli, 2005) can be added, recently was used with success a s a tool for vegetation survey, and in many applications of vegetation science. Main application of this survey method are:

1. evaluate and compare the about 30 ecological parameters of vegetation, e.g. through radar plots;
2. evaluate the ecological quality (Q_x) of each group of parameters (T,F,E,U);
3. verify and to estimate the biological territorial capacity of a tessera (BTC);
4. check the level of ecological maturity of a tessera (BTC/BTC*);
5. survey the mean BTC and the BTC classes composition of a landscape unit or one of its ecotopes (thus allowing the measure of other ecological indexes, Cfr. Ingegnoli 2002 and Giglio & Ingegnoli 2005);
6. estimate/quantify results of management intervention on some parameters effects on the whole tesserae or LU.

7. An example of application: the forests of the territory of Mori (Trento)

The Mori municipality is about 35 sq.Km, 53% covered by forest. It consists of 4 Landscape Units (LU) presenting different landscape types (Fig. 9):
LU1 (Mori-Talpina): valley floor *rural-suburban landscape,*
LU2 (Loppio): valley floor agricultural-protective landscape
LU3 (Val di Gresta): mountain agricultural-protective landscape
LU4 (Biaena Mount): mountain forest-agricultural landscape
Note that the first LU, Mori-Talpina, is the lowest one (from 200 to 550 m a.s.l.) and the more urbanised one: anyway 1/3 is covered by forest.

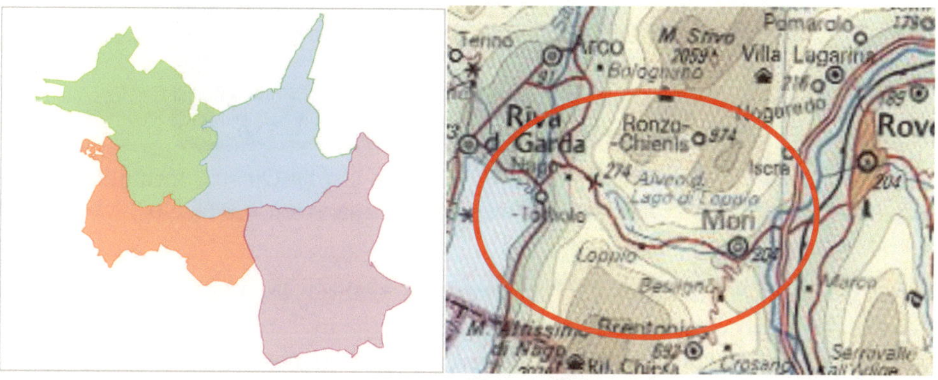

Fig. 9. The localization of the municipality of Mori, in the Southern part of Trentino, near the upper Garda Lake, and (right) the division of the territory in 4 landscape units of : (1) Mori-Talpina (violet), (2) Loppio (pink), (3) val Gresta (green) and (4) mount Biaena (pale blue).

7.1 Character of the forests

The distribution and types of forests lying on the territory of the municipality of Mori (TN) were surveyed in the year 2007 by Ingegnoli and Giglio, following the LaBISV Method. Mixed oak forests (*Ostrya* woods) are the most widespread formation (59.7%) followed, in the upper vegetation belt, by pine forests (*Pinus sylvestris* and *Pinus nigra*), spruce forests

(*Picea abies*) and beach forests (*Fagus sylvatica*), respectively 11.5, 8.7 and 5.4%. To have an idea of these forests, see Fig. 10, in which is shown *Ostrya* formations and Conifers ones.

Fig. 10. Picture of the Mori territory: (*left*) a vision of an *Ostrya-Quercus* formation, with some Pines on the slope, and (right) a view of the mount Biaena, from 700 to 1400, which presents spruce and beach formations.

The most impressive characteristic of forest vegetation in Mori Municipality is the considerable difference between the physiognomy of the investigated forest and woods and their proper ecological characters, due to human management and historical events: the phytosociological attribution to a proper association is often very difficult. For 11 forested tesserae dominated by spruce - the attribution of which to a certain phytosociological syntaxa was not clear- data concerning species have been elaborated following this formula:

$$TFC = [k\ SP/SP^*] \times DM^{1/3} \quad (13)$$

where: TFC = theoretical forest character; SP = surveyed species pertaining to a certain *Phytosociological Alliance;* SP* = possible species pertaining to the same Alliance; (SP/SP* in %); k = coefficient available to consider misbehaving and companion species (e.g: k= 1,1); DM = dominant in % plant biomass (elevated 1/3).

11	PHYSIONOMY	***Ostrya*-wood A**	ha 387,5
	SYSTEMIC CHARACTERS		
	PROPER ECOLOGICAL CHARACTERS		**Prealpine *Ostrya*-wood with *Quercus petraea***
	PHYTOSOCIOLOGICAL ATTRIBUTION		cl. *Querco-Fagetea*, ord. *Quercetalia pubescentis*, all. *Orno-Ostrenyon*, ass. *Buglossoidi-Ostryetum*
	medium SPATIAL STRUCTURE		Deciduous broad-leaves 95%, medium high 12m
	medium BIOLOGICAL TERRITORIAL CAPACITY		BTC = 5,03 Mcal/m^2/year
12	PHYSIONOMY	***Ostrya*-wood B**	ha 531,93
	SYSTEMIC CHARACTERS		
	PROPER ECOLOGICAL CHARACTERS		**Mixed wood with *Ostrya carpinifolia, Fraxinus ornus* and *Quercus pubescens***
	PHYTOSOCIOLOGICAL ATTRIBUTION		cl. *Querco-Fagetea*, ord. *Quercetalia pubescentis*, all. *Orno-Ostrenyon*, ass. *Orno-Ostryetum*
	medium SPATIAL STRUCTURE		Deciduous broad-leaves 97%, medium high 9,2m
	medium BIOLOGICAL TERRITORIAL CAPACITY		BTC = 4,93 Mcal/m^2/year
13	PHYSIONOMY	***Ostrya*-wood C**	ha 188,45
	SYSTEMIC CHARACTERS		
	PROPER ECOLOGICAL CHARACTERS		***Ostrya*-wood with *Quercus petraea* and *Quercus pubescens***
	PHYTOSOCIOLOGICAL ATTRIBUTION		cl. *Querco-Fagetea*, ord. *Quercetalia pubescentis*, all. *Orno-Ostrenyon*
	medium SPATIAL STRUCTURE		Deciduous broad-leaves 99%, medium high 10,6 m
	medium BIOLOGICAL TERRITORIAL CAPACITY		BTC = 5,4 Mcal/m^2/anno

Table 4. Physionomic-ecological map of forests: legend and surfaces

Three Alliances have been concerned: *Erico-Pinion, Piceion abietis, Fagion* (see Fig. 11). Results are shown in the figure. As you can see, tesserae n° 1-8 may be ecologically considered as Spruce coenosys, while the last two are Pine forests with spruce; n° 9 is a mixed one.
Under the physiognomy of *Ostrya*-wood, we gather at least three types of coppice woods, related to the two phytosociological associations of *Buglossoidi-Ostryetum* and *Orno-Ostyetum,* as shown in table 4.

Concerning Pine wood (*Pinus nigra austriaca*) on mount Garda (Mori), mainly planted by foresters about 60 years ago, it presents many characters of the *Fraxino orni-Pinetum nigrae* Martin Bosse (1967). This formation has been described by Pollini (1969) in the Karst near Trieste, with species like: *Amelanchier ovalis, Lembotropis nigricans, Erica carnea, Goodiera repens, Sesleria* sp., etc. The present site in Mori could represent the most Western site of this association in Italy.

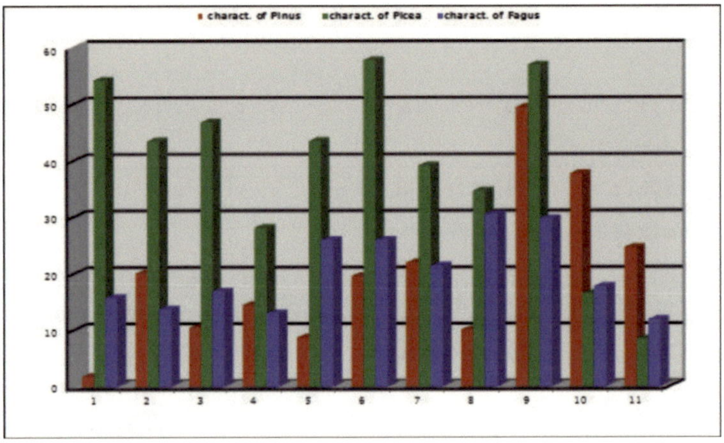

Fig. 11. The distribution of proper ecological characters of the alliance of Pinus (red), Picea (green) or Fagion (blue), following the above mentioned formula, within each surveyed tessera of spruce forest.

7.2 The thermophylous vegetation of Mori-Talpina

The results from the survey of 13 forested tesserae in the LU 1 of Mori-Talpina are shown in table 5, where: *pB* measure the plant biomass above ground; *BTC* is the biological territorial capacity of vegetation (Mcal/m^2/year); *Q* represent the four ecological qualities of the tessera (Ect = ecocenotope, LU = landscape unit, Ts = tessera, pB = plant biomass, B = % of coniferous species, BTC* maturity threshold, 85% of the model curve).

The average BTC of the forests of this LU 1 is quite low (about 4.9 Mcal/m^2/year) if compared with the values of the other 3 LU of Mori (see Tab. 6). Anyway, no one of the forest types reaches a hight mean of biological territorial capacity (e.g. BTC = 8-9 Mcal /m^2/yer). But the most evident difference among the 4 landscape units emerges in the chorological analysis, as we can see in Fig. 12, especially concerning the LU1 versus the others 3 regarding the Euri-Mediterranean, the Euro-Siberian and the Orophytae species. This analysis is based on 118-192 species per LU.

The Ellenberg indexes (*sensu* Pignatti, 2005) -resulted from the analysis of the species of the Mori-Talpina Landscape Unit- have been compared with 2 case study, the first in Menaggio (Lake of Como, Pre-Alpine climate), the second in Zoagli (near Genoa, Mediterranean climate). In figure 13, we may observe, despite the high presence of Euri-Mediterranean species, the good similarity with the other Pre-Alpine case and the differences versus the Ligurian landscape (true Mediterranean).

Rel. N°	Site	Height a.s.l.	*Dominant trees*	canopy height m	pB m³/ha	BTC Mcal /m²/a	% Q (Ts)	% Q (pB)	% Q (Ect)	% Q (LU)	B	BTC*
1	Zovo, p. 10	440 m	*Q. petraea Fraxinus ornus*	7,7	61,2	4,37	45,5	21,2	65	49	6	42,8
2	Besagno S	440	*Castanea sativa*	13,9	114,5	4,55	25	37,9	56,8	46,5	0	44,6
3	Talpina, p. 17a	410	*Q. petraea C. betulus*	12,1	126,7	4,52	32,6	37,9	57,3	45,5	0	44,3
4	Talpina, p. 17b	440	*Fagus sylatica*	17,2	255,1	6,41	51,5	59	65	52,5	0	62,8
5	N Corno	230	*Pinus nigra*	16,4	205,6	6,1	51,7	59	74,6	52,3	67	63,3
6	Le Coste	360	*Pinus strobus*	16,2	279,5	3,77	35,3	43,9	46	30,3	86	40,3
7	Talpina, Cava p-18	380	*Pinus nigra, Q. petraea C. betulus*	12,2	173,1	4,99	38,4	43,9	57,8	52,5	17	48,9
8	Coste di Tierno p-15	490	*Pinus nigra, Pinus strobus*	12,7	156,9	4,28	35,4	38,5	50,5	49,9	80	45,8
9	Santuario	320	*Pinus nigra*	16,6	238,1	5,48	39,3	44	70	62,6	97	58,6
10	Mori Vecchio W	280	*Pinus nigra, Ostrya carpin.*	11,3	143,3	4,57	34,3	53,3	60,4	43,3	72	47,4
11	Piede la Lasta	270	*Celtis australis, Q. pubescens*	8,7	117,4	5,00	40,2	37,9	54,1	59,1	0	49
12	Talpina vallecola	350	*Fraxinus excelsior, Fraxinus ornus*	18,6	200,1	4,84	41,6	43,9	61,2	30,1	23	48,8
13	Talpina Doss del Gal	430	*Pinus Nigra, Quercus sp. Carpinus betul.*	16.3	137	4.67	18.8	38	69.6	47.5	43	47.7
	Average values	**372**		**13.8**	**169.9**	**4.89**	**37.7**	**43.0**	**60.6**	**47,8**	**37,8**	**49,6**

Table 5. Landscape Unit 1 MORI forested area Km² 3,29 (27,7% LU)

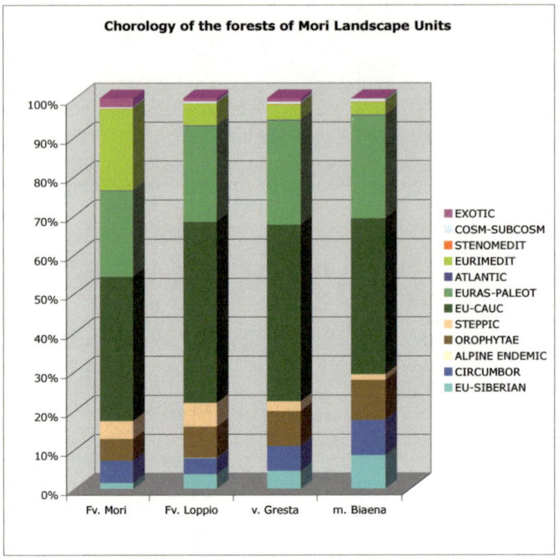

Fig. 12. The chorological spectrum of the forests of Mori LU shows the difference between the LU1 and the others, especially regarding the Euri-Mediterranean the Orophytae and the Euro-Siberian species.

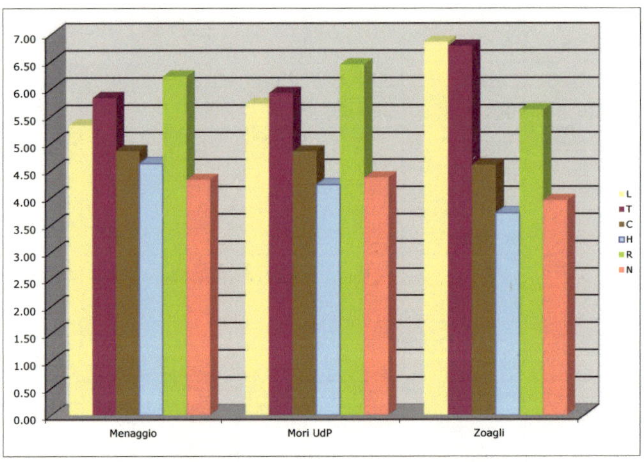

Fig. 13. The Ellenberg indexes resulted from the analysis of the species of the Mori-Talpina Landscape Unit have been compared with 2 case study, the first in Menaggio (Pre-Alpine conditions), the second in Zoagli (Mediterranean conditions). L= Light, T = temperature, C = continentality, H = humidity, R = soil reaction, N = soil nutrients.

7.3 Further applications of the LaBISV and their importance

It could be very important to remember that studying the landscape we can not measure and evaluate only the natural vegetation. Today, many of the European municipality- maybe the most parts of them- have few remnant patches of natural vegetation, and wide areas of human or near-human vegetation, *in primis* the agricultural one. Even in this case study of Mori, we expose table 6, in which some examples of survey of human vegetation are shown.

Tesserae	Sites	N°	Q Ts	Q pB	Q Ect	Q LU	BTC	pB	Hv
Vineyard I	Besagno	1	57,7	9,5	49,6	37,3	**1,93**	13,5	2,5
Vineyard II	Piantino/VGr.	2	28,1	9,5	47,8	31,3	**1,47**	10,6	2,3
Vineyard III	stadio/Mori	3	33,8	9,5	42,8	23,8	**1,35**	12	2,5
Vineyard IV	terrazzo/Mori	4	45,9	9,5	48,2	33,7	**1,71**	11	2,4
Vineyard V	Valle S. Felice	11	29,6	12,6	45	36,9	**1,63**	12,5	2,3
Vineyard VI	Valle S.F.	12	50,5	36,9	65,7	45,6	**2,36**	14	2,4
Potato field	Sud di Nomesino	5	17,4	7,6	65,8	50,2	**0,71**	0,9	0,7
Cabbage field I	Nagia/VGr.	6	34,2	37,6	74,8	53,9	**0,97**	2,5	Bare s.
Cabbage field II	Pannone/VGr.	7	44,5	26,9	62,2	41	**0,87**	2,5	0,4
Meadow II	Nagia/VGr.	10	27,7	21,9	61,9	39,2	**0,59**	0,7	0,7

BTC is the biological territorial capacity of vegetation (Mcal/m^2/year); *Q* represent the four ecological qualities of the tessera (Ect = ecocenotope, LU = landscape unit, Ts = tessera, pB = plant biomass as % of the maximum quality, Hv = high of vegetation.

Table 6. Example of survey through the LaBISV method of human vegetation (agricolture) in Mori.

We are now prepared to answer to crucial questions like these:

- how to consider the contribution of any tessera to the metastability of the landscape unit (LU)?
- how to compare the data of the forest patch with those of other vegetation elements in this LU?
- how to use the ecological characters of all the different types of vegetation, existing within a LU, to arrive to a diagnostic evaluation of the entire landscape?
- how to integrate the other main ecological parameters of the LU, like the ones related to animals and the ones related to human habitat or the carrying capacity[9] (SH/SH*) ?

The scientific diagnostic evaluation of the ecological state of a landscape unit allows a "physician of the environment" to change the present methodologies on territorial planning. As shown in Tab. 7, the LaBISV survey, allowed to elaborate interesting data on the ecological state of this territory, useful to avoid to consider the parameters pertaining to the entire municipality, in contrast with the bureaucratic procedure. In reality, it is possible to demonstrate that the sharp differences among the landscape units bring planning towards these ecological division of the territory, not towards the administrative ones.

[9] In landscape bionomics the ratio between the measured standard habitat per capita and the theoretical one (SH/SH*) gives the value of the carrying capacity of a landscape unit (see Ingegnoli, 1993, 2002; Ingegnoli & Giglio, 2005).

Landscape Unit	Area (ha)	Human Habitat (% LU)	Forest Cover (% LU)	BTC of the forests Mcal/m²/year	BTC of the LU Mcal/m²/year
LU1 (Mori-Talpina)	1.175	57.9	36.8	4.87	2.33
LU2 (Loppio)	602	45.5	43.8	5.08	3.04
LU3 (Gresta valley)	847	30.5	65.5	5.40	3.84
LU4 (mount Biaena)	836	23.3	72.0	5.90	4.47
Mori Municipality	3.460	40.7	52.5	5.28	3.34

Table 7. Differences among the ecological parameters of the entire municipality of Mori and the four landscape units.

8. Conclusion

At the end of this chapter, it is necessary to present another aspect of the application derived from the principles and methods proposed by Ingegnoli. Let us consider a case study, again in Mori, related to the EIS (Environmental Impact Statement) for a cave in the hill of Talpina.

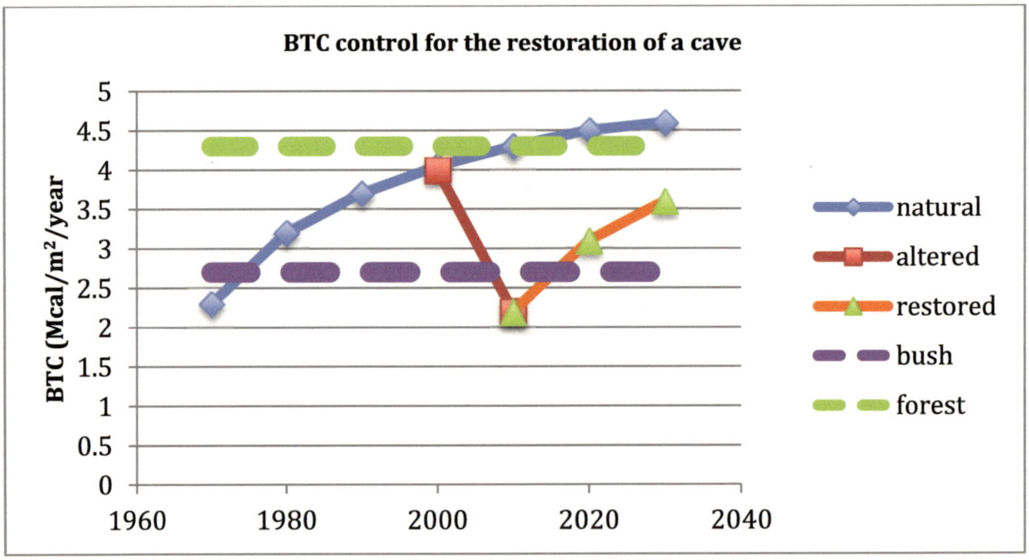

Fig. 14. Example of the ecological control of the restoration of a cave. The BTC function is available to evaluate the proposed opening of a cave after the comparison of the previewed restoration actions with the natural growth of the area and the thresholds indicating the main self-organisation structure of the ecocoenotope, from bush to forest.

The main model elaborated for the EIS, shown in Fig. 14, contributed to avoid the opening of a cave in the SCI area Talpina (Site of Community Importance, EU). The mentioned limits of the old concept of succession, due to non-equilibrium thermodynamics (Cfr. 5.1), eliminate the efficiency of environmental compensation, today based on restoration actions. This method of compensation does not consider the concept of "transformation deficit" (*sensu* Ingegnoli, 2002), which measure the lack of dissipation (of energy and related information) of a landscape system. In Fig. 14, this deficit concern the area between the lines of natural behaviour and the restored one, after the break of alteration. Moreover, the function of BTC allows to underline the thresholds indicating the main self-organisation structure of the ecocoenotope, from bush to forest.

In conclusion, the aim of this chapter is: (a) to demonstrate the possibility and the necessity to revise basic concepts of landscape ecology in the light of the new scientific theory, mainly derived from the non-equilibrium thermodynamics, concerning living systems and, consequently, (b) to revise the main concepts of vegetation science in the light of the new "Landscape Bionomics" and indicate the new methodological approach LaBISV (c) to underline the possibility to use the biological territorial capacity of vegetation (BTC) to evaluate landscape transformations.

Finally, note that human and animal coenosis have been investigated too, with analogous methodologies related to non equilibrium thermodynamics, trying to quantify the field of existence of about 12 temperate landscape types, with the help of a parametric diagnostic index.

9. Acknowledgement

The present evolution of my thinking has been influenced by deep discussions with colleagues and friends as Richard T.T. Forman, Zev Naveh, Sandro Pignatti, Roberto Canullo, Bruno Petriccione, and with my brother Alessandro Ingegnoli. A very special appreciation to Elena Giglio Ingegnoli, who reviewed the chapter with a good competence of the discipline.

10. References

Allen, T.F.H. & Starr, T.B. (1982). *Hierarchy: Perspective for Ecological Complexity*. University Chicago Press, USA.

Ashby, W.R. (1962). Principles of the Self-Orgainzation System. In: *Principles of Self-Organization*. Von Foerster, H. & Zopf, G.W. (Eds.), Pergamon Press, Oxford, UK

Bengtsson, J. Angelstam, P. Elmqvist, T. Emanuelsson, U. Folke, C. Ihse, M. Moberg, F. & Nistrom, M. (2003). Reserves, Resilience and Dynamic Landscapes. *Ambio*, Vol. 32 No. 6, pp 389-396

Bennet, M.D. & Smith, J.B. (1991). Nuclear DNA amounts in angiosperms. *Phil. Trans. R. Soc. Lond.* B. 334, pp 309-345

Box, E.O. (1987). Plant Life Form and Mediterranean Environments. *Annali di Botanica* XLV, pp 7-42, Rome, Italy, ISSN 0365-0812,

Braun Blanquet, J. (1928). *Pflanzensoziologie: Grundzüge der Vegetationskunde*. Berlin

Duvigneaud, P. (1977). Ecologia. In: *Enciclopedia del Novecento*. Vol. II, pp 237-250. Enciclopedia Italiana Treccani, Roma

Ellenberg, H. (1960). *Grundlagen der Pflanzenverbreitung*. Stuttgart, Ulmer.

Ellenberg, H. (1974). Zeigerwerte der Gefässpflanzen Mitteleuropas. In: *Scripta Geobotanica 9*, Ellenberg, H. et al. (Eds.). Göttingen

Ellenberg, H. (1978). Vegetation Mitteleuropas mit den Alpen. In: *Oekologischer Sicht*. Ulmer

Forman, R.T.T. & Godron, M. (1986). *Landscape Ecology*. John Wiley & Sons, ISBN 0-471-87037-4, New York, USA

Forman, R.T.T. (1995). *Land Mosaics, The ecology of Landscapes and Regions*. Cambridge University Press, ISBN 0 521 47462 0, New York

Gehu, J.M. (1988). L'analyse symphytosociologique et geosymphytosociologique de l'éspace. Tthéorie et métodologie. *Coll. Phytosoc.* 17, pp 11-46

Godron, M. (1984). *Ecologie de la végétation terrestre*. Masson, ISBN 2-225-80010-3, Paris, France

Haber, W. (1990). Basic concept of landscape ecology and their application in land management. In: *Ecology for tomorrow*. Kanawabe, H, Ohgushi, T & Higashi, M. (Eds.). *Phys. Ecol. Japan*. ISSN 0370-9612, pp 131-146

Ingegnoli, V. (1980). *Ecologia e progettazione*. Cooperativa Universitaria Studio e Lavoro, Milano

Ingegnoli, V. (1991). Human influences in landscape change: thresholds of metastability. In: *Terrestrial and aquatic ecosystems: perturbation and recovery*. Ravera O (Ed.), pp 303-309. Ellis Horwood Lmtd, ISBN 0-13-928201-7, Chichester, UK

Ingegnoli, V. (1993). *Fondamenti di Ecologia del paesaggio*. CittàStudi, ISBN 88-251-0032-9, Milano, Italy

Ingegnoli, V. (Ed.) (1997). *Esercizi di Ecologia del Paesaggio*. CittàStudi, ISBN 88-251-7178-1, Milano, Italy

Ingegnoli, V. (1999). Definition and evaluation of the BTC (biological territorial capacity) as an indicator for landscape ecological studies on vegetation. In: *Sustainable Landuse Management: The Challenge of Ecosystem Protection*. Windhorst W., Enckell P.H. (Eds.) *EcoSys: Beitrage zur Oekosystemforschung, Suppl* Bd 28, pp 109-118, Germany

Ingegnoli, V. (2002). *Landscape Ecology: A Widening Foundation*. Springer-Verlag, ISBN 3-540-42743-0, Berlin, Heidelberg, New York

Ingegnoli, V. (2006). Ecological transformations in the forest landscape unit at one alpine spruce (*Picea abies* K.) CONECOFOR plot, 1998-2004. *Ann.CRA – Centro Ric. Selv.*- Vol. 34, pp 49-56, ISSN 0390-0010

Ingegnoli, V. (2010). Ecologia del paesaggio. In: In: *XXI Secolo*. Gregory, T. (Ed.). Istituto della Enciclopedia Italiana Treccani, vol. IV. pp. 23-33, Roma

Ingenoli, V. (2011). *Bionomia del paesaggio. L'ecologia del paesaggio biologico-integrata per la formazione di un "medico" dei sistemi ecologici*. Springer-Verlag, Milano

Ingegnoli, V. & Pignatti, S. (Eds.) (1996). *L'ecologia del paesaggio ih Italia*. CittàStudi-UTET, ISBN 88-251-0110-4, Milano

Ingegnoli, V. & Giglio, E. (1999). Proposal of a synthetic indicator to control ecological dynamics at an ecological mosaic scale. Annali di Botanica LVII, pp 181-190, Rome, Italy, ISSN 0365-0812

Ingegnoli, V. & Giglio, E. (2005) *Ecologia del paesaggio. Manuale per conservare, gestire e pianificare l'ambiente*. Sistemi Editoriali SE, Esselibri-Spa, ISBN 88-513-0218-9, Napoli, Italy

Ingegnoli, V. & Pignatti, S. (2007). The impact of the widened landscape ecology on vegetation science: towards the new paradigm. Rendiconti Lincei Vol. 18, Issue 2, pp 89-122, ISSN 11206349, Retrieved from http: //www.springerlink.com/index/10.1007/BF02967218

Lorenz, K. (1978). *Vergleichende Verhaltensforschung: Grundlagen der Ethologie.* Springer-Verlag, Berlin, Wien

Manzelli, P. (1999). Il cervello, la percezione, il colore. In: *Didattica delle Scienze,* n. 202, Firenze

Manzelli, P. & Costa, M.G. (1994). *I segreti dell'acqua. L'opera scientifica di Giorgio Piccardi.* Roma

Martin-Bosse, H. (1967). Shwarzfahernwalden in Karnten. Springer-Verlag, Berlin, ISBN 3211807985

Naveh Z, Lieberman A (1984) *Landscape Ecology: theory and application.* Springer-Verlag, ISBN 0-387-90849-8, New York, Berlin, Heidelberg

Naveh Z, Lieberman A (1990) *Landscape Ecology: theory and application.* II Edition. Springer-Verlag, ISBN 0-387-94059-6, New York, Berlin, Heidelberg

Noss, R.F. (1983). A regional Landsclpe approach to maintain diversity. *BioScience,* 33, pp 700-706

Noss, R.F. (1997). Hierarchical indicators for monitoring changes in biodiversity. In: Principles of Conservation Biology. II ed.Meffe, GK, Carroll, C.R. & Contributors, Sinauer Associates, inc. ISBN 0-87893-521-5, Sunderland, Massachusetts, USA

Odum, E.P. (1971). *Fundamentals of Ecology.* 3rd ed. WB. Saunders Co., ISBN 0-7216-6941-7. Philadelphia

Odum, E.P. (1983). *Basic Ecology.* CBS College Publ. USA

Oldeman, R.A.A. (1990). *Forests: elements of sylvology.* Springer-Verlag, ISBN 3-540-51883-5. Berlin, Heidelberg, New York

O'Neill, R.V., DeAngelis, D.L., Waide, J.B. & Allen, T.F.H. (1986). *A Hierarchical Concept of Ecosystems.* Princeton University Press, ISBN 0-691-08437-8, Princeton, New Jersey, USA

Pattee, H.H. (1973). (Ed.) *Hierarchy Theory. The Challenge of Complex Systems.* New York, NY: Georges Braziller

Pignatti, S. (1979). I piani di vegetazione in Italia. *Giorn. Bot. It.* 113, pp 411-428

Pignatti, S. (1980). Reflections on the phytosociological approach and the epistemological basis of vegetation science. *Vegetatio* 42, pp 181-185

Pignatti, S. (1988). Ecologia del Paesaggio. In: *La vita delle piante.* Honsell, E., Giacomini, V. & Pignatti, S. pp 472-483, UTET, ISBN 88-02-04232-2. Torino

Pignatti, S. (1994). *Ecologia del Paesaggio.* UTET, ISBN 88-02-04671-9. Torino

Pignatti, S. (1996). Some Notes on Complexity in Vegetation. *Journ, of Vegetation Sc.* 7, pp 7-12

Pignatti, S., Dominici, E. & Pietrosanti, S. (1998). La biodiversità per la valutazione della qualità ambientale. *Atti dei Convegni Lincei* 145, pp 63-80

Pignatti, S. & Trezza, B. (2000). *Assalto al pianeta, Attività produttiva e crollo della biosfera.* Bollati Boringhieri, ISBN 88-339-1216-7. Torino

Pignatti, S., Box, E.O. & Fujiwara, K. (2002). A new paradigm for the XXIth Century. *Annali di Botanica,* ISSN 0365-0812 Vol. II, pp31-58

Popper, K.R. (1982). *The open universe : an argument for indeterminism.* WW Bartley III, London

Popper, K.R. (1990). *A World of Propensities.* Thoemmes, Bristol, UK

Prigogine, I. (1988). *La nascita del tempo.* Edizioni Teoria, Roma

Prigogine, I. (1996). *La fin dès certitudes: temps, chaos et les lois de la nature.* Editions Odile Jacob, ISBN 2-7382-0330-8. Paris

Prigogine, I., Nicolis, G. & Babloyatz, A. (1972). Thermodynamics of evolution. *Physics Today* Vol. 25, no11, pp 23-28; no12, pp 38-44

Rivas-Martinez, S. (1987). Nociones sobre Fitosociología, biogeografía y bioclimatología. In : *La vegetacion de España,* pp 19-45. Universidad de Alcalá de Henares, Madrid,

Ruelle, D. (1991). *Hasard et chaos.* Éditions Odile Jacob, Paris

Tüxen, R. (1956). Die heutige potentielle natürliche Vegetation als Gegenstand der Vegetationkartierung. Angew. *Pflanzensoziologie Stolzenau,* Weser13, pp 5-42

Tüxen, R. (Ed.) (1978). Assoziationkomplexe. *Ber. Intern. Symp. Veg.*, Rinteln. Cramer Verlag, Vaduz

U.S. Congress, Office of Technology Assessment (1986). *Grassroots Conservation of Bioiog- ical Diversity in the United States – Background Paper* #1, OTA-BP-F-38 (Washing- ton, DC: U.S. Government Printing Office)

Von Bertalanffy, L. (1968). *General System theory, Foundations, Development and Applications.* George Braziller New York

Walter, H. (1973). *Vegetation of the Earth in Relation to Climate and the Eco-Physiological Conditions.* Springer-Verlag. New York, Heidelberg, Berlin

Weiss, P.A. (1969). The living systems: determinism stratified. In: *Beyond Reductionism: New Perspectives in Life Sciences.* Koestler, A. & Smithies, J.R. (Eds.) Hutchinson of London, pp 3-55

Westhoff, V. (1970). Vegetation study as a branch of biological science. *Landbouwhogeschool Wageningen, Misc Pap 5,* pp 11-30

Whittaker, R.H. (1975). *Communities and ecosystems.* 2nd ed., Macmillan Pub, ISBN 0-02-427390-2. New York

Zanzi, L. (1999). Storia e cultura della conservazione. In: *Biodiversità, estinzione, conservazione.* Massa, R. & Ingegnoli, V. (Eds.), pp 19-49. Utet Libreria, Torino

Zonneveld, I.S. (1995). *Land ecology.* Academic Publishing, ISBN 90-5103-101-7. SPB Academic Publishing, Amsterdam

8

The Mean-Field Theory in the Study of Ferromagnets and the Magnetocaloric Effect

J. S. Amaral, S. Das and V. S. Amaral
Departamento de Física and CICECO, Universidade de Aveiro
Portugal

1. Introduction

List of Symbols

H	applied magnetic field	β	dependence of T_C in volume
λ	mean-field exchange parameter	K	compressibility
M	magnetization	α_1	thermal expansion
σ	reduced magnetization	T_0	ordering temperature (no volume coupling)
T	temperature	v	volume
χ	magnetic susceptibility	v_0	volume (no magnetic interaction)
N	number of spins	G	Gibbs free energy
J	spin	M_{sat}	saturation magnetization
g	gyromagnetic ratio	S	entropy
μ_B	Bohr magneton	p	pressure
k_B	Boltzmann constant	η	Bean-Rodbell model parameter
T_C	Curie temperature	$\mathcal{B}_J$	Brillouin function (spin J)
μ_{eff}	effective moment	H_c	critical field
C	Curie constant	x	fraction of ferromagnetic phase

Effective field theories, such as the molecular mean-field model (Coey, 2009; Kittel, 1996), are invaluable tools in the study of magnetic materials (Gonzalo, 2006). The framework of the molecular mean-field allows a description of the most relevant thermodynamic properties of a magnetic system, in a simplified way. For this reason, this century-old description of cooperative magnetic effects is still used in ongoing research for a wide range of magnetic materials, although its limitations are well known, such as neglecting fluctuation correlations near the critical temperature and low temperature quantum excitations (Aharoni, 2000).

In this work, we present methodologies and results of a mean-field analysis of the magnetocaloric effect (MCE) (Tishin & Spichin, 2003). The MCE is common to all magnetic materials, first discovered in 1881 by the German physicist Emil Warburg. The effect describes the temperature variation of a ferromagnetic material when subjected to an applied magnetic field change, in adiabatic conditions. In isothermal conditions, there occurs a change in magnetic entropy due to the magnetic field change, and heat is transferred. The first major application of the MCE was presented in the late 1920s when cooling via adiabatic demagnetization was independently proposed by Debye and Giauque. The application of the adiabatic demagnetization process made it possible to reach the very low temperature value of 0.25 K in the early 1930s, by using an applied field of 0.8 T and 61 g of the paramagnetic salt $Gd_2(SO_4)_3{\cdot}8H_2O$ as the magnetic refrigerant.

Pioneered by the ground-breaking work of G. V. Brown in the 1970's, the concept of room-temperature magnetic cooling has recently gathered strong interest by both the scientific and technological communities (Brück, 2005; de Oliveira & von Ranke, 2010; Gschneidner Jr. & Pecharsky, 2008; Gschneidner Jr. et al., 2005; Tishin & Spichin, 2003). The discovery of the giant MCE (Pecharsky & Gschneidner, 1997) resulted in this renewed interest in magnetic refrigeration, which, together with recent developments in rare-earth permanent magnets, opened the way to a new, efficient and environmentally-friendly refrigeration technology.

The development and optimization of magnetic refrigerator devices depends on a solid thermodynamic description of the magnetic material, and its properties throughout the steps of the cooling cycles. This work will present, in detail, the use of the molecular mean-field theory in the study of ferro-paramagnetic phase transitions, and the MCE. The dependence of magnetization on external field and temperature can be described, in a wide validity range. This description is also valid for both second and first-order phase transitions, which will become particularly useful in describing the magnetic and magnetocaloric properties of the so-called "giant" and "colossal" magnetocaloric materials.

An overview of the Weiss molecular mean-field model, and the inclusion of magneto-volume effects (Bean & Rodbell, 1962) is presented, providing the theoretical background for simulating the magnetic and magnetocaloric properties of second and first-order ferromagnetic phase transition systems. The numerical methods employed to solve the transcendental equation to determine the $M(H,T)$ (where M is magnetization, H applied magnetic field and T Temperature) dependence of a ferromagnetic material with a second-order phase transition are described. In the case of first-order phase transitions, the use of the Maxwell construction is shown in order to estimate the equilibrium solution from the two distinct metastable solutions and the single unstable solution of the state equation.

The generalized formulation of the molecular mean-field interaction leads to a novel mean-field scaling method (Amaral et al., 2007), that allows a direct estimation of the mean-field exchange parameters from experimental data. The application of this scaling method is explicitly shown in the case of simulated data, to exemplify its application and to highlight its robustness and general approach. Experimental magnetization data of second (La-Sr-Mn-O based) and first-order (La-Ca-Mn-O based) ferromagnetic manganites is then analyzed under this framework. We show how the Bean-Rodbell mean-field model can adequately simulate the magnetic properties of these complex magnetic systems, candidates for application for room-temperature magnetic refrigerant materials (Amaral et al., 2005; Gschneidner & Pecharsky, 2000; Phan & Yu, 2007).

An overview of the MCE is presented, focusing on the use of the Maxwell relations to estimate the magnetic entropy change of a magnetic phase transition. The thermodynamics of the molecular mean-field model presents us also a new method to estimate the MCE from magnetization data. Results of magnetic entropy variation values are compared, highlighting the difficulties of estimating the MCE in first-order phase transition systems.

The interest on the magnetocaloric properties of first-order phase transition systems, in terms of fundamental physics and also magnetic refrigeration applications, has opened debate on the validity of the use of Maxwell relations to estimate the MCE in these systems (Giguère et al., 1999). Using simulated data of a first-order mean-field system, we verify the consequences of the common use of the Maxwell relation to estimate the MCE from non-equilibrium magnetization data.

The recent reports of "colossal" values of magnetic entropy change of first-order phase transition systems (de Campos et al., 2006; Gama et al., 2004; Rocco et al., 2007) are also discussed, and are shown to be related to the mixed-phase characteristics of the system. We present a detailed description on how the misuse of the Maxwell relation to estimate the MCE of these systems justifies the non-physical results present in the bibliography (Amaral & Amaral, 2009; 2010).

Understanding the thermodynamics of a mixed-phase ferromagnetic system allows the construction of a new methodology to correct the results from the use of the Maxwell effect on magnetization data of these compounds. This methodology is theoretically justified, and its application to mean-field data is presented (Das et al., 2010a;b). In contrast to other suggestions in the bibliography (Tocado et al., 2009), this novel methodology permits a realistic estimative of the magnetic entropy change of a mixed-phase first-order phase transition system, with no need of additional magnetic or calorimetric measurements.

2. Molecular mean-field theory and the Bean-Rodbell model

2.1 Ferromagnetic order and the Weiss molecular field

A simplified approach to describing ferromagnetic order in a given magnetic material was put forth by Weiss, in 1907. This concept of a molecular field assumes the magnetic interaction between magnetic moments as equivalent to the existence of an additional internal interaction/exchange field that is a function of the bulk magnetization M:

$$H_{total} = H_{external} + H_{exchange} \text{ and } H_{exchange} = \lambda M, \tag{1}$$

where λ is the mean-field exchange parameter.

The general representation of the molecular mean-field model is then

$$\sigma = f\left[\frac{H+\lambda M}{T}\right]. \tag{2}$$

where f is the general function that applies in the paramagnetic system (e. g. the Brillouin function).

From a linear approximation of the susceptibility (Curie law):

$$\chi = \frac{M}{H} = \frac{NJ(J+1)g^2\mu_B^2}{3k_BT_C} = \frac{N\mu_{eff}^2}{3k_BT_C}; \tag{3}$$

where μ_{eff} is the effective magnetic moment: $\mu_{eff} = g[J(J+1)]^{1/2}\mu_B$.

We define the Curie temperature T_C as the temperature where the ferromagnetic to paramagnetic transition occurs, and there is a divergence in the susceptibility:

$$\chi = \frac{C}{T-T_C}, \text{ where } C = \frac{NJ(J+1)g^2\mu_B^2}{3k_B} \text{ and } T_C = C\lambda. \tag{4}$$

The exchange parameter can be estimated from the following relation, as long as N and J are known.

$$\lambda = \frac{3k_BT_C}{Ng^2J(J+1)\mu_B^2} \tag{5}$$

Typical values of λ correspond to molecular fields in the order of hundreds of Tesla.

2.2 Magneto-volume effects: The Bean-Rodbell model

The Bean-Rodbell model (Bean & Rodbell, 1962) adds a phenomenological description of magneto-volume effects to the classical molecular mean-field model of Weiss. The dependence of exchange interaction on interatomic spacing is then considered, taking into account three new parameters: β, which corresponds to the dependence of ordering temperature on volume, and also the volume compressibility, K and thermal expansion α_1. The formulation behind the model is as follows:

$$T_C = T_0 \left[1 + \beta \left(\frac{v - v_0}{v_0} \right) \right], \tag{6}$$

where T_C is the Curie temperature corresponding to a lattice volume of v, while v_0 is the equilibrium lattice volume in the absence of magnetic interactions, corresponding to a Curie temperature of T_0 if magnetic interactions are assumed, but with no magneto-volume effects. The free energy of the system can therefore be described, taking into account magnetic and volume interactions. For simplicity, we consider a purely ferromagnetic interaction. For a description including anti-ferromagnetic interactions, see Ref. (Bean & Rodbell, 1962).

$$G = G_{\text{field}} + G_{\text{exchange}} + G_{\text{volume}} + G_{\text{pressure}} + G_{\text{entropy}} \tag{7}$$

Considering first a spin 1/2 system, and the molecular field exchange interaction, we have that the Gibbs free energy per unit volume is:

$$\begin{aligned} G_v = & -HM_{sat}\sigma - \frac{1}{2}Nk_BT_c\sigma^2 + \frac{1}{2K}\left[\frac{v - v_0}{v_0}\right]^2 + p\left(\frac{v - v_0}{v_0}\right) \\ & - TNk_B\left[\ln 2 - \frac{1}{2}\ln\left(1 - \sigma^2\right) - \sigma \tanh^{-1}\sigma\right] - TS_{\text{lattice}}. \end{aligned} \tag{8}$$

where σ is the reduced magnetization, M_{sat} the saturation magnetization and N the number of particles for volume v_0. While the original description of Bean and Rodbell does not initially consider the lattice entropy, we will keep the generality of the calculations along our description of the model. The lattice entropy term is as follows:

$$S_{\text{lattice}} = 3Nk_B\left[\frac{x}{e^x - 1} - ln(1 - e^{-x})\right], \tag{9}$$

where $x \equiv h\nu/k_BT$ with ν being the phonon frequency. Eq. 9 can be expanded via the Debye approximation:

$$S_{\text{lattice}} = Nk_B\left[4 - 3\ln\Theta/T + (3/40)(\Theta/T^2) + ...\right] \tag{10}$$

where $\Theta \equiv h\nu_{\text{max}}/T$. From the previous expression we obtain:

$$\partial S/\partial v \cong -3Nk_Bd\ln(\nu_{\text{max}})/dv = \alpha_1/K \tag{11}$$

where α_1 is the thermal expansion coefficient ($\alpha_1 \equiv (1/v)(\partial v/\partial T)_p$) and K is the compressibility ($K \equiv -(1/v)(\partial v/\partial p)_T$).
By substituting Eq. 6 into Eq. 8, deriving in volume by using also Eq. 11, the relation between magnetization and volume that corresponds to the energy minimum is

$$\frac{v - v_0}{v_0} = \frac{1}{2} N K k_B T_0 \beta \sigma^2 + T v_o \alpha_1 - pK \tag{12}$$

By substituting the previous relation into the Gibbs free energy (Eq. 8), and minimizing in respect to volume, we obtain

$$\begin{aligned}(G_v)_{\min} = & -H M_{sat}\sigma - \frac{1}{2} N k_B T_0 \sigma^2 \left[1 - \beta(pK - \alpha_1 T)\right] \\ & - p^2 K/2 - \alpha_1^2 T^2/2K + \alpha_1 T p - \frac{1}{2K}\left(\frac{1}{2} N k_B T_0 \sigma^2 \beta\right)^2 \\ & - T N k_B \left[4 + \ln 2 - \frac{1}{2}\ln(1 - \sigma^2) - \sigma \tanh^{-1}\sigma\right].\end{aligned} \tag{13}$$

By minimizing as a function of σ, we obtain the implicit dependence of σ on temperature, for spin 1/2.

$$\frac{T}{T_0} = \frac{\sigma}{\tanh^{-1}\sigma}\left(1 - \beta(pK - \alpha_1 T) + \frac{\eta\sigma^2}{3} + M_{sat} H\right) \tag{14}$$

where the η parameter defines the order of the phase transition, if $\eta \leq 1$, the transition is second-order and if $\eta > 1$, the transition is first-order. The value of η is:

$$\eta = \frac{3}{2} N k_B K T_0 \beta^2 \ ; \text{(spin = 1/2)} \tag{15}$$

$$\eta_J = \frac{5}{2}\frac{[4J(J+1)]^2}{[(2J+1)^4 - 1]} N k_B K T_0 \beta^2 \ ; \text{(arbitrary J spin)}. \tag{16}$$

We can rewrite Eq. 14, in the more familiar molecular-mean field expression type, $M = f[(H + \lambda M)/T]$, since $\tanh^{-1}\sigma = (H + \lambda(M,T)M)/T$, (for spin = 1/2):

$$\tanh^{-1}\sigma = \frac{g\mu_B H/2k_B + (1 - \beta p K + \beta\alpha_1 T) T_0 \sigma + (\eta/3) T_0 \sigma^3}{T}. \tag{17}$$

We can therefore consider, in the absence of external pressure, and considering the lattice entropy change small, that the molecular field dependence in magnetization follows the simple form of $H_{exchange} = \lambda_1 M + \lambda_3 M^3$.
Considering a generalized spin system, with no applied pressure, nor the lattice entropy contribution, the implicit dependence of σ on temperature is

$$T(\sigma, H) = \frac{g\mu_B J H/k_B + aT_0\sigma + bT_0\sigma^3}{\mathcal{B}_J^{-1}(\sigma)}, \tag{18}$$

where

$$a = \frac{3J}{J+1}; \qquad b = \frac{9}{5}\frac{(2J+1)^4 - 1}{(2(J+1))^4}\eta_J = b'\eta_J \tag{19}$$

and $\mathcal{B}_J{}^{-1}(\sigma) = \partial S_J/\partial\sigma$, where $\mathcal{B}_J$ is the Brillouin function for a given J spin.

If the lattice entropy change is taken into consideration, the effect corresponds introducing the $\beta\alpha_1 T$ term into the first-order term of the exchange field, in the same way as the spin 1/2 system.
If we choose to describe the exchange field as $\lambda_1 M + \lambda_3 M^3$, it becomes practical to rewrite the conditions of the model explicitly in terms of the λ_1 and λ_3 parameters, bulk magnetization M, spin and the saturation magnetization, M_{sat}. This corresponds to the following expression, where the η parameter can be defined as:

$$\eta = \lambda_3 / \left[b' T_0 k_B / (g \mu_B J^2 M_{\text{sat}}^3) \right], \tag{20}$$

where the b' parameter is previously defined in Eq. 19. The λ_3 parameter includes the β (dependence of ordering temperature on volume) and K (compressibility) system variables. The direct consequence of the previous expression is that, by substituting the T_0 value, the ratio of λ_1 and λ_3, together with the system parameters define the nature of the transition, following the next simplified expression:

$$\eta = \frac{3J^2 M_{sat}^2}{b'} \frac{\lambda_3}{\lambda_1}. \tag{21}$$

2.3 Numerical approach

2.3.1 Second-order phase transitions

As shown in the previous section, the Bean-Rodbell model can describe a magnetovolume induced first-order phase transition. While the numerical approach to simulate first-order phase transitions in the Landau theory is straightforward (finding the roots of a polynomial and then which of the two local minima corresponds to the absolute free energy minimum), in the case of the Bean-Rodbell model the case is more complicated in computational terms. Even in the more simple second-order phase transition, solving the transcendental equation $M = f[(H + \lambda M)/T]$ cannot be done algebraically, and so numerical methods are employed. The classic visual representation of the numerical approach is presented in Fig. 1.

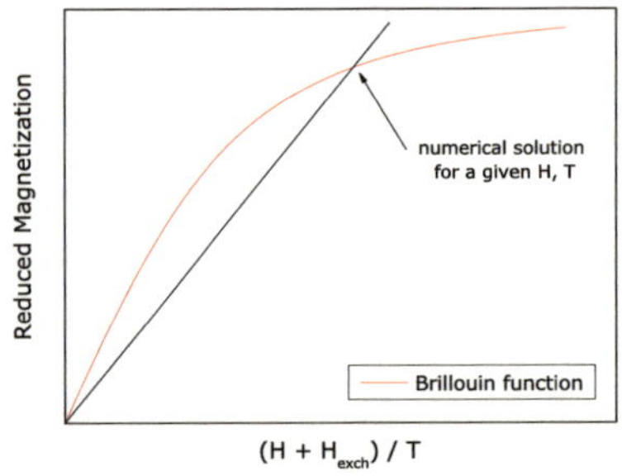

Fig. 1. Graphical solution of the mean-field state equation, adapted from Ref. (Kittel, 1996).

This graphical approach is easily converted into numerically finding the roots of the following function:

$$M_{\text{sat}} \mathcal{B}_J(J, \lambda_1, \lambda_3, M, H, T) - M(H, T); \tag{22}$$

Finding the roots of the above equation can be numerically achieved by using the optimized method suggested by T. Dekker, employing a combination of bisection, secant, and inverse quadratic interpolation methods (Forsythe et al., 1976).

2.3.2 First-order phase transitions

For the first-order phase transition, there are multiple solutions that need to be calculated, corresponding to the stable (equilibrium), metastable and unstable branches. Fig. 2(a) shows a representation of these solutions.

The methodology for obtaining the various M solutions in this situation is more numerically intensive than in a second-order system, apart from subdividing the interval of magnetization values into multiple sub-intervals to search for the multiple roots.

In order to calculate the critical field value H_c and consequently the full equilibrium solution (stable branch), the Maxwell construction (Callen, 1985) is applied, which consists of matching the energy of the two phases, in the so-called equal-area construction (Fig. 2(b)).

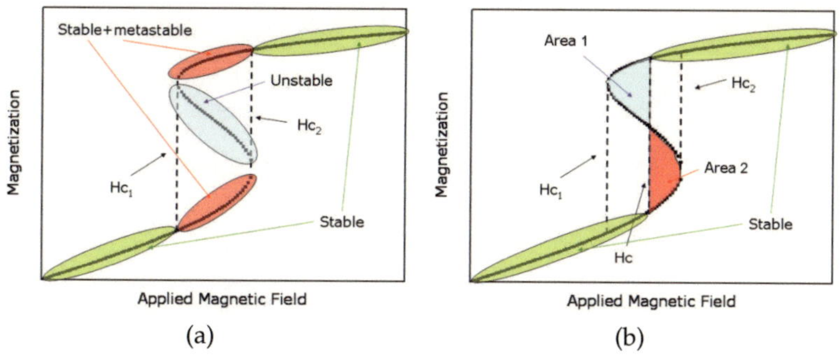

Fig. 2. a) The multiple solution branches from the roots of Eq. 22, for a first-order transition from the Bean-Rodbell model, and b) the Maxwell construction for determining the critical field H_c and the full equilibrium solution, for a first-order magnetic phase transition system.

In numerical terms, applying this graphical methodology becomes a matter of integrating the areas between the metastable and unstable solutions, between the Hc_1 and Hc_2 field values, until the value of area 1 is equal to area 2. This operation is numerically intensive, but manageable for realistic field interval values. The most important numerical concern is adequately reproducing all branches (solutions), in a way that the algorithm correctly integrates each area. In programming terms, this becomes a complicated problem, but becomes controllable by a careful definition of the various number of roots of the functions, and developing an optimized integration algorithm for each independent situation that can appear within this approach.

2.3.3 Estimating magnetic entropy change

Within the molecular field model, the relation between the magnetic entropy and the magnetic equation of state is simply defined. Let us consider that the magnetic equation of state is a generalized f function, and so $M = f[(H + \lambda(M,T)M)/T]$. We can then integrate the magnetic entropy relation:

$$S_M = \int f^{-1}(M)dM. \tag{23}$$

So to calculate the entropy change between two distinct field values H_1 and H_2:

$$-\Delta S_M(T)_{\Delta H} = \int_{M|_{H_1}}^{M|_{H_2}} f^{-1}(M)dM. \tag{24}$$

where $f^{-1}(M)$ is simply the argument of the state function for a given magnetization value:

$$f^{-1}(M) = \frac{H + \lambda(M,T)M}{T}. \tag{25}$$

We can generalize the previous result by considering an explicit dependence of the exchange field in temperature. We rewrite the previous equation as

$$f^{-1}(M) = \frac{H}{T} + \frac{\lambda(M,T)M}{T} \to H = Tf^{-1}(M) - \lambda(M,T)M \tag{26}$$

and using the following Maxwell relation (Callen, 1985):

$$\left(\frac{\partial S}{\partial M}\right)_T = -\left(\frac{\partial H}{\partial T}\right)_M, \tag{27}$$

entropy can be estimated by

$$\Delta S(T)_{H_1 \to H_2} = -\int_{M_{H1}}^{M_{H2}} \left(\frac{\partial H}{\partial T}\right)_M dM, \tag{28}$$

leading to

$$-\Delta S_M(T)_{H_1 \to H_2} = \int_{M|_{H_1}}^{M|_{H_2}} \left(f^{-1}(M) - \left(\frac{\partial \lambda}{\partial T}\right)_M M\right) dM. \tag{29}$$

Compared to Eq. 24, the derivative $\partial\lambda/\partial T$ directly affects the result. We shall explore the use of Eq. 24 to calculate the magnetic entropy change and compare it to the use of the Maxwell relation.

3. A molecular mean-field scaling method

3.1 Methodology

As presented in section 2.2, the molecular mean-field theory gives us a simple and often effective tool to describe a ferromagnetic system. If one is studying magnetization data from a given material, obtaining the mean-field parameters is not immediate. To do so, one usually needs to set the spin value and/or the number of ions N, and the mean-field state function is the Brillouin function or Langevin function (for a high spin value). From then on, the λ_1 parameter can be obtained from low-field M versus T measurements and a linear Curie-Weiss law fit of the inverse susceptibility. Subsequent fits to each $M(H)$ isotherm can then be performed. Such an approach can be quite complex, particularly if one considers a system where the magnetic ions can have different spin states, such as mixed-valence manganites, where the ratio between ions needs to be previously assumed (Szewczyk et al., 2000). Obtaining higher orders of the mean-field exchange parameter (λ_3, λ_5, etc.) can be done by performing simulations to describe experimental data, as done by Bean and Rodbell to describe MnAs (Bean & Rodbell, 1962).

A different approach to obtain the mean-field parameters from experimental magnetization data is presented here, based on data scaling. A summarized version of this work has been published in 2007 (Amaral et al., 2007). We consider the general mean field law, $M(H,T) = f((H + H_{exch})/T)$, where the state function f is not pre-determined, and that λ (as in $H_{exch} = \lambda M$) may depend on M and/or T. Then for corresponding values with the same M, $(H + H_{exch})/T)$ is also the same, the value of the inverse $f^{-1}(M)$ function:

$$\frac{H}{T} = f^{-1}(M) - \frac{H_{exch}}{T} \tag{30}$$

By taking H and T groups of values for a constant M and Eq. 30, the plot of H/T versus $1/T$ is linear if λ does not depend on T. The slope is then equal to H_{exch}, for each M value. Then, each isomagnetic line is shifted from the others, since its abscissa at $H/T = 0$ is simply the inverse temperature of the isotherm which has a the spontaneous magnetization equal to the M value (Fig. 3).

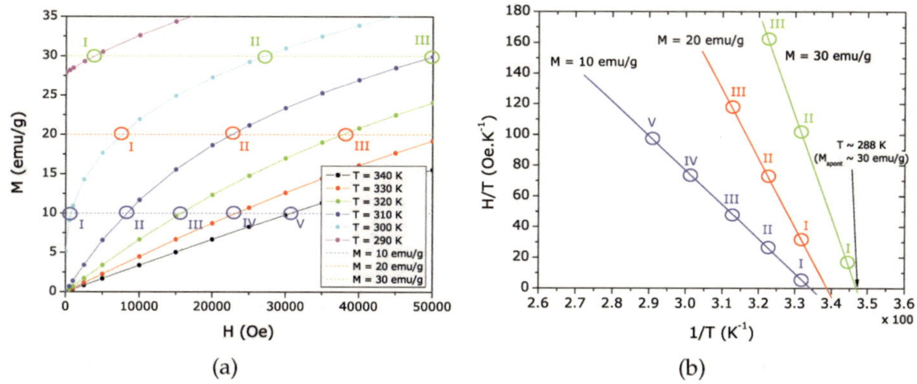

Fig. 3. a) Isomagnetic ($M = 10, 20$ and 30 emu/g) points from mean-field generated data in an M versus H plot (lines are eye-guides), and b) corresponding H/T versus $1/T$ plot (lines are linear fits to isomagnetic points).

In a similar fashion, a simple case of a constant λ (i. e. independent of M and T), a plot of H/MT versus $1/T$ will show parallel lines for all M values, with slope equal to H_{exch}/M, which in turn is equal to λ.
In a first-order phase transition, the discontinuity of $M(H,T)$ means that when interpolating data for constructing the isomagnetic curves, care should be taken not to interpolate the discontinuity in $M(H,T)$. This is shown in Fig. 4 and is a direct consequence of there being a region in the H/T versus $1/T$ plot that has no data, much like the preceding $M(H,T)$ plot.

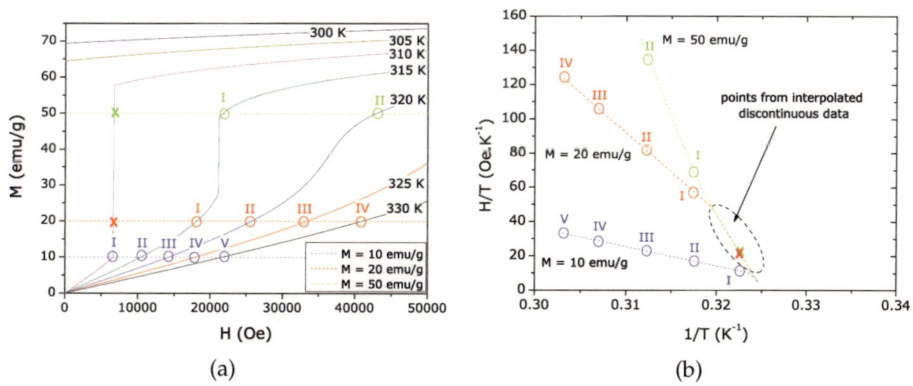

Fig. 4. a) Isomagnetic ($M = 10, 20$ and 30 emu/g) points from mean-field generated data in an M versus H plot (lines are eye-guides), and b) corresponding H/T versus $1/T$ plot (lines are linear fits to isomagnetic points).

Extrapolating this linear relation within this region will not present any real physical result, namely any relation to the spontaneous magnetization, which has a discontinuous jump. This point will be clearer in further simulation results.
From Eq. 30, the dependence of the exchange field on M is obtained directly. In principle, one can expect that the exchange field is given by a series of odd powers of M, $H_{exch} = \lambda M = \lambda_1 M + \lambda_3 M^3 + \ \ldots$. This follows from the frequently found expansion of the free energy in powers of M, e.g. when considering magnetovolume effects within the mean-field model by the Bean-Rodbell model as described in section 2.2. Note that the demagnetizing factor is intrinsically taken into account as a constant contribution to λ_1:

$$H_{total} = H_{applied} + H_{exch} - DM = H_{applied} + (\lambda_1 - D)M + \lambda_3 M^3 + \ldots \tag{31}$$

where D is the demagnetizing factor, in the simple assumption of an uniform magnetization. After obtaining H_{exch}, the second step of this method consists on building the scaling plot of M versus $(H + H_{exch})/T$, where data should collapse to the one curve that describes the system, the f function. Analyzing the f function is a further important step to study magnetic systems and to compare the results of theoretical microscopic models.
The above mentioned collapse on the scaling plot can be used to evaluate the validity of the mean-field analysis. In this sense the method is self-consistent: only if H_{exch} has been properly evaluated, will the points collapse into a single curve.

3.1.1 Second-order phase transition

As a first immediate example of this methodology, let us consider mean-field generated data, for a spin 2 system, with saturation magnetization of 100 emu g^{-1} and $T_C \sim 300$ K. No dependence of λ on T was considered. M versus H data, from 290 to 330 K, at a 1 K temperature step and 100 Oe feld step, are shown in Fig. 5(a).

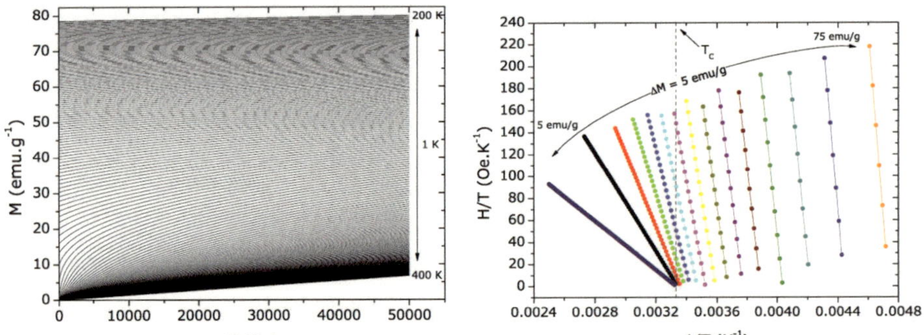

Fig. 5. a) Isothermal magnetization versus applied magnetic field, from 200 to 400 K, at a 1 K temperature step and 100 Oe field step and b) Isomagnetic H/T versus $1/T$ plot, of data from the molecular mean-field model, from $M = 5$ emu/g (dark blue line) to $M = 75$ emu/g (orange line), with a 5 emu/g step.

We then plot H/T versus $1/T$ at constant values of magnetization, following Eq. 30 (Fig. 5(b)). Since λ does not depend on T, there is a linear behavior of the isomagnetic curves, which are progressively shifted into higher $1/T$ values. From Eq. 30, the slope of each isomagnetic line of Fig. 5(b) will then give us the dependence of the exchange field in M (Fig. 6(a)).

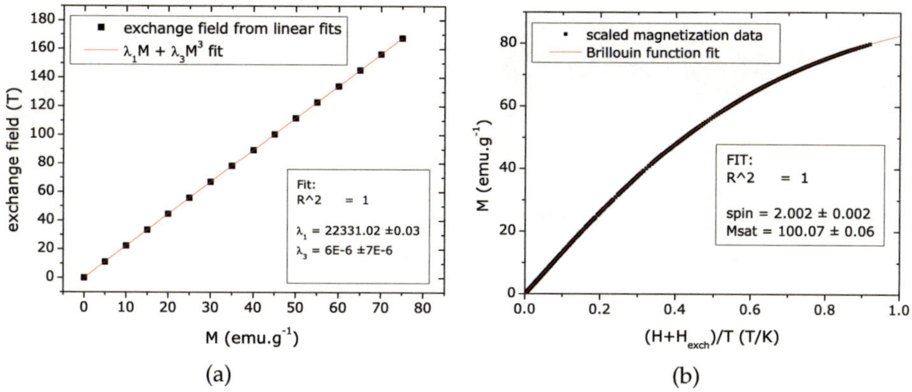

Fig. 6. a) Fit of the exchange field dependence on M. Solid squares represent the slope of each isomagnetic curve, from Figure 5(b) and b) Brillouin function fit of scaled data from the mean-field model, from Figure 5(a).

Having determined the $\lambda(M)$ dependence, we can now proceed to scale all the magnetization data, to determine the mean-field state function (Fig. 6(b)).
As expected, the scaled data closely follows a Brillouin function, with spin 2, and a saturation magnetization of 100 emu g^{-1}. We can then describe, interpolate and extrapolate $M(H,T)$ at will, since the full mean-field description is complete (exchange parameters and state function).

3.1.2 First-order phase transition

As shown previously, this approach is also valid if a first-order magnetic phase transition in considered. There is no fundamental difference on the methodology, apart from the expected higher order terms of $\lambda(M)$. Care must be taken when interpolating $M(H)$ data within the irreversibility zone, so that no values of M correspond to the discontinuities. We simulate a first-order magnetic phase transition by adding a λ_3 dependence of the molecular exchange field, equal to 1.5 (Oe emu^{-1} g)^3, to the previous second-order transition parameters. Isothermal magnetization data is shown in Fig 7(a). The discontinuity in magnetization values is visible, and we can estimate that the critical field is around 2.5 T, for this simulation parameters.
From the $M(H,T)$ data, we plot the corresponding isomagnetic H/T versus $1/T$ plot (Fig. 7(b)).
As shown previously in Fig. 4 b), if interpolations in $M(H,T)$ are done within the discontinuities, points that do not follow the expected linear behavior appear. These points should not be included for the linear fits to determine $\lambda(M)$. In the rest of the plot, the linear relation between H/T and $1/T$ is kept, as expected. Linear fits are then easily made to each isomagnetic line, and we obtain the exchange field dependence on magnetization (Fig. 8(a)). The $\lambda_1 M + \lambda_3 M^3$ dependence of the mean-field exchange parameter is well defined. We obtain λ_1 and λ_3 values that are, within the fitting error, equivalent to the initial parameters. This shows us that the first-order nature of the transition and the associated discontinuities should not affect this mean-field scaling methodology.
We can then construct the scaling plot, using the obtained λ_1 and λ_3 parameters (Fig. 8(b)). From the scaling plot and the subsequent fit with the Brillouin function, we obtain values of spin and saturation magnetization close to the the initial parameters of the simulation.

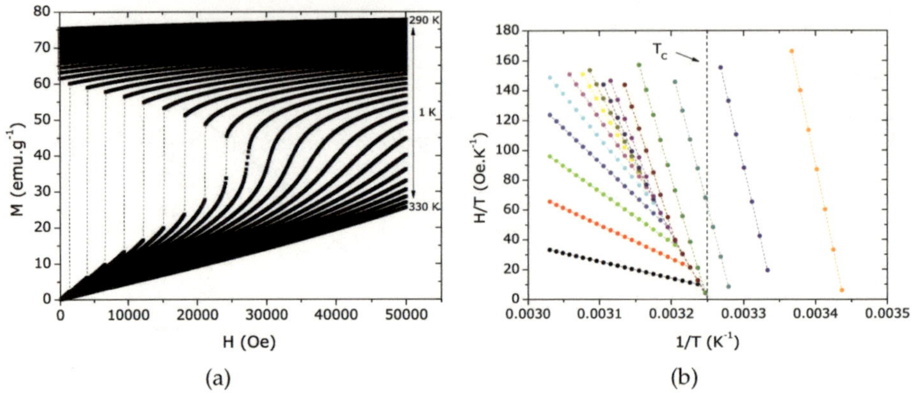

Fig. 7. a) Isothermal M versus H data of a first-order magnetic phase transition, from the Bean-Rodbell model and b) corresponding isomagnetic H/T versus $1/T$ plot, for a first-order mean-field system, and a 5 emu g^{-1} step.

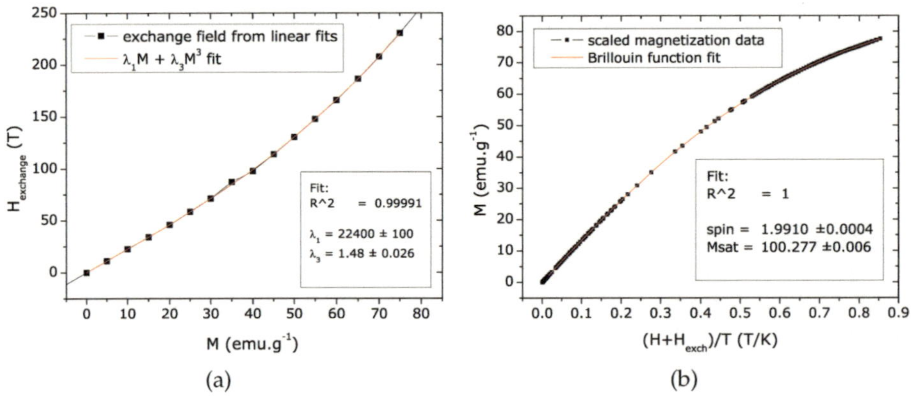

Fig. 8. a) Exchange field fit for a first-order mean-field system, with the $\lambda_1 M + \lambda_3 M^3$ law, and b) corresponding mean-field scaling plot and Brillouin function fit.

3.2 Applications

In the previous section, we have shown how it is possible to obtain directly from bulk magnetization data, and only considering the mathematical properties of the general mean-field expression $M = f[(H + \lambda M)/T]$, a direct determination of the molecular field exchange parameter λ and its dependence on M, and the mean-field state function f, which will contain information on the magnetic entities in play, and their interactions.

One immediate application for this method is to use this description of the magnetic properties of the system as a way to interpolate/extrapolate experimental data, and/or as a smoothing criteria to noisy $M(H,T)$ and corresponding $\Delta S_M(T)$ curves. It is worth mentioning that while this can be also performed within Landau theory, since the mean-field theory is not limited to small M values, the mean-field description of the system can have a broader application range: lower T and higher H values, up to saturation. Still, the methodology presented here is time-consuming, even with optimized numerical data analysis programs. When considering experimental magnetization data for $T < T_C$, care must be taken to adequately disregard data from the magnetic domain region (low fields). Still, the ability to determine the mean-field parameters directly from experimental data becomes attractive taking in mind that one can estimate magnetic entropy (and consequently magnetic entropy change) within the mean-field model, by using Eq. 29, reproduced here for convenience.

$$-\Delta S_M(T)_{H_1\rightarrow H_2} = \int_{M|_{H_1}}^{M|_{H_2}} \left(f^{-1}(M) - \left(\frac{\partial\lambda}{\partial T}\right)_M M\right) dM.$$

And so not only can the $M(H,T)$ values be interpolated/extrapolated, the entropy curves and their dependence in field and temperature can also be easily interpolated and extrapolated as well. This becomes particularly appealing if one wishes to make thermal simulations of a magnetic refrigeration device, and, within a physical model (and not by purely numerical approximations), the magnetocaloric response of the material, at a certain temperature and a certain field change is directly calculated. As an example of this approach, bulk isothermal magnetization data of two ferromagnetic manganite systems will be analyzed in this section. Fig. 9(a) shows the magnetization data of the ferromagnetic, second-order phase transition $La_{0.665}Er_{0.035}Sr_{0.30}MnO_3$ manganite, obtained by SQUID measurements. Fig. 9(b) shows the isomagnetic H/T versus $1/T$ plot, up to 50 emu/g in a 5 emu/g step, which could be reduced in order to have more points.

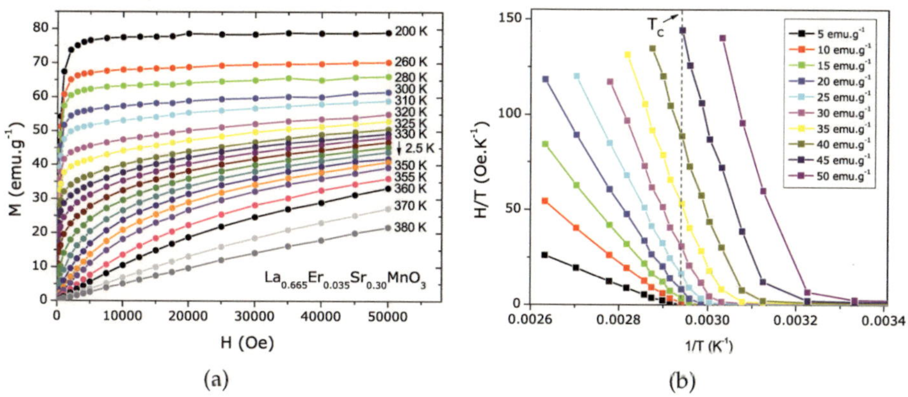

Fig. 9. a) Magnetization data of $La_{0.665}Er_{0.035}Sr_{0.30}MnO_3$ and b) corresponding isomagnetic H/T versus $1/T$ plot. Lines are eye-guides.

Each point at constant M is obtained from data interpolation os the isothermal $M(H)$ data. From linear fits to the H/T versus $1/T$ plot, the dependence of the exchange field in magnetization is directly obtained (Fig. 10(a)). The exchange field is fitted to a $\lambda_1 M + \lambda_3 M$ function. The scaling plot is then constructed (Fig. 10(b)).
For calculation purposes, the scaling function of Fig. 10(b) was described as an odd-terms polynomial function. The Fig. shows some data point that are clearly deviated from the scaling function. These points correspond to the magnetic domain region (low fields, $T < T_C$). With the exchange field and mean-field state function described, the magnetic behavior of this material can then be simulated. Also, magnetic entropy change can be calculated from the mean-field relation of Eq. 29. Result from these calculations, together with the experimental $M(H,T)$ data and $\Delta S_M(H,T)$ results from Maxwell relation integration are shown in Fig. 11. A good agreement between the experimental $M(H,T)$ curves and the mean-field generated curves with the obtained parameters is obtained. The entropy results show some deviations, particularly near T_C. While the mean-field theory does not consider fluctuations near

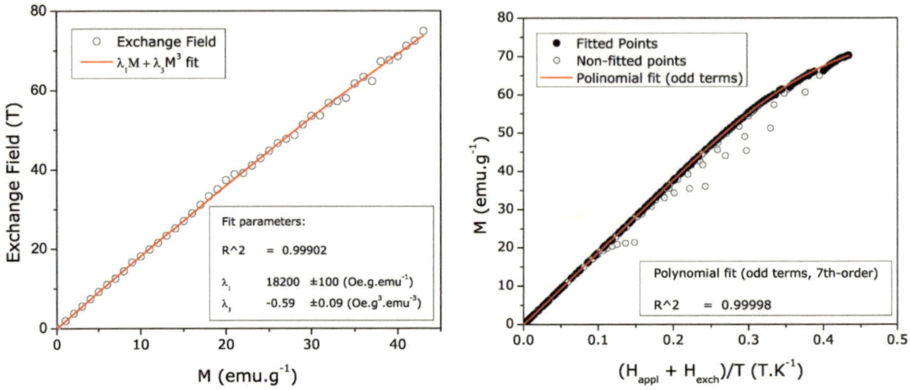

(a) (b)

Fig. 10. Interpolating a) experimental $M(H,T)$ data and b) magnetic entropy change results by mean-field simulations for the second-order phase transition manganite $La_{0.665}Er_{0.035}Sr_{0.30}MnO_3$.

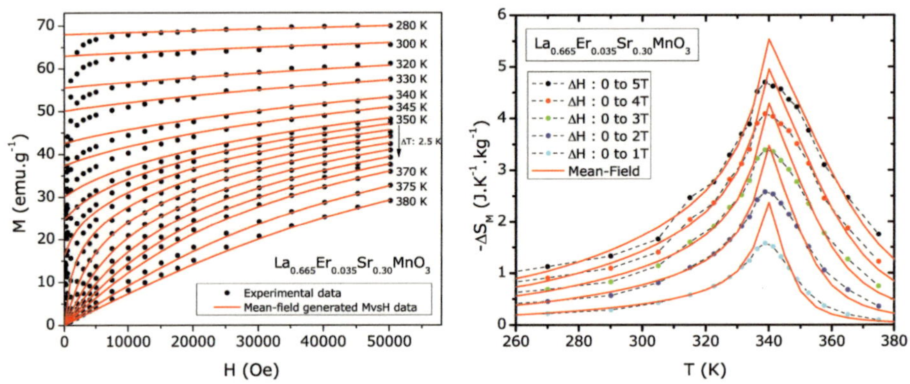

(a) (b)

Fig. 11. Interpolating a) experimental $M(H,T)$ data and b) magnetic entropy change results by mean-field simulations for the second-order phase transition manganite $La_{0.665}Er_{0.035}Sr_{0.30}MnO_3$.

T_C, these deviations can be attributed to that fact. Still, by considering disorder effects (chemical/structural inhomogeneity), a better description of magnetocaloric properties can be obtained (Amaral et al., 2008).

We now consider bulk magnetization data of a the first-order ferromagnetic phase transition $La_{0.638}Eu_{0.032}Ca_{0.33}MnO_3$ manganite. Fig. 12(a) shows isothermal magnetization data obtained from SQUID measurements, and Fig. 12(b) shows the corresponding isomagnetic H/T versus $1/T$ plot.

The exchange field H_{exch} dependence on magnetization (Fig. 13(a)) and the mean-field state function (Fig. 13(b)) are then obtained.

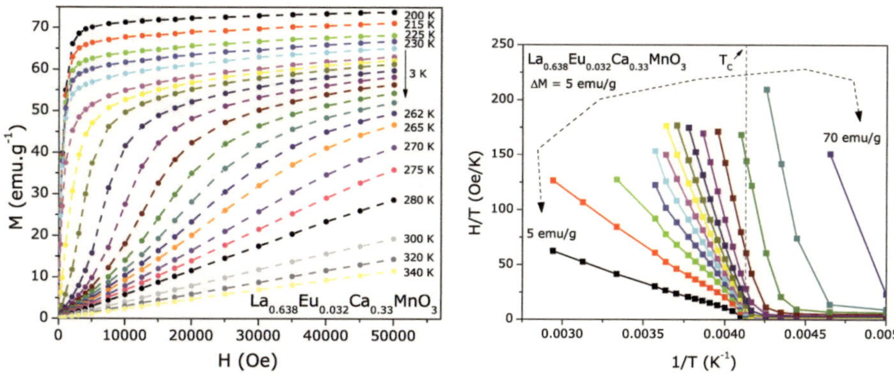

(a) (b)

Fig. 12. a) Magnetization data of $La_{0.638}Eu_{0.032}Ca_{0.33}MnO_3$ and b) corresponding isomagnetic H/T versus $1/T$ plot. Lines are eye-guides.

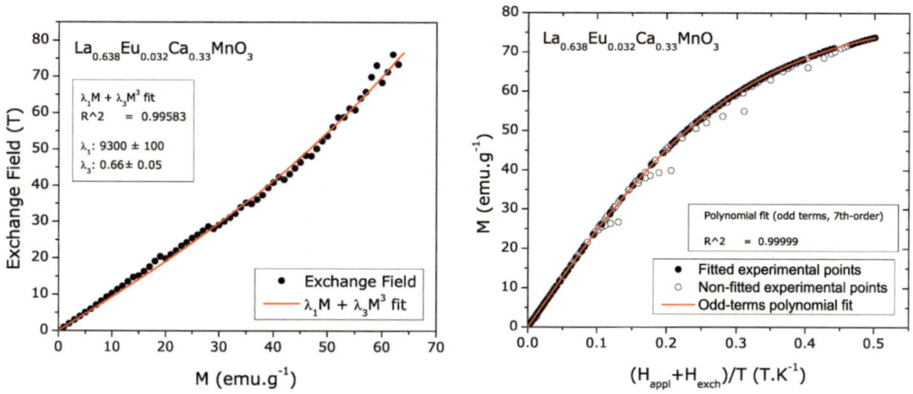

(a) (b)

Fig. 13. Interpolating a) experimental $M(H,T)$ data and b) magnetic entropy change results by mean-field simulations for the second-order phase transition manganite $La_{0.665}Er_{0.035}Sr_{0.30}MnO_3$.

Like the previous example of the second-order manganite, the mean field state function f is fitted to a polynomial function, for calculation purposes. With the λ_1 and λ_3 exchange parameters and the f function described, $M(H,T)$ simulations can be performed, and compared to the experimental values. Also, magnetic entropy change can be estimated from the mean-field relation of Eq. 29 and compared to results form the use of the Maxwell relation. Results are shown in Fig. 14.

The results of this mean-field scaling method are also very promising for this first-order phase transition system. The insight that can be gained from the use of this methodology for a given magnetic system can be of great interest. In a simplistic approach, we can say that if this scaling method does not work, then the system does not follow a molecular mean-field behavior, and other methods need to be pursued in order to interpret the magnetic behavior of the system. It is important to emphasize that this scaling analysis is global, in the sense that it encompasses the consistency of the whole set of magnetization data.

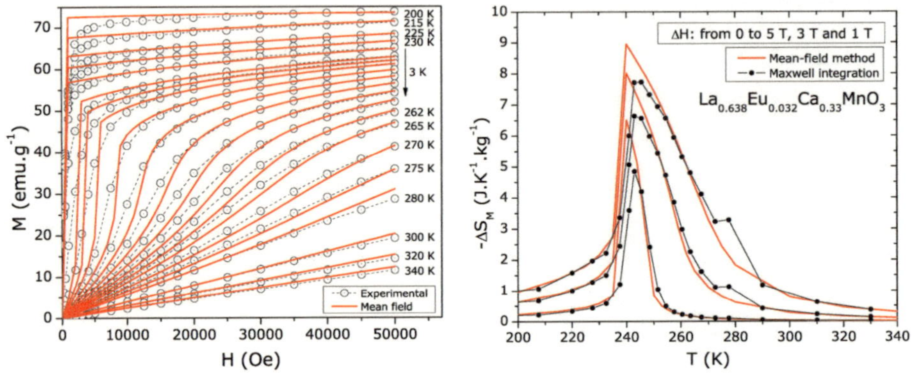

(a) (b)

Fig. 14. Interpolating a) experimental $M(H,T)$ data and b) magnetic entropy change results by mean-field simulations, of the first-order phase transition manganite $La_{0.638}Eu_{0.032}Ca_{0.33}MnO_3$.

3.3 Limitations

Of course, there are limitations to the use of this method, even if one is successful in determining the exchange field parameters and, from what appears to be a good scaling plot, determine the mean-field exchange function. For extensive and smooth $M(H,T)$ data, interpolating isomagnetic data should not pose a real problem, but choosing which points in the H/T versus $1/T$ to fit or to disregard (due to magnetic domains or from the discontinuities of first-order transitions) can remove the confidence on the final scaling plot, and consequently on the mean-field state function.

This simple approach also does not take into account any potencial explicit dependence of the exchange field on temperature. While this dependence is possible, it is generally not considered in the molecular mean-field framework. On the examples we have shown earlier, no such $\lambda(T)$ dependence was considered.

Nevertheless, the best way to evaluate if the mean-field model and obtained parameters are able to describe experimental data is to compare simulations to experiment.

4. The magnetocaloric effect in first-order magnetic phase transitions

4.1 Estimating magnetic entropy change from magnetization measurements

The most common way to estimate the magnetic entropy change of a given magnetic material is from isothermal bulk magnetization measurements. To this effect, one has to simply integrate the Maxwell relation. However, the validity of this approach has been questioned for the case of a first-order magnetic phase transition. The first argument comes from purely numeric considerations, since the discontinuities of the thermodynamic parameters, common to first-order transitions, will make the usual numerical approximations less rigorous in their vicinity. Since the first reports of materials presenting the giant MCE, anomalous 'spikes' in the $\Delta S_M(T)$ plots are commonly seen in literature, for first-order systems. This so-called magnetocaloric peak effect, is present in results form magnetization measurements, but does not appear in calculations using specific heat data.

Indeed, the most immediate culprit for these peaks to occur would be the numerical approximations, which become less rigorous near the transition (Wada & Tanabe, 2001). The peak effect has also been discussed in other perspectives, most notably by Pecharsky and Gschneidner (Pecharsky & Gschneidner, 1999) and Giguère *et al.* (Giguère et al., 1999), for the case of $Gd_5Si_2Ge_2$. Their arguments behind the presence of the entropy peak (and its absence in results for calorimetric measurements) differed considerably. Pecharsky and Gschneidner argue that (Pecharsky & Gschneidner, 1999):

> "The obvious sharp peak observed in $\Delta S_M(T)$ calculated from magnetization data significantly exceeds that calculated from heat capacity and is most likely associated with the fact that the magnetic transition occurs simultaneously with the crystal structure change in this alloy system. The magnetization data reflect only the changes in the magnetic entropy and are insensitive to the overlapping changes in the lattice entropy, while heat capacity data reflect the change in the combined entropies (lattice, electronic and magnetic), thus providing more reliable magnetocaloric effect values."

Giguère *et al.* in contrast argue that (Giguère et al., 1999):

> "The sudden, discontinuous entropy change is related to the phase transition itself, and is approximately independent of the applied field. The field shifts the transition only to higher temperatures. This entropy change cannot be calculated from the Maxwell relations, for two reasons: (i) It is not a magnetic entropy change, and (ii) M(T) or M(H) is not a continuous, derivable function. For first order transitions, the Clausius-Clapeyron (CC) equation offers a way to calculate the entropy change."

The arguments are in almost total disagreement with each other. The only point in common is that the direct use of the Maxwell relation on magnetization data would only report on the change of magnetic entropy, and not the change in 'non-magnetic' (lattice/electronic) entropy. Our analysis of this subject starts exactly in this point. What is 'magnetic entropy' and 'non-magnetic' entropy change, and why would non-magnetic entropy changes be invisible in magnetization measurements.

4.1.1 Thermodynamics

Let us go back to some basics. To estimate entropy change from specific heat measurements, one needs to measure C_p in both zero and non-zero applied field. The difference between curves, integrated in T, will then correspond to the entropy difference between the $H = 0$ and $H \neq 0$ conditions. This is rigorously the MCE, as seen by isothermal entropy change. On the other hand, magnetic measurements give us the bulk magnetization value, which is then used as a thermodynamic variable. If the dependence of M on H and T is known, then the magnetic entropy change is easily calculated. The only possible entropy change that is invisible in magnetization measurements would then be the non-magnetically coupled entropy change in lattice/electronic degrees of freedom. Now, magnetization is defined thermodynamically as

$$M(H, T, p) = \left(\frac{\partial G}{\partial H}\right)_{T,p}, \tag{32}$$

so any change of the Gibbs free energy of system due to a change in applied field will then result on a change of the M value. In other words, any change within the thermodynamic system that occurs due to a change in applied field has repercussions in M. And so the full magnetic entropy change is obtainable from $M(H, T)$. This point of view will become clearer upon investigating a magnetovolume coupling induced first-order phase transition. As we will show, the use of the CC relation does not allow us to calculate 'non-magnetic entropy' variations, as the entropy change due to the lattice volume change is directly calculated from the use of the Maxwell relation. It is helpful to have a visual sense of the application of the Maxwell relation on magnetization data to obtain entropy change, as we will discuss in the following section. A summarized version of the following section is given in (Amaral & Amaral, 2010).

4.1.2 Visual representation

Let us consider a second-order phase transition system. M is a valid thermodynamic parameter, i.e., the system is in thermodynamic equilibrium and is homogeneous. Numerically integrating the Maxwell relation corresponds to integrating the magnetic isotherms in field, and dividing by the temperature difference:

$$\Delta S_M = \sum_0^{H'} \left(\frac{M_{i+1} - M_i}{T_{i+1} - T_i} \right) \Delta H_i = \frac{\int_0^{H'} [M(T_{i+1}, H) - M(T_i, H)]\, dH}{T_{i+1} - T_i} \tag{33}$$

which has a direct visual interpretation, as seen in Fig. 15(a).
If the transition is first-order, there is an 'ideal' discontinuity in the M vs. H plot. Still, apart from expected numerical difficulties, the area between isotherms can be estimated, (Fig. 15(b)).

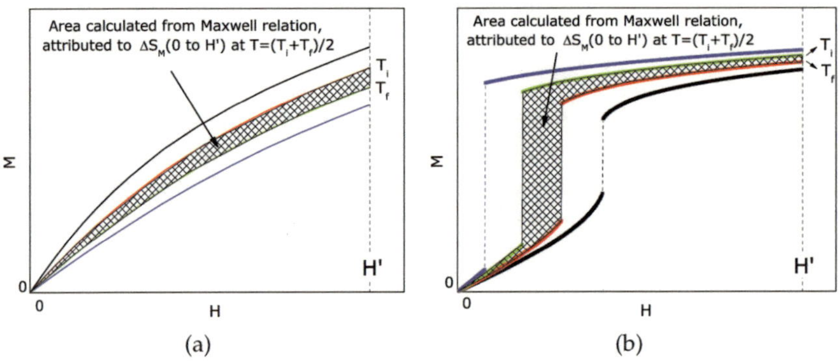

Fig. 15. Schematic diagrams of a a) second-order and b) first-order M vs. H plots, showing the area between magnetic isotherms. From Eq. 33 these areas directly relate to the entropy change.

The CC relation is presented in Eq. 34

$$\left| \frac{\Delta T}{\Delta H_C} \right| = \left| \frac{\Delta M}{\Delta S} \right|, \tag{34}$$

where ΔM is the difference between magnetization values before and after the discontinuity for a given T, ΔH_C is the shift of critical field from ΔT and ΔS is the difference between the entropies of the two phases.
The use of the CC relation to estimate the entropy change due to the first-order nature of the transition also has a very direct visual interpretation (Fig. 16(a)):
From comparing Figs. 15(b) and 16(a), we can see how all the magnetic entropy variation that can be accounted for with magnetization as the order parameter is included in calculations using the Maxwell relation (Fig. 16(b)).

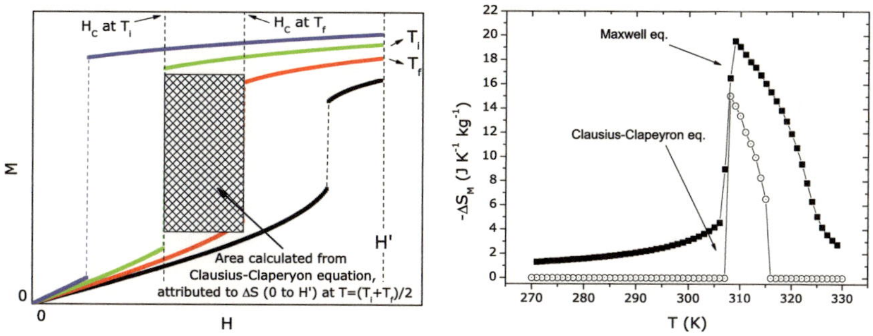

(a) (b)

Fig. 16. a) schematic diagram of the area for entropy change estimation from the Clausius-Clapeyron equation, from a M vs. H plot of a magnetic first-order phase transition system, and b) magnetic entropy change versus temperature, estimated from the Maxwell relation (full symbols) and corresponding entropy change estimated from the Clausius-Clapeyron relation (open symbols).

All the magnetic entropy change is accounted for in calculations using the Maxwell relation. So there is no real gain nor deeper understanding of the systems to be had from the use of the CC relation to estimate magnetic entropy change. The 'non-magnetic entropy' is indeed accounted for by the Maxwell relation. The argument that the entropy peak exists, but specific heat measurements measure the lattice and electronic entropy in a way that conveniently smooths out this peak, is in contrast with the previously shown results. The entropy peak effect does not appear in calculations on purely simulated magnetovolume first-order transition systems, which seems to conflict with the arguments from Pecharsky and Gschneidner.

Of course, all of this reasoning and arguments have a common presumption: M is a valid thermodynamic parameter. In truth, for a first-order transition, the system can present metastable states, and so the measured value of M may not be a good thermodynamic parameter, and also the Maxwell relation is not valid. In the following section, the consequences of using non-equilibrium magnetization data on estimating the MCE is discussed.

4.2 Irreversibility effects

We consider simulated mean field data of a first-order phase transition system, with the same initial parameters as used for the $M(H,T)$ data shown in Fig. 7(a), now considering the metastable and stable solutions of the transcendental equation. Results are shown in Fig. 17(a).

To assess the effects of considering the non-equilibrium solutions of $M(H,T)$ as thermodynamic variables in estimating the magnetic entropy change via the Maxwell relation, we use the three sets of $M(H,T)$ data. The result is presented in Fig. 17(b).

The use of the Maxwell relation on these non-equilibrium data produces visible deviations, and in the case of metastable solution (2), the obtained peak shape is quite similar to that reported by Pecharsky and Gschneidner for $Gd_5Si_2Ge_2$ (Pecharsky & Gschneidner, 1999). In this case $\Delta S_M(T)$ values from caloric measurements follow the half-bell shape of the equilibrium solution, but from magnetization measurements, an obvious sharp peak in $\Delta S_M(T)$ appears. Similar deviations have been interpreted as a result of numerical artifacts

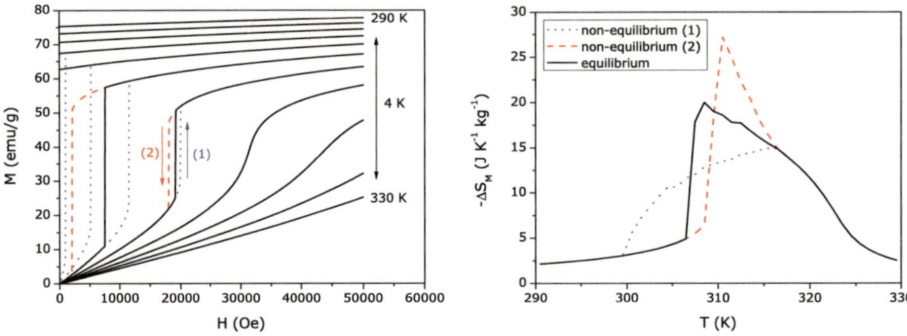

(a) (b)

Fig. 17. a) M versus H isotherms from Landau theory, for a first-order transition, with equilibrium (solid lines) and non-equilibrium (dashed and dotted lines), and b) estimated ΔS_M versus T for equilibrium and non-equilibrium solutions, from the use of the Maxwell relation.

(Wada & Tanabe, 2001), but are not present in a first-order system with no visible hysteresis (Hu et al., 2001).
For the considered model parameters, the overestimation of ΔS_M from using the Maxwell relation in nonequilibrium can be as high as 1/3 of the value obtained under equilibrium, for an applied field change of 5 T.
For large values of H, where M is near saturation in the paramagnetic region, the upper limit to magnetic entropy change, $\Delta S_M(\text{max}) = Nk_B \ln(2J+1)$, is reached, which for the chosen model parameters is ~ 60 J.K^{-1}.kg^{-1}. However, this is exceeded by around 10% by the use of the Maxwell relation to non-equilibrium values. If a stronger magneto-volume coupling is considered (λ_3 = 8 Oe (emu/g)$^{-3}$), the limit can be exceeded by $\sim$ 30 J.K^{-1}kg^{-1}, clearly breaking the thermodynamic limit of the model, falsely producing a colossal MCE (Fig. 18).

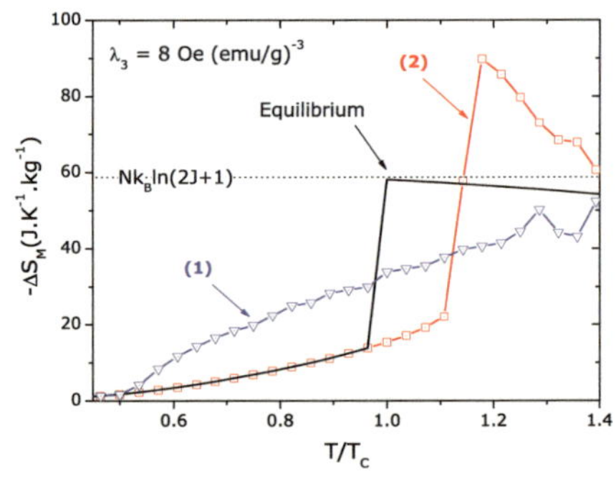

Fig. 18. $-\Delta S_M(T)$, obtained from the use of the Maxwell relation on equilibrium (black line) and metastable (colored lines) magnetization data from the Bean-Rodbell model with a magnetic field change of 1000 T.

The mean-field model also allows the study of mixed-state transitions, by considering a proportion of phases (high and low magnetization) within the metastability region. Magnetization curves are shown in the inset of Fig. 19, for λ_3 = 2 Oe (emu/g)$^{-3}$, corresponding to a critical field $\sim$ 10T. The mixed-phase temperature region is from 328 to

329 K, where the proportion of FM phase is set to 25% at 329 K, 50% at 328.5 K and 75% at 328 K.

The deviation resulting from using the mixed-state M vs. H curves and the Maxwell relation to estimate ΔS_M is now larger compared to the previous results (Fig. 19), since now the system is also inhomogeneous, further invalidating the use of the Maxwell relation. The thermodynamic limit to entropy change is again falsely broken. Note how the temperatures that exceed the limit of entropy change are the ones that include mixed-phase data to estimate ΔS_M.

This result shows how the estimated value of ΔS_M can be greatly increased solely as a consequence of using the Maxwell relation on magnetization data from a mixed-state transition, which is the case of materials that show a colossal MCE (Liu et al., 2007). It is worth noting that, at this time, there are no calorimetric measurements that confirm the existence of the colossal MCE, and its report came from magnetization data and the use of the Maxwell relation.

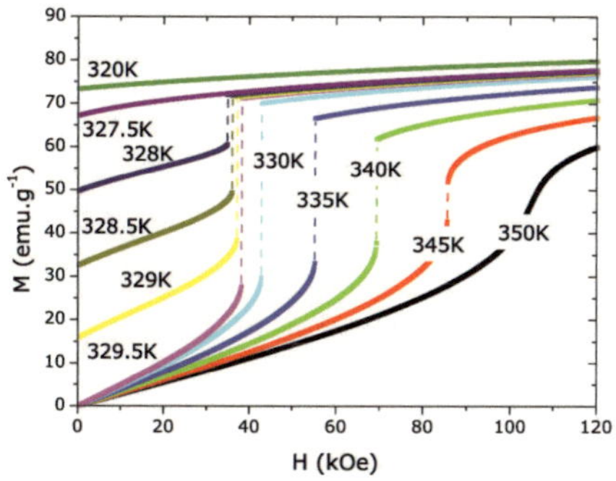

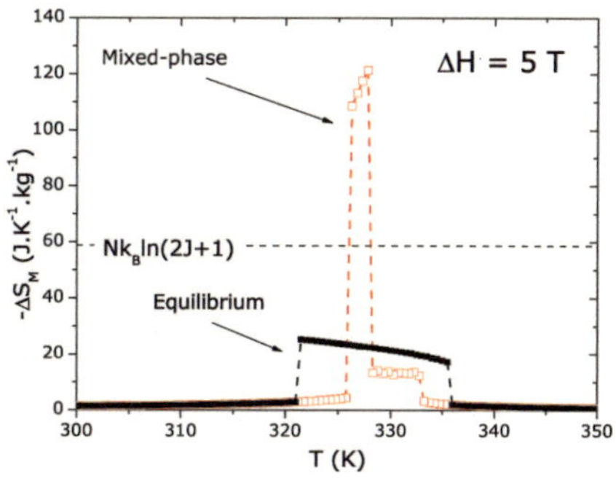

Fig. 19. a) M vs. H isotherms of a mixed-phase system from the mean-field model and b) corresponding $\Delta S_M(T)$ for ΔH=5T from Maxwell relation (open symbols), and of the equilibrium solution (solid symbols).

In the next section, an approach to make a realistic MCE estimation from mixed-phase magnetization data is presented.

4.3 Estimating the magnetocaloric effect from mixed-phase data

It is possible to describe a mixed phase system, by defining a percentage of phases x, where one phase has an $M_1(H,T)$ magnetization value and the other will have an $M_2(H,T)$ magnetization value. In a coupled magnetostructural transition, one of the phases will be in the ferromagnetic state (M_1) and the other (M_2) will be paramagnetic. By changing the temperature, the phase mixture will change from being in a high magnetization state (ferromagnetic) to a low magnetization state (paramagnetic), and so the fraction of phases (x) will depend on temperature. Explicitly, this corresponds to considering the total magnetization of the system as $M_{\text{total}} = x(T)M_1 + (1 - x(T))M_2$, for $H < H_c(T)$ and $M = M_1$ for $H > H_c(T)$, where x is the ferromagnetic fraction in the system (taken as a function of temperature only), M_1 and M_2 are the magnetization of ferromagnetic and paramagnetic phases, respectively and H_c is the critical field at which the phase transition completes.

So if we substitute the above formulation in the integration of the Maxwell relation, used to estimate magnetic entropy change, we can establish entropy change up to a field H as

$$\Delta S_{\text{cal}} = \frac{d}{dT}\int_0^H \left[xM_1 + (1-x)M_2\right] dH' = \frac{\partial x}{\partial T}\int_0^H (M_1 - M_2)dH' + \Delta S_{\text{avg}} \qquad (35)$$

for $H < H_c$, where

$$\Delta S_{\text{avg}} = x\int_0^H \frac{\partial M_1}{\partial T}dH' + (1-x)\int_0^H \frac{\partial M_2}{\partial T}dH'. \qquad (36)$$

Out of these terms, ΔS_{avg} is due to the weighted contribution of the ferro- and paramagnetic phase in the system while the first term results from the phase transformation that occurred in the system during temperature and field variation. In order to obtain the entropy change up to a field above the critical magnetic field H_c, its temperature dependence plays an important role (latent heat contribution) and total entropy change can be formulated as

$$\begin{aligned}\Delta S_{\text{cal}} &= \frac{\partial}{\partial T}\int_0^{H_c(T)} \left[xM_1 + (1-x)M_2\right] dH' + \frac{\partial}{\partial T}\int_{H_c(T)}^{H} M_1 dH' \\ &= \frac{\partial x}{\partial T}\int_0^{H_c(T)} (M_1 - M_2)dH' + (1-x)\frac{\partial}{\partial T}H_c\left[M_1 - M_2\right]_{CT} + \Delta S_{\text{avg}} \\ &+ \int_{H_c(T)}^{H} \frac{\partial M_1}{\partial T}dH'. \end{aligned} \qquad (37)$$

The first term in the previous expression represents the contribution of phase transformation, while the second term represents the fraction (1-x) of the latent heat contribution which is measured in the calorimetric experiment in the region of mixed state (since part of the sample is already in the ferromagnetic state, at zero field) and the last two terms are solely from the magnetic contribution.

For both $H < H_c$ and $H > H_c$ cases, the contribution from the temperature dependence of mixed phase fraction ($\partial x/\partial T$) represents the main effect from non-equilibrium in the thermodynamics of the system and therefore creates major source of error in the entropy calculation.

So, by estimating magnetic entropy change using the Maxwell relation and data from a mixed-phase magnetic system adds a non-physical term, which, as we will see later, can be estimated from analyzing the magnetization curves and the $x(T)$ distribution. Let us use mean-field generated data and a smooth sigmoidal $x(T)$ distribution (Fig. 20).

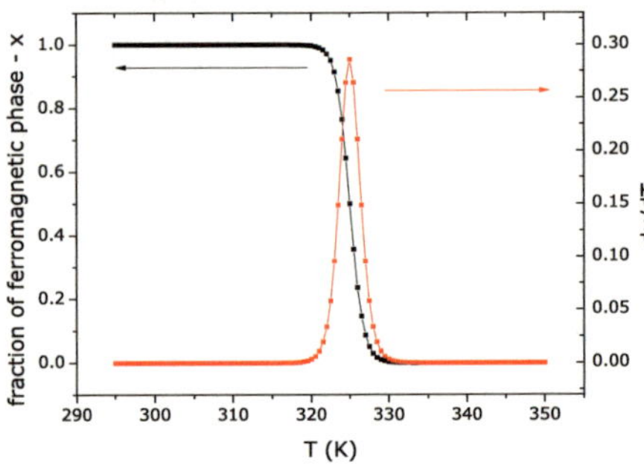

Fig. 20. Distribution of ferromagnetic phase of system, and its temperature derivative.

Such a wide distribution will then produce M versus H plots that strongly show the mixed-phase characteristics of the system, since the step-like behavior is well present (Fig.

21(a)). Using the Maxwell relation to estimate magnetic entropy change, we obtain the peak effect, exceeding the magnetic entropy change limit (Fig. 21(b)).

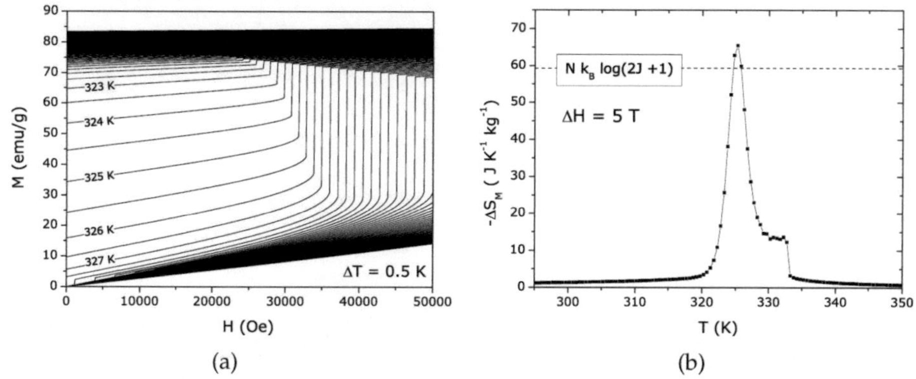

Fig. 21. a) Isothermal M versus H plots of a simulated mixed-phase system, from 295 to 350 K (0.5 K step) and b) magnetic entropy change values resulting from the direct use of the Maxwell relation.

As the entropy plot shows us, the shape of the entropy curve and the $\partial x/\partial T$ function (Fig. 20) share a similar shape. This points us to Eqs. 35 or 37. It seems that the left side of the entropy plot may just be the result of the presence of the mixed-phase states, while for the right side of the entropy plot, there is some 'true' entropy change hidden along with the $\partial x/\partial T$ contribution. By using Eqs. 35 or 37, we present a way to separate the two contributions, and so estimate more trustworthy entropy change values. We plot the entropy change values obtained directly from the Maxwell relation, as a function of $\partial x/\partial T$. This is shown in Fig. 22(a), for the data shown in Figs. 21(a) and 20.

Plotting entropy change as a function of the temperature derivative of the phase distribution gives us a tool to remove the false $\partial x/\partial T$ contribution to the entropy change. As we can see in Fig. 22(a), there is a smooth dependence of entropy in $\partial x/\partial T$, which allows us to extrapolate the entropy results to a null $\partial x/\partial T$ value, following the approximately linear slope near the plot origin (dashed lines of Fig. 22(a)). This slope is constant as long and the magnetization difference between phases ($M_1 - M_2$) is approximately constant, which is observed in strongly first-order materials. The results of eliminating the $\partial x/\partial T$ contribution to the Maxwell relation result are presented in Fig. 22(b).

By eliminating the contribution of the temperature derivative of the mixed-phase fraction, the entropy 'peak' effect is eliminated, in a justified way. The resulting entropy curve resembles the results obtained from specific heat measurements when compared to results from magnetic measurements, as seen in Refs. (Liu et al., 2007) and (Tocado et al., 2009), among others.

However, this corrected entropy is always less than the value in equilibrium condition. This is because we deal with a fraction (1-x) of the phase M_2 remaining to transform which will give a fraction of latent heat entropy (Eq. 37) since part (x) of phase is already transformed at zero field. This average entropy change weighted by the fraction of each phase present, can be measured in calorimetric experiments. We regard $x(T)$ and $\partial x/\partial T$ as parameters that can be externally manipulated by changing the measurement condition/sample history and should therefore be carefully handled to obtain the true entropy calculation.

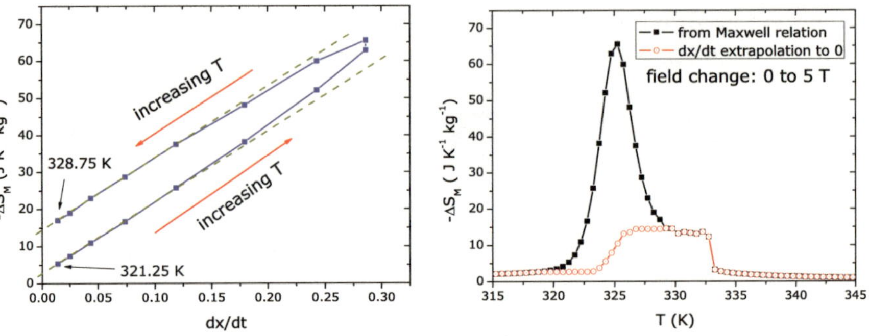

(a) (b)

Fig. 22. a) Entropy change, as obtained from the use of the Maxwell relation of mixed-phase magnetization data, versus a) $\partial x/\partial T$ and b) versus T, with values extrapolated to $\partial x/\partial T \rightarrow 0$.

We can conclude that, for a first-order magnetic phase transition system, estimating magnetic entropy change from the Maxwell relation can give us misleading results. If the system presents a mixed-phase state, the entropy 'peak' effect can be even more pronounced, clearly exceeding the theoretical limit of magnetic entropy change.

5. Acknowledgements

We acknowledge the financial support from FEDER-COMPETE and FCT through Projects PTDC/CTM-NAN/115125/2009, PTDC/FIS/105416/2008, CERN/FP/116320/2010, grants SFRH/BPD/39262/2007 (S. Das) and SFRH/BPD/63942/2009 (J. S. Amaral).

6. References

Aharoni, A. (2000). *Introduction to the Theory of Ferromagnetism*, Oxford Science Publications.

Amaral, J. S. & Amaral, V. S. (2009). The effect of magnetic irreversibility on estimating the magnetocaloric effect from magnetization measurements, *Appl. Phys. Lett.* 94: 042506.

Amaral, J. S. & Amaral, V. S. (2010). On estimating the magnetocaloric effect from magnetization measurements, *J. Magn. Magn. Mater.* 322: 1552.

Amaral, J. S., Reis, M. S., Amaral, V. S., Mendonça, T. M., Araújo, J. P., Sá, M. A., Tavares, P. B. & Vieira, J. M. (2005). Magnetocaloric effect in Er- and Eu-substituted ferromagnetic La-Sr manganites, *J. Magn. Magn. Mater.* 290: 686.

Amaral, J. S., Silva, N. J. O. & Amaral, V. S. (2007). A mean-field scaling method for first- and second-order phase transition ferromagnets and its application in magnetocaloric studies, *Appl. Phys. Lett.* 91(17): 172503.

Amaral, J. S., Tavares, P. B., Reis, M. S., Araújo, J. P., Mendonça, T. M., Amaral, V. S. & Vieira, J. M. (2008). The effect of chemical distribution on the magnetocaloric effect: A case study in second-order phase transition manganites, *J. Non-Cryst. Solids* 354: 5301.

Bean, C. P. & Rodbell, D. S. (1962). Magnetic disorder as a first-order phase transformation, *Phys. Rev.* 126(1): 104.

Brück, E. (2005). Developments in magnetocaloric refrigeration, *J. Phys D: Appl. Phys.* 38(23): R381.

Callen, H. B. (1985). *Thermodynamics and an introduction to thermostatistics*, 2nd edn, John Wiley and Sons, New York, USA.

Coey, J. (2009). *Magnetism and Magnetic Materials*, Cambridge University Press, Cambridge.

Das, S., Amaral, J. S. & Amaral, V. S. (2010a). Handling mixed-state magnetization data for magnetocaloric studies – a solution to achieve realistic entropy behaviour, *J. Phys D: Appl. Phys.* 43(15): 152002.

Das, S., Amaral, J. S. & Amaral, V. S. (2010b). Prediction of realistic entropy behavior from mixed state magnetization data for first order phase transition materials, *J. Appl. Phys.* 107(9): 09A912.

de Campos, A., Rocco, D. L., Carvalho, A. M. G., Caron, L., Coelho, A. A., Gama, S., da Silva, L. M., Gandra, F. C. G., dos Santos, A. O., Cardoso, L. P., Von Ranke, P. J. & de Oliveira, N. A. (2006). Ambient pressure colossal magnetocaloric effect tuned by composition in $Mn_{1-x}Fe_xAs$, *Nature Materials* 5(10): 802.

de Oliveira, N. A. & von Ranke, P. J. (2010). Theoretical aspects of the magnetocaloric effect, *Physics Reports-Review Section of Physics Letters* 489(4-5): 89.

Forsythe, G. E., Malcolm, M. A. & Moler, C. B. (1976). *Computer Methods for Mathematical Computations*, Prentice-Hall.

Gama, S., Coelho, A. A., de Campos, A., Carvalho, A. M. G., Gandra, F. C. G., von Ranke, P. J. & de Oliveira, N. A. (2004). Pressure-induced colossal magnetocaloric effect in MnAs, *Phys. Rev. Lett.* 93(23): 237202.

Giguère, A., Foldeaki, M., Gopal, B. R., Chahine, R., Bose, T. K., Frydman, A. & Barclay, J. A. (1999). Direct measurement of the "giant" adiabatic temperature change in $Gd_5Si_2Ge_2$, *Phys. Rev. Lett.* 83(11): 2262.

Gonzalo, J. A. (2006). *Effective Field Approach to Phase Transitions and Some Applications to Ferroelectrics*, World Scientific, Singapore.

Gschneidner Jr., K. A. & Pecharsky, V. K. (2008). Thirty years of near room temperature magnetic cooling: Where we are today and future prospects, *International Journal of Refrigeration* 31(6): 945.

Gschneidner Jr., K. A., Pecharsky, V. K. & Tsokol, A. O. (2005). Recent developments in magnetocaloric materials, *Reports on Progress in Physics* 68(6): 1479.

Gschneidner, K. A. & Pecharsky, V. K. (2000). Magnetocaloric materials, *Annual Review of Materials Science* 30: 387.

Hu, F. X., Shen, B. G., Sun, J. R., Cheng, Z. H., Rao, G. H. & Zhang, X. X. (2001). Influence of negative lattice expansion and metamagnetic transition on magnetic entropy change in the compound $LaFe_{11.4}Si_{1.6}$, *Appl. Phys. Lett.* 78(23): 3675.

Kittel, C. (1996). *Introduction to Solid State Physics*, 7th edn, John Wiley and Sons, New York.

Liu, G. J., Sun, J. R., Shen, J., Gao, B., Zhang, H. W., Hu, F. X. & Shen, B. G. (2007). Determination of the entropy changes in the compounds with a first-order magnetic transition, *Appl. Phys. Lett.* 90(3): 032507.

Pecharsky, V. K. & Gschneidner, K. A. (1997). Giant magnetocaloric effect in $Gd_5Si_2Ge_2$, *Phys. Rev. Lett.* 78(23): 4494.

Pecharsky, V. K. & Gschneidner, K. A. (1999). Heat capacity near first order phase transitions and the magnetocaloric effect: An analysis of the errors, and a case study of $Gd_5(Si_2Ge_2)$ and Dy, *J. Appl. Phys.* 86(11): 6315.

Phan, M. H. & Yu, S. C. (2007). Review of the magnetocaloric effect in manganite materials, *J. Magn. Magn. Mater.* 308(2): 325.

Rocco, D. L., de Campos, A., Carvalho, A. M. G., Caron, L., Coelho, A. A., Gama, S., Gandra, F. C. G., dos Santos, A. O., Cardoso, L. P., von Ranke, P. J. & de Oliveira, N. A. (2007). Ambient pressure colossal magnetocaloric effect in $Mn_{1-x}Cu_xAs$ compounds, *Appl. Phys. Lett.* 90(24): 242507.

Szewczyk, A., Szymczak, H., Wisniewski, A., Piotrowski, K., Kartaszynski, R., Dabrowski, B., Kolesnik, S. & Bukowski, Z. (2000). Magnetocaloric effect in $La_{1-x}Sr_xMnO_3$ for x = 0.13 and 0.16, *Appl. Phys. Lett.* 77(7): 1026.

Tishin, A. M. & Spichin, Y. I. (2003). *The Magnetocaloric Effect and its Applications*, IOP Publishing, London.

Tocado, L., Palacios, E. & Burriel, R. (2009). Entropy determinations and magnetocaloric parameters in systems with first-order transitions: Study of MnAs, *J. Appl. Phys.* 105: 093918.

Wada, H. & Tanabe, Y. (2001). Giant magnetocaloric effect of $MnAs_{1-x}Sb_x$, *Appl. Phys. Lett.* 79(20): 3302.

9

Entropy Generation in Viscoelastic Fluid Over a Stretching Surface

Saouli Salah and Aïboud Soraya
University Kasdi Merbah, Ouargla, Algeria

1. Introduction

Due to the increasing importance in processing industries and elsewhere when materials whose flow behavior cannot be characterized by Newtonian relationships, a new stage in the evolution of fluid dynamics theory is in progress. An intensive effort, both theoretical and experimental, has been devoted to problems of non-Newtonian fluids. The study of MHD flow of viscoelastic fluids over a continuously moving surface has wide range of applications in technological and manufacturing processes in industries. This concerns the production of synthetic sheets, aerodynamic extrusion of plastic sheets, cooling of metallic plates, etc.

(Crane, 1970) considered the laminar boundary layer flow of a Newtonian fluid caused by a flat elastic sheet whose velocity varies linearly with the distance from the fixed point of the sheet. (Chang, 1989; Rajagopal et al., 1984) presented an analysis on flow of viscoelastic fluid over stretching sheet. Heat transfer cases of these studies have been considered by (Dandapat & Gupta, 1989, Vajravelu & Rollins, 1991), while flow of viscoelastic fluid over a stretching surface under the influence of uniform magnetic field has been investigated by (Andersson, 1992).

Thereafter, a series of studies on heat transfer effects on viscoelastic fluid have been made by many authors under different physical situations including (Abel et al., 2002, Bhattacharya et al., 1998, Datti et al., 2004, Idrees & Abel, 1996, Lawrence & Rao, 1992, Prasad et al., 2000, 2002). (Khan & Sanjayanand, 2005) have derived similarity solution of viscoelastic boundary layer flow and heat transfer over an exponential stretching surface.

(Cortell, 2006) have studied flow and heat transfer of a viscoelastic fluid over stretching surface considering both constant sheet temperature and prescribed sheet temperature. (Abel et al., 2007) carried out a study of viscoelastic boundary layer flow and heat transfer over a stretching surface in the presence of non-uniform heat source and viscous dissipation considering prescribed surface temperature and prescribed surface heat flux.

(Khan, 2006) studied the case of the boundary layer problem on heat transfer in a viscoelastic boundary layer fluid flow over a non-isothermal porous sheet, taking into account the effect a continuous suction/blowing of the fluid, through the porous boundary. The effects of a transverse magnetic field and electric field on momentum and heat transfer characteristics in viscoelastic fluid over a stretching sheet taking into account viscous dissipation and ohmic dissipation is presented by (Abel et al., 2008). (Hsiao, 2007) studied the conjugate heat transfer of mixed convection in the presence of radiative and viscous dissipation in viscoelastic fluid past a stretching sheet. The case of unsteady magnetohydrodynamic was carried out by (Abbas et al., 2008). Using Kummer's funcions, (Singh, 2008) carried out the study of heat source and radiation effects on magnetohydrodynamics flow of a viscoelastic fluid past a stretching sheet with prescribed power law surface heat flux. The effects of non-uniform heat source, viscous dissipation and thermal radiation on the flow and heat transfer in a viscoelastic fluid over a stretching surface was considered in (Prasad et al., 2010). The case of the heat transfer in magnetohydrodynamics flow of viscoelastic fluids over stretching sheet in the case of variable thermal conductivity and in the presence of non-uniform heat source and radiation is reported in (Abel & Mahesha, 2008). Using the homotopy analysis, (Hayat et al., 2008) looked at the hydrodynamic of three dimensional flow of viscoelastic fluid over a stretching surface. The investigation of biomagnetic flow of a non-Newtonian viscoelastic fluid over a stretching sheet under the influence of an applied magnetic field is done by (Misra & Shit, 2009). (Subhas et al., 2009) analysed the momentum and heat transfer characteristics in a hydromagnetic flow of viscoelastic liquid over a stretching sheet with non-uniform heat source. (Nandeppanavar et al., 2010) analysed the flow and heat transfer characteristics in a viscoelastic fluid flow in porous medium over a stretching surface with surface prescribed temperature and surface prescribed heat flux and including the effects of viscous dissipation. (Chen, 2010) studied the magneto-hydrodynamic flow and heat transfer characteristics viscoelastic fluid past a stretching surface, taking into account the effects of Joule and viscous dissipation, internal heat generation/absorption, work done due to deformation and thermal radiation. (Nandeppanavar et al., 2011) considered the heat transfer in viscoelastic boundary layer flow over a stretching sheet with thermal radiation and non-uniform heat source/sink in the presence of a magnetic field

Although the forgoing research works have covered a wide range of problems involving the flow and heat transfer of viscoelastic fluid over stretching surface they have been restricted, from thermodynamic point of view, to only the first law analysis. The contemporary trend in the field of heat transfer and thermal design is the second law of thermodynamics analysis and its related concept of entropy generation minimization.

Entropy generation is closely associated with thermodynamic irreversibility, which is encountered in all heat transfer processes. Different sources are responsible for generation of entropy such as heat transfer and viscous dissipation (Bejan, 1979, 1982). The analysis of entropy generation rate in a circular duct with imposed heat flux at the wall and its extension to determine the optimum Reynolds number as function of the Prandtl number and the duty parameter were presented by (Bejan, 1979, 1996). (Sahin, 1998) introduced the second law analysis to a viscous fluid in circular duct with isothermal boundary conditions. In another paper, (Sahin, 1999) presented the effect of variable viscosity on entropy generation rate for heated circular duct. A comparative study of entropy generation rate inside duct of different shapes and the determination of the optimum duct shape subjected to isothermal boundary condition were done by (Sahin, 1998). (Narusawa, 1998) gave an analytical and numerical analysis of the second law for flow and heat transfer inside a rectangular duct. In a more recent paper, (Mahmud & Fraser, 2002a, 2002b, 2003) applied the second law analysis to fundamental convective heat transfer problems and to non-Newtonian fluid flow through channel made of two parallel plates. The study of entropy generation in a falling liquid film along an inclined heated plate was carried out by (Saouli & Aïboud-Saouli, 2004). As far as the effect of a magnetic field on the entropy generation is concerned, (Mahmud et al., 2003) studied the case of mixed convection in a channel. The effects of magnetic field and viscous dissipation on entropy generation in a falling film and channel were studied by (Aïboud-saouli et al., 2006, 2007). The application of the second law analysis of thermodynamics to viscoelastic magnetohydrodynamic flow over a stretching surface was carried out by (Aïboud & Saouli 2010a, 2010b).

The objective of this paper is to study the entropy generation in viscoelastic fluid over a stretching sheet with prescribed surface temperature in the presence of uniform transverse magnetic field.

2. Formulation of the problem

In two-dimensional Cartesian coordinate system (x, y) we consider magneto-convection, steady, laminar, electrically conduction, boundary layer flow of a viscoelastic fluid caused by a stretching surface in the presence of a uniform transverse magnetic field and a heat source. The x-axis is taken in the direction of the main flow along the plate and the y-axis is normal to the plate with velocity components u, v in these directions.

Under the usual boundary layer approximations, the flow is governed by the following equations:

$$\frac{\partial u}{\partial x}+\frac{\partial v}{\partial y}=0 \tag{1}$$

$$u\frac{\partial u}{\partial x}+v\frac{\partial u}{\partial y}=\nu\frac{\partial^2 u}{\partial y^2}-k_0\left(u\frac{\partial^3 u}{\partial x\partial y^2}+v\frac{\partial^3 u}{\partial y^3}-\frac{\partial u}{\partial y}\frac{\partial^2 u}{\partial x\partial y}+\frac{\partial u}{\partial y}\frac{\partial^2 u}{\partial y^2}\right)-\frac{\sigma B_0^2}{\rho}u \tag{2}$$

The constant $k_0=-\frac{\alpha_1}{\rho}$ is the viscoelastic parameter.

The boundary conditions are given by

$$y=0, u=u_p=\lambda x, v=0 \tag{3a}$$

$$y=\infty, u=0, \frac{\partial u}{\partial y}=0 \tag{3b}$$

The heat transfer governing boundary layer equation with temperature-dependent heat generation (absorption) is

$$\rho C_p\left(u\frac{\partial T}{\partial x}+v\frac{\partial T}{\partial y}\right)=k\frac{\partial^2 T}{\partial y^2}+Q(T-T_\infty) \tag{4}$$

The relevant boundary conditions are

$$y-0, T=T_p=A\left(\frac{x}{l}\right)^2+T_\infty \tag{5a}$$

$$y=\infty, T=T_\infty \tag{5b}$$

3. Analytical solution

The equation of continuity is satisfied if we choose a dimensionless stream function $\Psi(x,y)$ such that

$$u=\frac{\partial\Psi}{\partial y},v=-\frac{\partial\Psi}{\partial x} \tag{6}$$

Introducing the similarity transformations

$$\eta=y\sqrt{\frac{\lambda}{\nu}},\Psi(x,y)=x\sqrt{\nu\lambda}f(\eta) \tag{7}$$

Momentum equation (2) becomes

$$f'^2(\eta)-f(\eta)f''(\eta)=f'''(\eta)-\frac{k_0\lambda}{\nu}\left[2f'(\eta)f'''(\eta)-f(\eta)f^{IV}(\eta)-f''^2(\eta)\right]-Mnf'(\eta) \tag{8}$$

where $Mn=\dfrac{\sigma B_0^2}{\rho\lambda}$

Now let us seek a solution of Eq. (6) in the form

$$f'(\eta)=e^{-\alpha\eta}\,(\alpha\succ 0 \tag{9}$$

which is satisfied by the following boundary conditions:

$$\eta=0,f(0)=0,f'(0)=1 \tag{10a}$$

$$\eta=\infty,f'(\infty)=0,f''(\infty)=0 \tag{10b}$$

On substituting (7) into (6) and using boundary conditions (10a) and (10b) the velocity components take the form

$$u=\lambda xf'(\eta) \tag{11}$$

$$v=-\sqrt{\nu\lambda}f(\eta) \tag{12}$$

Where $k_1=\dfrac{k_0\lambda}{\nu}$ is the viscoelastic parameter, and

$$\alpha=\sqrt{\frac{1+Mn}{1-k_1}} \tag{13}$$

Defining the dimensionless temperature

$$\Theta(\eta)=\frac{T-T_\infty}{T_p-T_\infty} \tag{14}$$

and using (9), (11), (12), Eq. (14) and the boundary conditions (5a) and (5b) can be written as

$$\Theta''(\eta)+\frac{\Pr}{\alpha}\left(1-e^{-\alpha\eta}\right)\Theta'(\eta)-\left(2\Pr e^{-\alpha\eta}-\beta\right)\Theta(\eta)=0 \tag{15}$$

$$\eta=0,\Theta(0)=1 \tag{16a}$$

$$\eta=\infty,\Theta(\infty)=0 \tag{16b}$$

Where $\Pr=\frac{\mu C_P}{k}$ and $\beta=\frac{Qv}{\lambda k}$ are respectively the Prandtl number and the heat/sink parameter.
Introducing the variable

$$\xi=\frac{\Pr}{\alpha^2}e^{-\alpha\eta} \tag{17}$$

And inserting (17) in (15) we obtain

$$\xi\Theta''(\xi)+\frac{\Pr}{\alpha}\left(1-\frac{\Pr}{\alpha^2}+\xi\right)\Theta'(\xi)-\left(2-\frac{\beta}{\alpha^2\xi}\right)\Theta(\xi)=0 \tag{18}$$

And (16a) and (16b) transform to

$$\xi=\frac{\Pr}{\alpha^2},\Theta\left(\frac{\Pr}{\alpha^2}\right)=1 \tag{19a}$$

$$\xi=0,\Theta(0)=0 \tag{19b}$$

The solution of Eq. (18) satisfying (19a) and (19b) is given by

$$\Theta(\xi)=\left(\frac{\alpha^2}{\Pr}\xi\right)^{a+b}\frac{M(a+b-2,2b+1,-\xi)}{M\left(a+b-2,2b+1,-\frac{\Pr}{\alpha^2}\right)} \tag{20}$$

The solution of (20) in terms of η is written as

$$\Theta(\xi)=e^{-\alpha(a+b)\eta}\frac{M\left(a+b-2,2b+1,-\frac{\Pr}{\alpha^2}e^{-\alpha\eta}\right)}{M\left(a+b-2,2b+1,-\frac{\Pr}{\alpha^2}\right)} \tag{21}$$

where $a=\frac{\Pr}{2\alpha^2}$, $b=\frac{\sqrt{\Pr^2-4\alpha^2\beta}}{2\alpha^2}$ and $M\left(a+b-2,2b+1,-\frac{\Pr}{\alpha^2}e^{-\alpha\eta}\right)$ is the Kummer's function.

4. Second law analysis

According to (Woods, 1975), the local volumetric rate of entropy generation in the presence of a magnetic field is given by

$$S_G = \frac{k}{T_\infty^2}\left[\left(\frac{\partial T}{\partial x}\right)^2 + \left(\frac{\partial T}{\partial y}\right)^2\right] + \frac{\mu}{T_\infty}\left(\frac{\partial u}{\partial y}\right)^2 + \frac{\sigma B_0^2}{T_\infty}u^2 \tag{22}$$

Eq. (22) clearly shows contributions of three sources of entropy generation. The first term on the right-hand side of Eq. (22) is the entropy generation due to heat transfer across a finite temperature difference; the second term is the local entropy generation due to viscous dissipation, whereas the third term is the local entropy generation due to the effect of the magnetic field. It is appropriate to define dimensionless number for entropy generation rate N_S. This number is defined by dividing the local volumetric entropy generation rate S_G to a characteristic entropy generation rate S_{G0}. For prescribed boundary condition, the characteristic entropy generation rate is

$$S_{G0} = \frac{k(\Delta T)^2}{l^2 T_\infty^2} \tag{23}$$

therefore, the entropy generation number is

$$N_S = \frac{S_G}{S_{G0}} \tag{24}$$

using Eq. (9), (21) and (22), the entropy generation number is given by

$$N_S = \frac{4}{X^2}\Theta^2(\eta) + \mathrm{Re}_l\,\Theta'^2(\eta) + \mathrm{Re}_l\,\frac{Br}{\Omega} f''^2(\eta) + \frac{BrHa^2}{\Omega} f'^2(\eta) \tag{25}$$

where Re_l and Br are respectively the Reynolds number and the Brinkman number. Ω and Ha, are respectively the dimensionless temperature difference and the Hartman number. These number are given by the following relationships

$$\mathrm{Re}_l = \frac{u_l\, l}{\nu}, \quad Br = \frac{\mu\, u_P^2}{k\Delta T}, \quad \Omega = \frac{\Delta T}{T_\infty}, \quad Ha = B_0\, l\sqrt{\frac{\sigma}{\mu}} \tag{26}$$

5. Results and discussion

The flow and heat transfer in a viscoelastic fluid under the influence of a transverse uniform magnetic field has been solved analytically using Kummer's functions and analytic expressions of the velocity and temperature have been used to compute the entropy generation. Figs. 1 and 2 show the variations of the longitudinal velocity $f'(\eta)$ and the transverse velocity $f(\eta)$ as function of η for several values of magnetic parameter Mn. It can be observed that $f'(\eta)$ decreases with η and $f(\eta)$ increases with η asymptotically for Mn keeping constant. For a fixed position η, both $f'(\eta)$ and $f(\eta)$ decreases with Mn, thus the presence of the magnetic field decreases the momentum boundary layer thickness and increase the power needed to stretch the sheet.

The effects of the viscoelastic parameter k_1 on the longitudinal velocity $f'(\eta)$ and the transverse velocity $f(\eta)$ are illustrated on figs. 3 and 4. As it can be seen, for a fixed value of η, both $f'(\eta)$ and $f(\eta)$ decrease as viscoelastic parameter rises. This can be explained by the fact that, as the viscoelastic parameter increases, the hydrodynamic boundary layer adheres strongly to the surface, which in turn retards the flow in the longitudinal and the transverse directions.

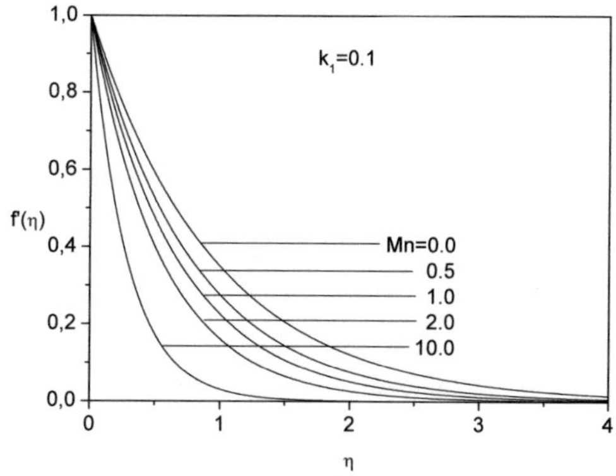

Fig. 1. Effect of the magnetic parameter on the longitudinal velocity.

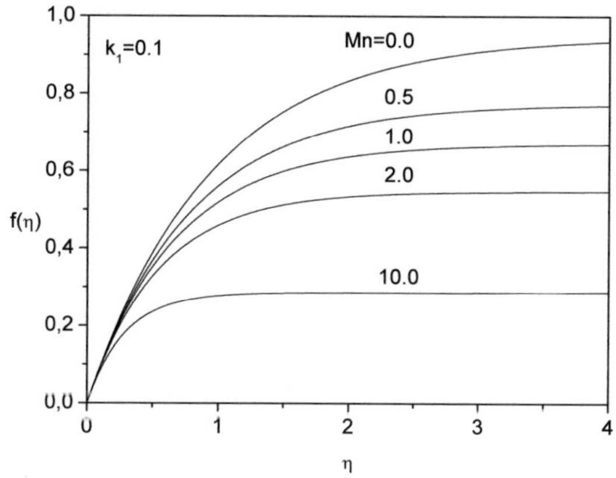

Fig. 2. Effect of the magnetic parameter on the transverse velocity.

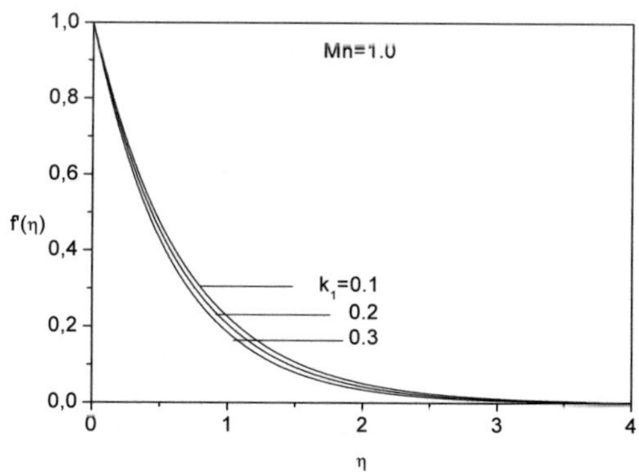

Fig. 3. Effect of the viscoelastic parameter on the longitudinal velocity.

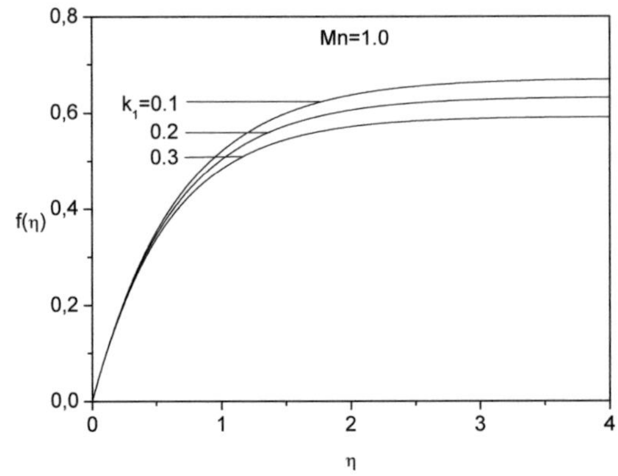

Fig. 4. Effect of the viscoelastic parameter on the transverse velocity.

Fig. 5 depicts the temperature profiles $\Theta(\eta)$ as function of η for different values of the Prandtl number $\Pr$. As it can be noticed, $\Theta(\eta)$ decreases with η whatever is the value of the Prandtl number, for a fixed value of η, the temperature $\Theta(\eta)$ decreases with an increase in Prandtl number which means that the thermal boundary layer is thinner for large Prandtl number.

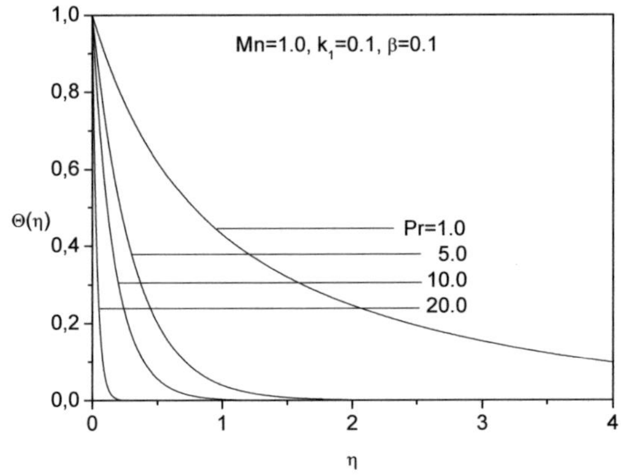

Fig. 5. Effect of the Prandtl number on the temperature.

The temperature profiles $\Theta(\eta)$ as function of η for different values of the magnetic Mn are plotted in fig. 6. An increase in the magnetic parameter Mn results in an increase of the temperature; this is due to the fact that the thermal boundary layer increases with the magnetic parameter. Fig. 7 represents graphs of temperature profiles $\Theta(\eta)$ as function of η for various values of the heat source/sink parameter β. For fixed value of η, the temperature $\Theta(\eta)$ augments with the heat source/ sink parameter β. This is due to the fact that the increase of the heat source/sink parameter means an increase of the heat generated inside the boundary layer leading to higher temperature profile.
The influence of the magnetic parameter Mn on the entropy generation number N_S is shown on fig. 8. The entropy generation number N_S decreases with η for Mn keeping constant. For

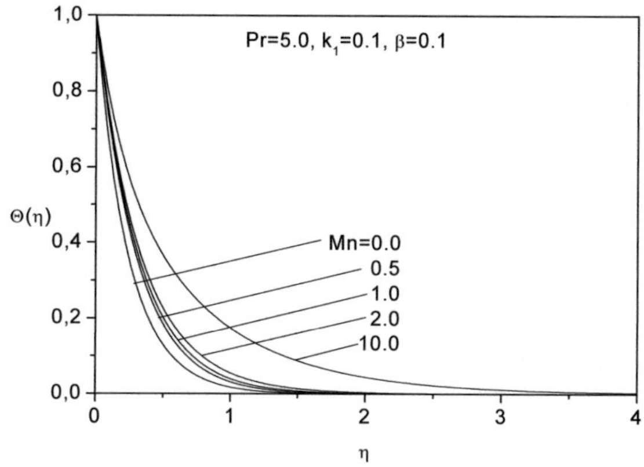

Fig. 6. Effect of the magnetic parameter on the temperature.

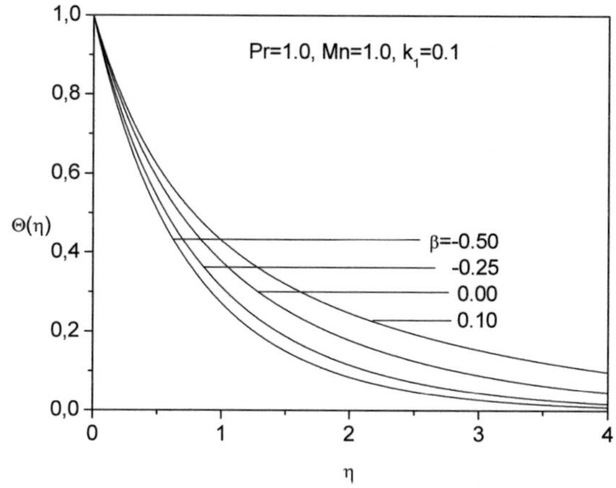

Fig. 7. Effect of the heat source/sink parameter on the temperature.

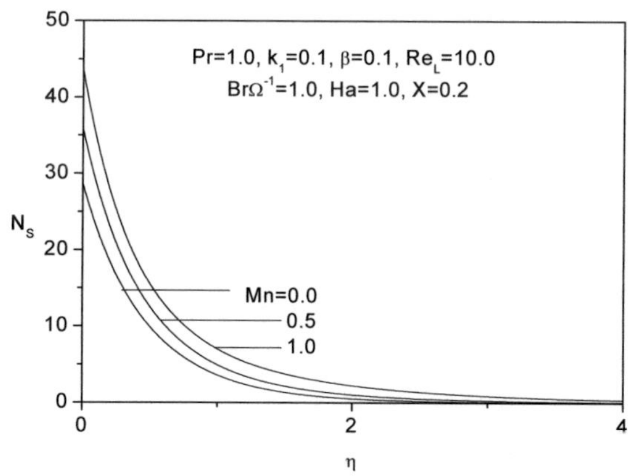

Fig. 8. Effect of the magnetic parameter on the entropy generation number.

fixed value of η , the entropy generation number increases with the magnetic parameter, because the presence of the magnetic field creates more entropy in the fluid. Moreover, the stretching surface acts as a strong source of irreversibility.

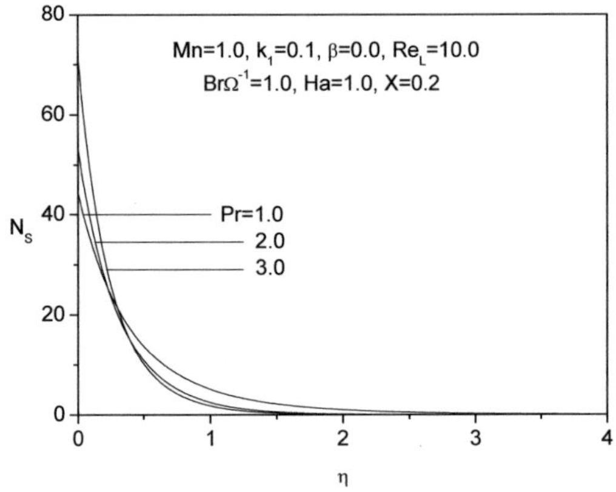

Fig. 9. Effect of the Prandtl number on the entropy generation number.

Fig. 9 illustrates the effect of the Prandtl number Pr on the entropy generation number N_S . The entropy generation number is higher for higher Prandtl number near the surface, but as η increases, the entropy generation number shows different variation. This is due to the fact that according to fig. 6, the temperature profiles decrease sharply with the increase of the Prandlt number.

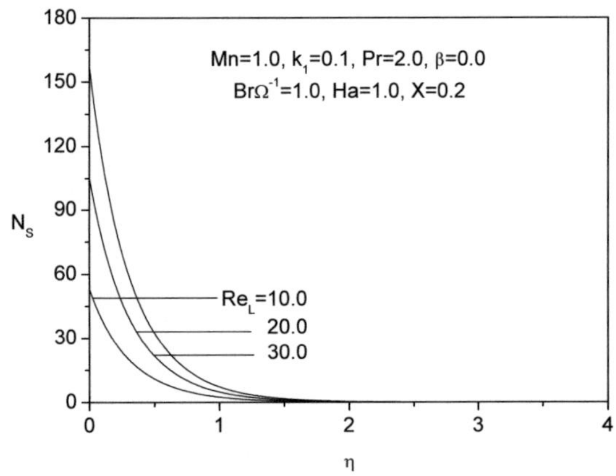

Fig. 10. Effect of the Reynolds number on the entropy generation number.

The influence of the Reynolds number Re_l on the entropy generation number is plotted on fig. 10. For a given value of η, the entropy generation number increases as the Reynolds number increases. The augmentation of the Reynolds number increases the contribution of the entropy generation number due to fluid friction and heat transfer in the boundary layer.

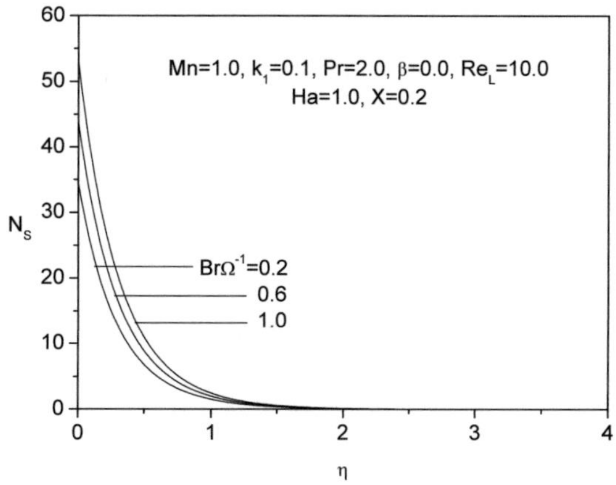

Fig. 11. Effect of the dimensionless group on the entropy generation number.

The effect of the dimensionless group parameter $Br\Omega^{-1}$ on the entropy generation number N_S is depicted in fig. 11. The dimensionless group determines the relative importance of viscous effect. For a given η, the entropy generation number is higher for higher dimensionless group. This is due to the fact that for higher dimensionless group, the entropy generation numbers due to the fluid friction increase.

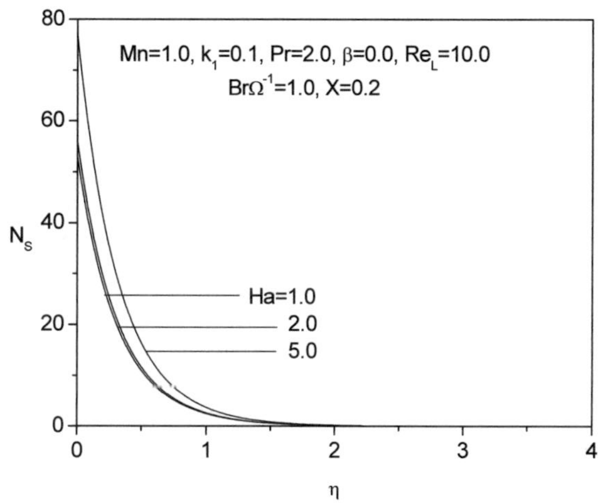

Fig. 12. Effect of the Hartman number on the entropy generation number.

The effect of the Hartman number Ha on the entropy generation number N_S is plotted in fig. 12. For a given η, as the Hartman number increases, the entropy generation number increases. The entropy generation number is proportional to the square of Hartman number which proportional to the magnetic field. The presence of the magnetic field creates additional entropy.

6. Conclusion

The velocity and temperature profiles are obtained analytically and used to compute the entropy generation number in viscoelastic magnetohydrodynamic flow over a stretching surface

The effects of the magnetic parameter and the viscoelastic parameter on the longitudinal and transverse velocities are discussed. The influences of the Prandtl number, the magnetic parameter and the heat source/sink parameter on the temperature profiles are presented. As far as the entropy generation number is concerned, its dependence on the magnetic parameter, the Prandlt, the Reynolds, the Hartmann numbers and the dimensionless group are illustrated and analyzed.

From the results the following conclusions could be drawn:

a. The velocities depend strongly on the magnetic and the viscoelastic parameters.
b. The temperature varies significantly with the Prandlt number, the magnetic parameter and the heat source/sink parameter.
c. The entropy generation increases with the increase of the Prandlt, the Reynolds, the Hartmann numbers and also with the magnetic parameter and the dimensionless group.
d. The surface acts as a strong source of irreversibility.

7. Nomenclature

A	constant, K
$\vec{B}_0$	uniform magnetic field strength, Wb.m^{-2}
Br	Brinkman number, $Br = \frac{\mu u_0^2}{k\Delta T}$
C_P	specific heat of the fluid, J.kg^{-1}.K^{-1}
f	dimensionless function
Ha	Hartman number $Ha = B_0 l \sqrt{\frac{\sigma}{\mu}}$
k	thermal conductivity of the fluid, W.m^{-1}.K^{-1}
k_1	viscoelastic parameter, $k_1 = \frac{k_0 \lambda}{\nu}$
k_0	viscoelastic parameter, m^2
l	characteristic length, m
M	Kummer's function
Mn	magnetic parameter, $Mn = \frac{\sigma B_0^2}{\rho \lambda}$
N_S	entropy generation number, $N_S = \frac{S_G}{S_{G0}}$
Pr	Prandlt number, $\Pr = \frac{\mu C_P}{k}$
Q	rate of internal heat generation or absorption, W.m^{-3}.K^{-1}

Re_l	Reynolds number based on the characteristic length, $\mathrm{Re}_l = \frac{u_l\, l}{\nu}$
S_G	local volumetric rate of entropy generation, W.m-3.K-1
S_{G0}	characteristic volumetric rate of entropy generation, W.m-3.K-1
T	temperature, K
u	axial velocity, m.s-1
u_l	plate velocity based on the characteristic length, m.s-1
u_p	plate velocity, m.s-1
v	transverse velocity, m.s-1
x	axial distance, m
X	dimensionless axial distance, $X = \frac{x}{l}$
y	transverse distance, m
α	positive constant
β	heat source/sink parameter, $\beta = \frac{Q\nu}{\lambda k}$
λ	proportional constant, s-1
η	dimensionless variable, $\eta = y\sqrt{\frac{\lambda}{\nu}}$
ξ	dimensionless variable, $\xi == \frac{\mathrm{Pr}}{\alpha^2} e^{-\alpha\eta}$
μ	dynamic viscosity of the fluid, kg.m-1.s-1
ν	kinematic viscosity of the fluid, m2.s-1
ΔT	temperature difference, $\Delta T = T_p - T_\infty$
Ω	dimensionless temperature difference, $\Omega = \frac{\Delta T}{T_\infty}$
Θ	dimensionless temperature, $\Theta = \frac{T - T_\infty}{T_p - T_\infty}$
ρ	density of the fluid, kg.m-3
σ	electric conductivity, Ω-1.m-1

subscripts

P	plate
∞	far from the sheet

8. References

Abbas, Z.; Wang, Y.; Hayat, T. & Oberlack M. (2008). Hydromagnetic flow of a viscoelastic fluid due to the oscillatory stretching surface, *Int. J. Non-Linear Mech.*, 43, pp. 783-793, ISSN 0020-7462.

Abel, M. S.; Khan, S. K. & Prasad, K. V. (2002). Study of viscoelastic fluid flow and heat transfer over stretching sheet with variable viscosity, *Int. J. Non-Linear Mech.* 37, pp. 81-88, ISSN 0020-7462.

Abel, M. S.; Siddheshwar, P. G. & Nandeppanavar, M. M. (2007). Heat transfer in a viscoelastic boundary layer low over a stretching sheet with viscous dissipation and non-uniform heat source, *Int. J. Heat Mass Transfer,* 50, pp. 960-966, ISSN 0017-9310.

Abel, M. S. & Mahesha, N. (2008). Heat transfer in MHD viscoelastic fluid over a stretching sheet with variable thermal conductivity, non-uniform heat source and radiation, *Appl. Math. Modelling,* 32, pp. 1965-1983, ISSN 0307-904X.

Abel, M. S.; Sanjayanand, E. & Nandeppanavar; M. M. (2008). Viscoelastic MHD flow and heat heat transfer over a stretching sheet with viscous and ohmic dissipation, *Comm. Nonlinear. Sci. and Num. Simulation,* 13, pp. 1808-1821, ISSN 1007-5704.

Abel, M. S. & Nandeppanavar, M. M. (2009). Heat transfer in MHD viscoelastic boundary layer flow over a stretching sheet with non-uniform heat source/sink, *Comm. NonLinear Sci. and Num. Simu.,* 14, pp. 2120-2131, ISSN 1007-5704.

Aïboud-Saouli, S.; Saouli, S.; Settou, N. & Meza, N. (2006). Thermodynamic analysis of gravity-driven liquid film along an inclined heated plate with hydromagnetic and viscous dissipation effects, *Entropy,* 8, pp. 188-199, ISSN 1099-4300.

Aïboud-Saouli, S.; Settou, N.; Saouli, S. & Meza, N. (2007). Second-law analysis of laminar fluid flow in a heated channel with hydromagnetic and viscous dissipation effects, *Applied Energy,* 84, pp. 279-289, ISSN 0306-2619.

Aïboud, S., & Saouli, S. (2010). Second law analysis of viscoelastic fluid over a stretching sheet subject to a transverse magnetic field with heat and mass transfer, Entopy, 12, pp. 1867-1884, ISSN 1099-4300.

Aïboud, S., & Saouli, S. (2010). Entropy analysis for viscoelastic magnetohydrodynamics flow over a stretching surface, *Int. J. Non-Linear Mech.* 45, pp. 482-489, ISSN 0020-7462.

Andersson, H. D. (1992). MHD flows of a viscoelastic fluid past a stretching surface, *Acta Mech.,* 95, pp. 227-230, ISSN 0001-5970.

Bejan, A. (1982). Second-law analysis in heat transfer and thermal design, Adv. Heat Transfer, 15, pp. 1-58, ISBN 10-0-12-020021.

Bejan, A. (1996). Entropy generation minimization. *CRC Press, Boca Raton,* ISBN-10:9780849396519, New York, USA.

Bejan, A. (1979). A study of entropy generation in fundamental convective heat transfer, J. Heat Transfer, 101, pp. 718-725, ISSN 0022-1481.

Bhattacharya, S.; Pal, A. & Gupta, A. S. (1998). Heat transfer in the flow of a viscoelastic fluid over a stretching surface, *Heat mass Transfer,* 34, pp. 41-45, ISSN 0947-7411.

Chang, W. D. (1989). The non-uniqueness of the flow of viscoelastic fluid over a stretching sheet, *Q. Appl. Math.,* 47, pp. 365-366, ISSN 0033-569X.

Chen, C. H. (2010). On the analytic solution of MHD flow and heat transfer for two types of viscoelastic fluidc over a stretching sheet with energy dissipation internal heat source and thermal radiation, *Int. J. Heat mass Transfer,* 53, pp. 4264-4273, ISSN 0017-9310.

Cortell, R. (2006). A note on flow and heat transfer of a viscoelastic fluid over a stretching sheet, *Int. J. Non-Linear Mech.,* 41, pp. 78-85, ISSN 0020-7462.

Crane, L. J. (1970). Flow past a stretching sheet, *Z. Angew. Math. Phys.,* 21, pp. 645-647, ISSN 0044-2275.

Dandapat, B. S. & Gupta, A. S. (1998). Flow and heat transfer in a viscoelastic fluid over a stretching sheet, *Int. J. Non-Linear Mech.*, 24, pp. 215-219, ISSN 0020-7462 .

Datti, P. S.; Prasad, K. V.; Abel, M. S. & Joshi, A. (2004). MHD viscoelastic fluid flow over a non-isothermal stretching sheet, *Int. J. Eng. Sci.,* 42, pp. 935-946, ISSN 0020-7225.

Hayat, T.; Sajid, M. & Pop, I. (2008). Three-dimensional flow over a stretching sheet in a viscoelastic fluid, *Nonlinear. Ana. Real World Appl.*, 9, pp. 1811-1822, ISSN 1468-1218.

Hsiao, K. L. (2007). Conjugate heat transfer of magnetic mixed convection with viscous dissipation effects for second-grade viscoelastic fluid past a stretching sheet, *Appl. Therm. Eng.*, 27, pp. 1895-1903, ISSN 1359-4311.

Idrees, M.K. & Abel, M. S. (1996). Viscoelastic flow past a stretching sheet in porous meadia and heat transfer with internal heat source, *Indian J. Theor. Phys.*, 44, pp. 233-244, ISSN 0019-5693.

Khan, S. K. & Sanjayanand, E. (2005). Viscoelastic boundary layer flow and heat transfer over an exponential stretching sheet, *Int. J. Heat Mass Transfer*, 48, pp. 1534-1542, ISSN 0017-9310.

Khan, S. K. (2006). Heat transfer in a viscoelastic fluid over a stretching surface with source/sink, suction/blowing and radiation, *Int. J. Heat Mass Transfer*, 49, pp. 628-639, ISSN 0017-9310.

Lawrence, P.S. & Rao, B. N. (1992). Heat transfer in the flow of viscoelastic fluid over stretching sheet, Acta Mech., 93, pp. 53-61, ISSN 0001-5970.

Mahmud, S. & Fraser, R. A. (2003). The second law analysis in fundamental convective heat transfer problems, *Int. J. Therm. Sci.*, 42, pp. 177-186, ISSN 1290-0729.

Mahmud, S. & Fraser, R. A. (2002). Thermodynamic analysis of flow and heat transfer inside channel with two parallel plates, *Exergy* 2, pp. 140-146, ISSN 1164-0235.

Mahmud, S. & Fraser, R. A. (2002). Inherent irreversibility of channel and pipe flows for non-Newtonian fluids, *Int. Comm. Heat Mass Transfer,* 29, pp. 577-587, ISSN 0947-7411.

Mahmud, S.; Tasnim, S. H. & Mamun, H. A. A. (2003). Thermodynamic analysis of mixed convection in a channel with transverse hydromagnetic effect, *In. J. Therm. Sci.*, 42, pp. 731-740, ISSN 1290-0729.

Misra, J. C., & Shit, G. C. (2009). Biomagnetic viscoelastic fluid flow over a stretching sheet, *Appl. Math. And Compu*, 210, pp. 350-361, ISSN 0096-3003.

Nandeppanavar, M. M.; Abel, M. S. & Vajravelu, K. (2010). Flow and heat transfer characteristics of a viscoelastic fluid in a porous medium over an impermeable stretching sheet with viscous dissipation, *Int. J. Heat Mass Transfer*, 53, pp. 4707-4713, ISSN 0017-9310.

Nandeppanavar, M. M.; Vajravelu, K. & Abel, M. S. (2011). Heat transfer in MHD viscoelastic boundary layer flow over a stretching sheet with thermal radiation and non-uniform heat source/sink, *Comm. Nonlinear. Sci. and Num. Simulation,* 16, pp. 3578-3590, ISSN 1007-5704.

Narusawa, U. (1998). The second-law analysis of mixed convection in rectangular ducts, *Heat Mass Transfer,* 37, pp. 197-203, ISSN 0947-7411.

Prasad, K.V.; Abel, M. S.; Khan, S.K. & Datti, P. S. (2002). Non-Darcy forced convective heat transfer in a viscoelastic fluid flow over a non-Isothermal stretching sheet, *J. Porous Media*, 5, pp. 41-47, ISSN 1091-028X.

Prasad, K.V.; Abel, M. S. & Khan, S. K. (2000). Momentum and heat transfer in viscoelastic fluid flow in a porous medium over a non-isothermal stretching sheet, *Int. J. Numer. Method Heat flow*, 10, pp. 786-801, ISSN 0961-5539.

Prasad, K.V.; Pal, D.; Umesh, V. & Prasanna Rao, N. S. (2010). The effect of variable viscosity on MHD viscoelastic fluid flow and heat transfer over a stretching sheet, *Comm. Nonlinear Sci. and Num. Simulation*, 15, pp. 331-344, ISSN 1007-5704.

Rajagopal, K. R. ; Na, T. Y. & Gupta, A. S. (1984). Flow of a viscoelastic sheet over a stretching sheet, *Rheo. Acta*, 23, pp. 213-221, ISSN 0035-4511.

Sahin, A. Z. (1998). Second law analysis of laminar viscous flow through a duct subjected to constant wall temperature, *J. Heat Transfer*, 120, pp. 76-83, ISSN 0022-1481.

Sahin, A. Z. (1999). Effect of variable viscosity on the entropy generation and pumping power in a laminar fluid flow through a duct subjected to constant heat flux, *Heat Mass Transfer*, 35, pp. 499-506, ISSN 0947-7411.

Sahin, A. Z. (1998). A second law comparison for optimum shape of duct subjected to constant wall temperature and laminar flow, *Heat Mass Transfer*, 33, pp. 425-430, ISSN 0947-7411.

Saouli, S. & Aïboud-Saouli, S. (2004). Second law analysis of laminar falling liquid film along an inclined heated plate, *Int. Comm. Heat Mass Transfer*, 31, pp. 879-886, ISSN 0947-7411.

Singh, A. K. (2008). Heat source and radiation effects on magneto-convection flow of a viscoelastic fluid past a stretching sheet: Analysis with Kummer's functions, *Int. Comm. Heat Mass Transfer*, 35, pp. 637-642, ISSN 0947-7411.

Vajravelu, K. & Rollins, D. (1991). Heat transfer in a viscoelastic fluid over a stretching sheet, *J. Math. Anal. Appl.* 158, pp. 241-255, ISSN 0022-247X.

Woods, L. C. (1975). Thermodynamics of fluid systems, *Oxford University Press*, ISBN-10: 0198561806, Oxford, UK.

10

From Particle Mechanics to Pixel Dynamics: Utilizing Stochastic Resonance Principle for Biomedical Image Enhancement

V.P. Subramanyam Rallabandi and Prasun Kumar Roy
National Brain Research Centre, Manesar, Gurgaon, India

1. Introduction

There is a noteworthy analogy between the statistical mechanical systems and the digital image processing systems. We can make pixel gray levels of an image correspondence to a discrete particles under thermodynamic noise (Brownian motion) that transits between binary state transition from a weak- signal state to a strong-signal state whereas a noisy signal to the enhanced signal in digital imaging systems. One such phenomenon in the physical systems is stochastic resonance (SR) where the signal gets enhanced by adding a small amount of mean-zero Gaussian noise. A local change is made in the image based upon the current values of pixels and boundary elements in the immediate neighborhood. However, this change is random, and is generated by the sampling from a local conditional probability distribution. These local conditional distributions are dependent on the global control parameter called "temperature" in physical systems (Geman & Geman, 1984). At low temperature the coupling between the particles is tighter means that the images appear more regular and whereas at higher temperature induce a loose coupling between the neighboring pixels and the image appears noisy or blurred image. At particular optimum temperature these particles comes much closer fashion and similarly the pixels of an image got arranged in much closer and leads to noise degradation and further enhances the signal. In this chapter, we discuss the application of the physical principle of stochastic resonance in biomedical imaging systems. Some of the applications of stochastic resonance are signal detection and signal transmission, image restoration, enhancement of noisy or blurred images and image segmentation.

Stochastic resonance (SR) is a phenomenon of certain nonlinear systems in which the synchronization between the input signal and the noise occurs when an optimal amount of additional noise is inserted into the system (Gammaitoni et al., 1998). Stochastic resonance is a ubiquitous and conspicuous phenomenon. The climatic model addressing the apparently periodic occurrences of the ice ages by the weak, periodic external signal was thought to be the first theoretical model of stochastic resonance phenomenon, from which the concept of stochastic resonance was put forward (Benzi et al., 1981). Since after the discovery by Benzi, there has been increasingly attracting applications of stochastic resonance in various fields like physics (Gammaitoni et al., 1998), (Anishchenko et al., 1999), chemistry (Horsthemke & Lefever, 2006), biology and neurophysiology (Moss et al., 2004), biomedical (Morse & Evans,

1996), engineering systems (Hongler et al., 2003), and signal processing applications (Badzey & Mohanty, 2005). Usually noise is the hindrance to any system but in some cases, a little extra amount of noise will help, rather than hinder, the performance improvement of the system by maximizing or minimizing the chosen performance measure, such as output signal-to-noise ratio (SNR) (Gammaitoni et al., 1998), or mutual information (Deco & Schrmann, 1998).

Stochastic resonance can be characterized as a resonant synchronization phenomenon, resulting from the combined action of noise and forcing signals. If the noise intensity and the system parameters are tuned properly, synchronization will happen between the noise and the signal, yielding the "enhancement" of the signal (Gammaitoni et al., 1998). The basic components required for SR phenomenon is the input signal, threshold and the system outputs with different noise intensities (Marks et al., 2002). In stochastic resonance systems, noise can be converted into a positive fact in the improvement of system performance when the synchronization between the input signal and noise occurs. Usually, there are two approaches to realize this synchronization between the input signal and noise. The first one is the traditional stochastic resonance. It realizes the stochastic resonance effect by adding an optimal amount of additional noise into the systems. The second approach is called parameter-induced stochastic resonance. It is discovered that the synchronization can also be realized by tuning the parameters of stochastic resonance systems without adding noise (Xu et al., 2004).

The plot between input noise intensity versus signal-to-noise ratio is shown in figure 1. From figure 1, we can notice that the output signal-to-noise ratio will be maximized or stochastic resonance phenomenon occurs for optimal noise intensity. It is obvious that the output signal will start to change at the same frequency as the input signal when an optimal amount of noise is inserted into the system. One way of showing the SR phenomenon is the frequency domain, where the information can be recovered from the response recording using Fourier analysis. First, we compute the discrete Fourier transform of the recording at discrete values of the frequency. The power spectral density (PSD) at each frequency can be calculated as twice the square of the Fourier transform at that frequency. The PSD provides the distribution of power over frequency in the recorded response. If a periodic signal is detected it will show as a peak in the PSD at the frequency of the signal.

2. Types of stochastic resonance models

2.1 Nonlinear systems

Many kinds of nonlinear systems have demonstrated stochastic resonance phenomena, such as static systems (Chapeau-Blondeau & Godivier, 1997), dynamic systems (Gammaitoni et al., 1998), (Wellens et al., 2004), discrete systems (Zozor & Amblard, 1999), and coupled systems (Jung et al., 1992). The traditional stochastic resonance requires the information-carrying signal to be weak and periodic (Gammaitoni et al., 1998). Now, aperiodic (Barbay et al., 2001) and suprathreshold signals can also be the input of certain stochastic resonance systems, in terms of aperiodic stochastic resonance (Park et al., 2004), (Sun et al., 2008) and suprathreshold stochastic resonance (Stocks, 2001) respectively.

The stochastic resonance paradigm is compatible with single-neuron models or synaptic and channels properties and applies to neuronal assemblies activated by sensory inputs and perceptual processes as well. In literature, the landmark experiments including psychophysics, electrophysiology, functional MRI, human vision, hearing and tactile functions, animal behavior, single/multiunit activity recordings have been explored. Models and experiments show a peculiar consistency with known neuronal and brain physiology (Moss et al., 2004). A number of naturally occurring 'noise' sources in the brain (e.g. synaptic transmission, channel gating, ion concentrations, membrane conductance) possibly accounting for stochastic resonance phenomenon.

2.2 Suprathreshold systems

Suprathreshold stochastic resonance can operate with signals of arbitrary amplitude and has been reported in the transmission of random aperiodic signals (Stocks, 2001). Noise is an essential part of stochastic resonance systems and will improve the system performance when synchronization between noise and input signals happens. The most common and extensively studied noise is the additive zero-mean white Gaussian noise (Wang, 2008). The noise, however, is no longer limited to white Gaussian noise and even it can be colored (Nozaki et al., 1999), or non-Gaussian noise (Kosko & Mitaim, 2001), (Rousseau, et al., 2006). In some cases, chaotic signals can replace the stochastic noise and generate the stochastic resonance effect. In order to describe SR phenomena quantitatively and reveal the synchronization between signals and noise, different manners to characterize stochastic resonance phenomena have been advanced over the years. For periodic signals, the most commonly used quantifier is signal-to-noise ratio (Gammaitoni et al., 1998). For aperiodic signals, cross-correlation measures (Collins et al., 1996), and information-based measures, such as mutual information (Deco & Schrmann, 1998), can be used instead. The theoretical analysis of stochastic resonance systems is often very difficult, because of the complexity of the systems. Approximation models and approaches have been adopted in these cases. Some of the useful tools for the theoretical analysis are two-state model (Ginzburg, & Pustovoit, 2002), Fokker-Planck equation (Hu et al., 1990), and linear-response theory (Casado-Pascual et al., 2003). The noise-enhanced feeding behavior of the paddle fish is an example of stochastic resonance phenomena in biological systems and Schmitt trigger in engineering systems (Gammaitoni et al., 1998).

2.3 Excitable systems

Another example of a system, often found in neuronal circuits, that exhibits SR is an excitable system. Unlike the double well bistable system discussed below, this system has a single rest state and an unstable excited state that is reached by crossing a barrier. An excitable system behavior of SR is shown in figure 2. The system has an inbuilt threshold and monitors (over time) whether an input crosses this threshold. If, when the receiver is looking at the input it lies above the threshold, a pulse is emitted figure 2(b) and (c). If, on the other hand, the input lies below the threshold, no pulse is emitted. The pattern of pulses can be used by the detector to determine frequency information about the signal. Again, when the whole signal lies below the threshold, no pulses are emitted and it will not be detected. If noise is added to this sub-threshold signal it may push the input above the threshold, this is most likely to happen at the peaks of the signal (Rousseau et al., 2005). Information about the signal frequency is contained in the emitted pulse train and can be recovered by the detector.

2.4 Bistable systems

Another typical example of the stochastic resonance system is the nonlinear bistable double-well dynamic system, which describes the overdamped motion of a Brownian particle in a symmetric double-well potential in the presence of noise and periodic forcing as shown in figure 3(a) and the particle in the double-well potential crossing the barrier from a weak-signal state to a strong-signal state as shown in figure 3(b). The bistable double-well systems have found several applications in signal processing (Leng et al., 2007) and fault diagnosis (Tan et al., 2009). It has been used to amplify the coherent signals (Badzey & Mohanty, 2005). We can make pixel gray levels of an image correspondence to a discrete particles under Brownian motion that transits between binary state transition whereas a noisy image to an enhanced image in digital imaging systems. The assignment of an energy function in the states of atoms or molecules in the physical system is determined by its Boltzmann's or Gibbs distribution. Because of the Gibbs distribution, markov random field (MRF) equivalence, this assignment also determines MRF image model (Geman & Geman, 1984). Similarly, the threshold-crossing rate of the stochastic resonator occurs only at the Kramer's frequency. In physical systems, at low temperature the coupling between the particles is tighter means that the images appear more regular and whereas at higher temperature induce a loose coupling between the neighboring pixels and the image appears noisy or blurred image. At particular optimum temperature these particles comes much closer and analogous the pixels of an image got arranged in much closer and leads to noise reduction and enhances the signal.

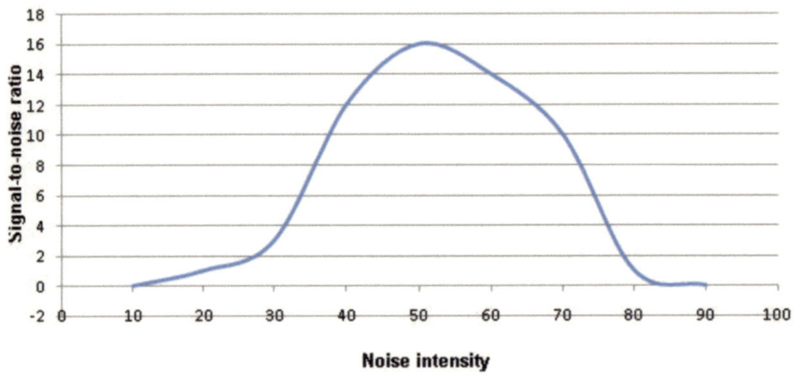

Fig. 1. Signal-to-noise ratio maximum peak occurs at an optimum level of noise intensity

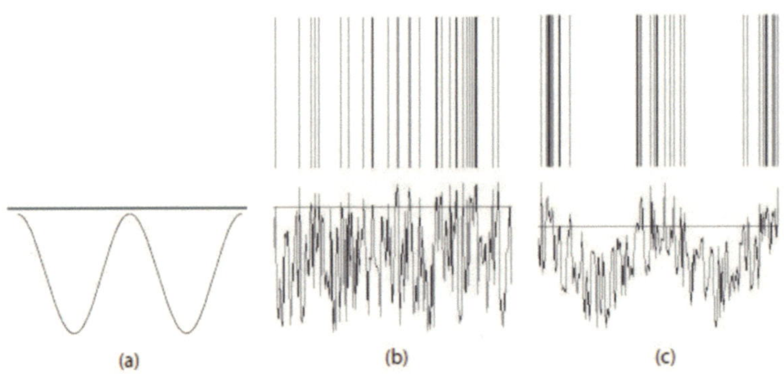

Fig. 2. An excitable system (a) A periodic signal lying below the threshold (b) If only noise is added to the system, threshold crossings are random and no information is contained in the pulse train, (c) If both the noise and signal are added to the system, the threshold crosses and hence the pulse train corresponds to peak of signal and information can be recovered.

In this chapter, we focuses on the phenomenon of stochastic resonance application in various medical imaging systems like computed tomography (CT) and magnetic resonance imaging (MRI).We investigate the applications of stochastic resonance techniques in medical image processing based on systematic and theoretical analysis, rather than only based on simulations. We develop a totally new formulation of two-dimensional parameter-induced stochastic resonance for nonlinear image processing. We reveal it is feasible to extend the concept of one-dimensional parameter-induced stochastic resonance to two-dimensional and use it for image processing. Compared with current SR-based methods, the current approach based on two-dimensional SR technique can eliminate the noise on the addition of noise into images, which can be used as a nonlinear filter for image processing. Here, we first propose a new two dimensional bistable stochastic resonance system in their respective integral transforms such as Radon and Fourier transforms respectively for CT and MR imaging.

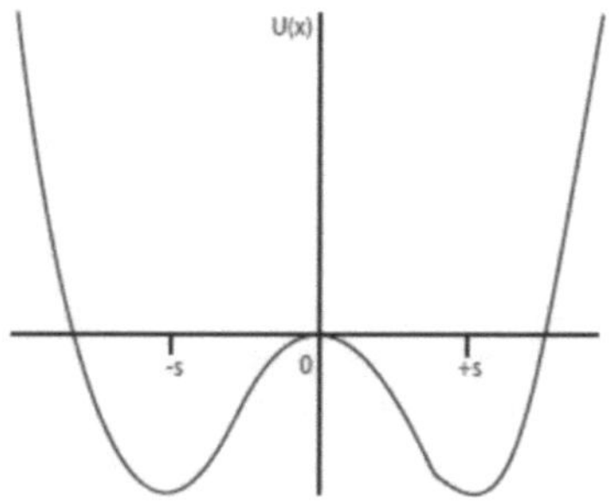

Fig. 3. (a) Bistable double well potential

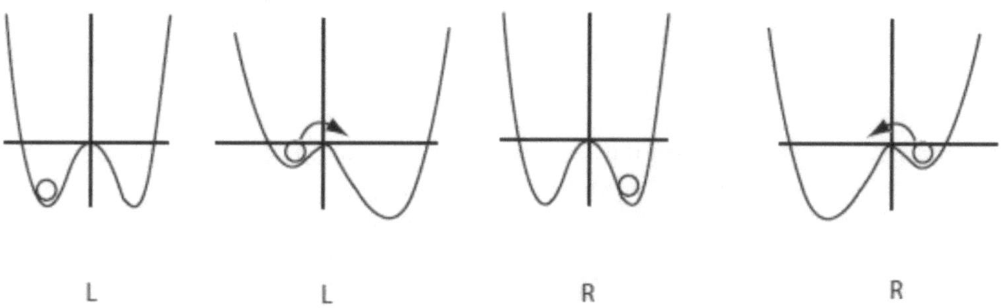

Fig. 3. (b) Particle in double well potential crossing the barrier when signal reaches peak

3. Mathematical framework

We now elaborate the bistable SR model in the theoretical form that is conventionally used by the physicists. We now ask how an image pixel would transform if mean-zero Gaussian fluctuation noise $\eta(t)$ is added, so that the pixel is transferred from a weak-signal state to a strong-signal state, i.e. a binary-state transition occurs. Actually, such a discrete image pixel under noise can be modeled by a discrete particle under Brownian motion, the particle transits between two binary states L and R, separated by a threshold (figure 3b). The theory of stochastic Brownian model is well known in statistical physics and thermodynamics, and the initial investigations on stochastic transition by (Kramers, 1940) and on the bistability theory of stochastic resonance by (McNamara, 1989). The transition of a Brownian particle between two-states (Gammaitoni et al., 1998), having a bistable potential, *U(x)*, is given by

$$U(x) = \frac{-a}{2}x^2 + \frac{b}{4}x^4 \tag{1}$$

where x is the particle's normalized position in the state parameter axis centred on the origin at $x = 0$ (figure 3a). We can obtain the equation of motion of the particle by delineating that its velocity $\dot{x}(t)$ as the algebraic resultant of the two causative factors of motion, namely the sinusoidal signal force term and the damping force term, the latter being the (negative) first differential of the potential, $U'(x)$ and hence given by:

$$\dot{x}(t) = -U'(x) + A_0 \cos(\Omega t + \phi) \tag{2}$$

where A_0, Ω and ϕ are respectively the signal amplitude, modulation frequency and phase. In order to occur the stochastic resonance phenomenon, we need to add small amount of mean-zero white Gaussian noise $\eta(t)$ to the particle, which causes the particle to move from one state to the other state, jumping and crossing over the threshold that has a threshold potential, ΔU as shown in figure 3b. As already mentioned earlier, each particle of the physical system above, corresponds to a pixel of the image, from a signal processing perspective. Note that $\eta(t)$ is the stochastic noise administered, having the mean or expected value of zero, i.e.

$$\xi[\eta(t)] = 0$$

with the autocorrelation function $\eta(t)$ being that of a Gaussian white noise, given by

$$\langle \eta(t)\eta(0) \rangle = 2\, D\, \delta(t)$$

Here $\delta(\tau)$ and D are the delta function and noise intensity respectively.
Mathematically, one can represent the random motion of the particle in a bistable potential in the presence of noise and periodic forcing can be given by:

$$\dot{x}(t) = -U'(x) + A_0 \cos(\Omega t + \phi) + \eta(t) \tag{3a}$$

where $U'(x) = -ax + bx^3$.
Since our aim is to obtain a maximal signal, we let the cosine term attain its maximum value i.e. unity, and substitute $U'(x)$ as obtained by differentiating eq. (1), we get from eq. (3a):

$$\dot{x}(t) = ax(t) - bx^3(t) + A_0 + \eta(t) \tag{3b}$$

The threshold-crossing rate of the stochastic resonator occurs at the Kramer's frequency

$$r_k = \frac{a}{\sqrt{2\pi}} \exp\left(-\frac{\Delta U}{D}\right) \tag{4}$$

Being reciprocal of Kramer's frequency, the periodicity or waiting time of the stochastic transition between two noise-induced inter-well transition which is given by $T_k(D) = 1/r_k$.
If we input a small periodic forcing term to the particle, stochastic switching and jumping occurs between the potential wells and the switching may become synchronized with the input. This stochastic synchronization happens if the mean waiting time satisfies the time-scale matching requirement (Gammaitoni et al., 1998)

$$T_\Omega = 2\ T_k(D)$$

where T_Ω is the period of the input periodic forcing term.

Stochastic resonance occurs if the signal-to-noise level of a system increases with the values of noise intensity. For lower noise intensities, the signal does not affect the system to cross threshold, so little signal is passed through it. For large noise intensities, the output is dominated by the noise, also leading to a low signal-to-noise ratio. For moderate optimal intensity level, the noise allows the signal to reach threshold, and increases the signal-to-noise ratio of a system. SR occurs at the maximum response of the signal i.e. signal-to-noise ratio. (SNR) and the alteration of the response of the signal due to stochastic resonator is given by

$$SNR = \frac{4a}{\sqrt{2}(\sigma_0\sigma_1)^2}\exp\left(\frac{-a}{2\sigma_0{}^2}\right) \tag{5}$$

With respect to figure 3a, the potential minima are located at $s = \pm\sqrt{(a/b)}$, while the height of the threshold potential barrier between the two states is $\Delta U = \left(a^2/4b\right)$. Considering the image enhancement scenario, one can posit that the x-axis corresponds to the normalized pixel intensity value with respect to the detector threshold value that is defined as $x = 0$, where it is analogous to noisy image to enhanced image.
Based on the power spectral density of a one dimensional signal or the coefficient of variance (CV) of an image, which is the contrast enhancement index defined as the performance measure of nonlinear bistable dynamic systems with fluctuating potential functions can be further enhanced by adding noise and tuning system parameters at the same time, if the input signal is Gaussian-distributed. Then, we extend these results to hazy or noisy images. The relative enhancement of the contrast of an image means the ratio of the coefficient of variance between the input noisy image and the output SR enhanced image. Therefore, we suggest a potential application of this mechanism in the recovery of weak signals corrupted by noise to biomedical imaging.

4. Application of stochastic resonance in biomedical imaging

4.1 SR-based Integral transform

In this section, we discuss the application of the bistability stochastic resonance model for the enhancement of commonly used medical images such as computed tomography and magnetic resonance imaging. Due to the fact that CT image reconstructed using Radom transform (Deans, 1983), whereas MR image formation corresponds to the Fourier transform (Lauterbur & Liang, 2001), we propose a bistable SR system operating in the spatial domain of the two-dimensional integral transforms. Let us consider the 2D spatial representation of an object as a function $\psi(x,y)$, which can be the image intensity or a 2D projection of a CT image, pixel gray value in T_1-weighted MR image where the pixel brightness respectively depend on the tissue relaxation rate or the spin density. The generalized MR or CT imaging equation in projective imaging case can be given by

$$I(x,y) = \int_{z=-\infty}^{\infty} \psi(x,y,z)dz$$

Since we consider a single slice of 3-D volume, and the 2-D image $\hat{I}(x,y)$ can be formed using respective Fourier integral transform (eq. 6a) and Radon transform (eq. 6b) which is given by (Rallabandi & Roy, 2008):

$$\hat{I}(k_x,k_y) = \int_{x=-\infty}^{\infty} \int_{y=-\infty}^{\infty} I(x,y).e^{-i2\pi(k_x x + k_y y)} dk_x dk_y \tag{6a}$$

$$\hat{I}(\rho,\theta) = \int_{x=-\infty}^{\infty} \int_{y=-\infty}^{\infty} I(x,y).\delta(x\cos\theta + y\sin\theta - \rho)d\rho d\theta \tag{6b}$$

where δ(.) is a dirac-delta function given for the plane of projection which is equal to 1 if x=0 and 0 otherwise.

We now derive a transformed image $I'(k_x,k_y)$ by subtracting the mean-zero noise image $I(k_x,k_y)$ image from the original image $\hat{I}(k_x,k_y)$ such that

$$I'(k_x,k_y) = \hat{I}(k_x,k_y) - < I(k_x,k_y) > \tag{7}$$

where < > denotes the spatial average value of pixel intensity of the original image $\hat{I}(k_x,k_y)$ .Now convoluting the stochastic resonator *SR* on the transformed image $I'(k_x,k_y)$, thereby obtaining the stochastically enhanced image $I''(k_x,k_y)$ which is given by:

$$I''(k_x,k_y) = \int_{x=-\infty}^{\infty} \int_{y=-\infty}^{\infty} SR\left[I'(k_x,k_y).e^{2\pi i(k_x x + k_y y)} dk_x dk_y \right] \tag{8}$$

Here *SR* is operated on the magnetic resonance image *I* as given in eq.3 (b) such that SR phenomenon occurs at maximum SNR given in eq.(5).

Now we need to solve the stochastic differential equation given in eq. (3b) using stochastic version of Euler-Maruyama's method using the iterative method as follows [Gard, 1998]:

$$x_{n+1} = x_n + k(ax_n - bx_n^3 + s_n) \tag{9}$$

in which $s_n = A_0 + \sigma\, w_n$, denotes the sequence of input signal and noise with the initial condition being x_0 = x (0), i.e. the initial value of x being 0. Observe that the zero-mean stochastic noise sequence $\{w_n\}$ has unit variance, σ_w^2 = 1. We discretize the stochastic simulation in terms of '*k*' steps as shown in eq. (9).

4.2 Selection of optimal parameters

Note that it is necessary to select the optimal bistability parameters of '*a*' and '*b*', we consider the output SNR as a function of noise intensity given in eq. (5) such that the pixel maps (Ω_0, σ_0) and (Ω_1, σ_1) have the relationship (Ye et al., 2003):

$$\frac{a}{b} = \left(\frac{\sigma_1^{\ 2}}{\sigma_0^{\ 2}} \right) \tag{10}$$

where (Ω_0, Ω_1) are respectively the signal frequencies of the input image and SR-enhanced image, while (σ_0, σ_1) are respectively the standard deviation of noise in the input image and SR-enhanced image. Our approach has been adapted and modified from the usual methodology of using the bistability-based stochastic resonance effect to enhance input noisy image based on the integral transform of the input image (Rallabandi & Roy, 2010). In our case, we fix one of the bistability parameters '*a*' at a particular value, and estimate the other parameter '*b*' according to the relation given in eq.(10). However, the choice of parameters '*a*' and '*b*' are selected for CT and MRI using the relationship given in eq. (10).

To furnish a readily obtainable quantitative index of image upgradation, we plot the gray-level histograms of the input image and the optimal enhanced image. As a ready approximation, it is known that as an image is enhanced and there is more finer or clearer heterogenous structuration obtained, this enhancement can be characterized by an increase in the image quality contrast parameter, which is the coefficient of variance (CV) of an image, that is, the ratio of variance to the mean of the image histogram given by $Q = \left(\sigma^2 / \mu \right)$.

Further, we can estimate the relative image enhancement factor due to SR by means of the ratio of the pre-enhancement (Q_A) and post-enhancement (Q_B), values of image quality index given by (Rallabandi & Roy, 2010)

$$F = \left(\sigma_B^{\ 2} \mu_A / \sigma_A^{\ 2} \mu_B \right) \tag{11}$$

The general illustration of using SR approach for CT/MRI images is shown in figure 4. We consider the noisy CT axial image so that the image became indistinct, which caused the obliteration of the lesion and its edema, and the midline falx cerebri (figure 5a). To this indistinct image, the SR-based Radon transform is applied (the resultant output image is shown in Figure 5b). Note that the noise in the image has been reduced, whereas clearer visibility has been attained by the representation of the edema, falx, and lesion, with an inner central core reminiscent of a calcified scolex blob inside (arrow; figure 5b).

We consider the T_1-weighted MR image of the malignant brain tumor, glioblastoma multiforme having mass effect in both the hemispheres, contraction of the ventricles and involvement of the splenium of the corpus callosum. Noise was added to this image so that it becomes indistinct; the gray matter, white matter and the lesion region cannot be distinguished and the sulci and gyri become obliterated (figure 6a). We then apply the SR enhancement process in Fourier domain and the resultant enhanced image is given in figure 6b. One may easily observe that the noise in the image has reduced, while the representation of the lesion, sulci, gyri, white and gray matter has appreciably restored with clearer demarcation. To enable a quantitative comparison, the image histograms are constructed, and are displayed to the right of the respective images. Figures 6c and 6d are the image histograms of figures 6a and 6b respectively.

The stochastic resonance imaging approach has advantages like that it can recover the image from noise and also enhance the selected region of tumor image. The proposed method can be used to distinguish boundaries between gray matter, white matter, and CSF and also delineate edematous zones, vascular lesions and proliferative tumor regions. This method would be of considerable use to clinicians since SR enhanced images, under a suitable choice of '*a*' and '*b*' parameters. One can reiterate that the advantage of SR procedure is that the process can adapt to the local image texture by altering these stochastic bistability parameters, so that the enhancement process is suitably optimized.

4.3 Contrast sensitivity

Stochastic resonance inherently is a process that is well tuned to enhance the contrast sensitivity and decrease the neurophysiological threshold of the human visual system, which have been well demonstrated experimentally when stochastic fluctuation of pixel intensity is administered to visual images on a computer screen observed by a subject (Simonotto et al; 1997). In other words, it may be emphasized that the development of a high performance contrast enhancement algorithm must hence attempt to enhance the contrast in the image, based not only on the local characteristics of the image but also on some basic human visual characteristics, especially those properties related to contrast. The development of a high-performance contrast enhancement algorithm must thus attempt to enhance the contrast in the image based not only on the local characteristics of the image but also on some basic human visual characteristics, especially those properties related to

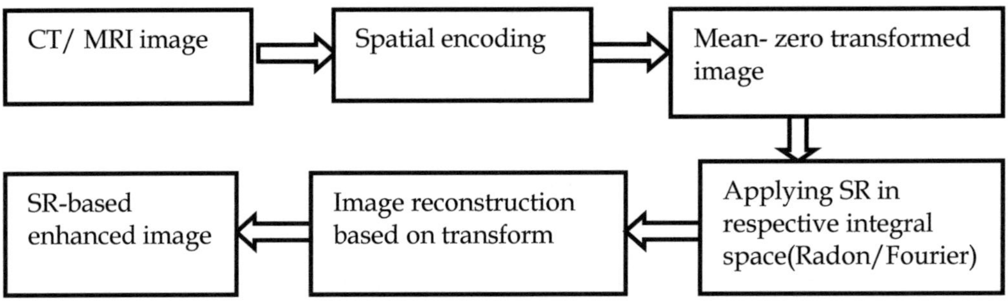

Fig. 4. Illustration of Stochastic Resonance in Radon/Fourier integral domain

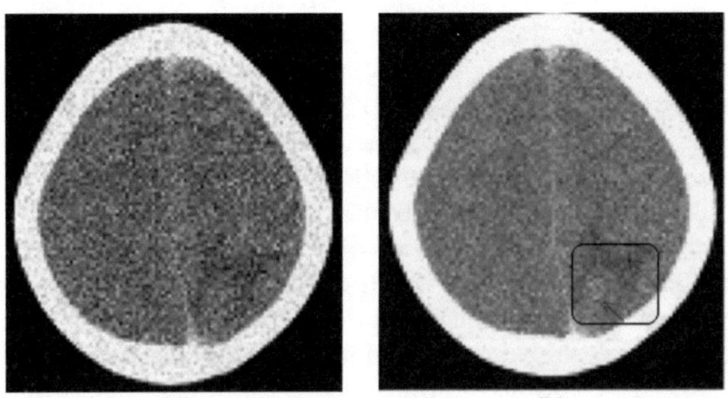

(a) (b)

Fig. 5. (a) Noisy or hazy CT image (b) SR-enhanced output image using Radon transform contrast (Piana et al., 2000). Nevertheless, the majority of enhancement procedures are neither tissue-selective nor tissue-adaptive, since in general the various texture properties in the image are enhanced evenly together. From an ergonomics perspective, the SR approach can be taken to enhance the performance of both aspects of the image visualization process, the radiological image processing device, and the human neurophysiological visual characteristics.

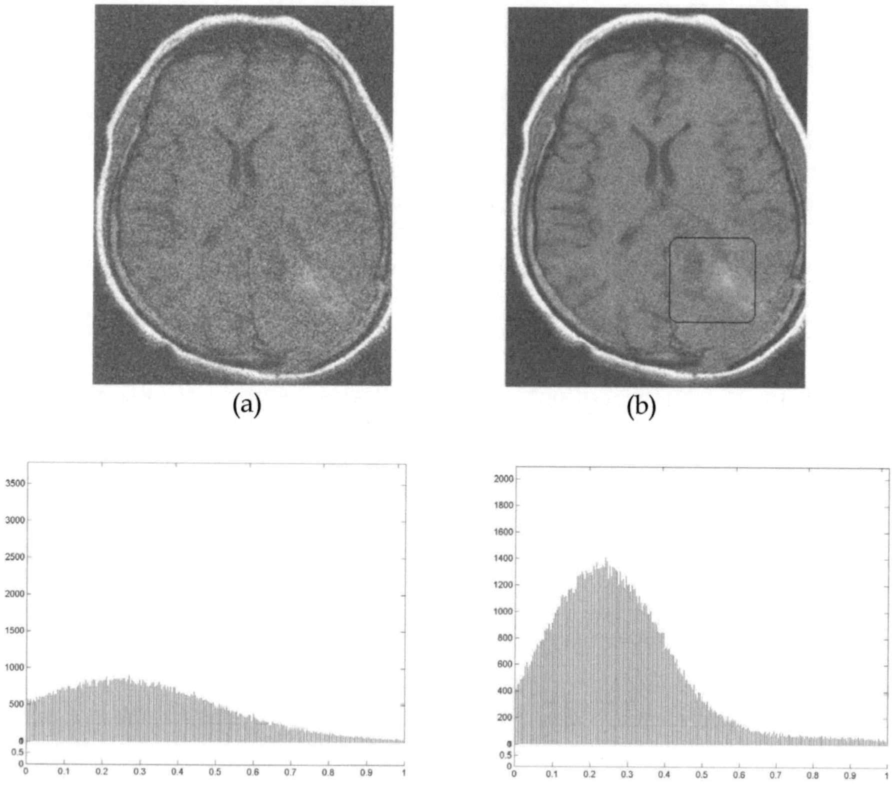

Fig. 6. (a) Noisy MR image (b) SR-enhanced image where the lesion, sulci and gyri are visible (c) & (d) Image histograms of input image of fig.6a and SR-enhanced image of fig.6b.

5. Conclusion

In this chapter, we discuss the phenomenon of stochastic resonance applicable to biomedical image processing, where the discrete image pixels are treated as discrete particles, whereby the gray value of an image pixel corresponds to a specific kinetic parameter of a physical particle in Brownian motion. For real-time applications, we can extend our approach for enhancing images which are poor in spatial resolution like positron emission tomography images and low signal-to-noise ratio images like functional MRI. Additionally, we aver that much appreciable scope exists in utilizing the stochastic resonance technique for enhancing higher order noisy images due to various operational conditions during scanning such as electronic device noise, thermal noise or nyquist frequency noise.

6. References

Anishchenko, V.S., Neiman, A.B., Moss, F. & and Schimansky-Geier, L. (1999). Stochastic Resonance: noise-enhanced order, *Physics-Uspekhi*, vol.42 issue 1, pp.7-36.

Badzey, R.L. & Mohanty, R (2005). Coherent signal amplification in bistable nanomechanical oscillators by stochastic resonance, *Nature*, vol. 437, pp. 995-998.

Barbay, S., Giacomelli, G. & Marin, F. (2000). Noise-assisted transmission of binary information: Theory and experiment, *Phys. Rev. E*, vol. 63, 051110.

Benzi, R., Sutera, A. & Vulpiani A. (1981). The mechanism of stochastic resonance, *J. Phys.A*, vol. 14, no. 11, pp.L453-L457.

Casado-Pascual, J., Denk, C., Gomez-Ordonez, J., Morillo, M. & Hanggi P. (2003). Gain in stochastic resonance: precise numerics versus linear response theory beyond the two-model approximation, *Phys. Rev. E*, vol. 67, 036109.

Chapeau-Blondeau, F. & Godivier, X. (1997). Theory of stochastic resonance in signal transmission by static nonlinear systems, *Phys. Rev. E*, vol. 55, no. 2, pp. 1478-1495.

Collins, J.J., Chow, C.C., Capela, A.C. & Imhoff, T.T. (1996). Aperiodic stochastic resonance, *Phys. Rev. E*, vol. 54, no. 5, pp. 5575-5584.

Deans, S. R. (1983).*The Radon Transform and Some of Its Applications*. New York: John Wiley & Sons.

Deco, G. & Schrmann, B. (1998). Stochastic resonance in the mutual information between input and output spike trains of noisy central neurons, *Phys. D*, vol. 117, pp. 276-282.

Gammaitoni, L., Hanggi, P., Jung, P. & Marchesoni, F. (1998). Stochastic resonance, *Reviews of Modern Physics*, vol. 70, no. 1, pp. 223-287.

Gard, T.C. (1998). *Introduction to Stochastic Differential Equations*. New York: Marcel-Dekker.

Geman, S. & Geman, D. (1984). Stochastic relaxation, Gibbs distributions, and the Bayesian restoration of images, *IEEE Trans. Pattern Analysis and Machine Intelligence*, vol. 6, no. 6, pp. 721-741.

Ginzburg, S.L. & Pustovoit, M.A. (2002). Stochastic resonance in two-state model of membrane channel with comparable opening and closing rates, *Phys. Rev. E*, vol.66, 021107.

Hongler, M., Meneses, Y., Beyeler, A. & Jacot, J. (2003). The resonant retina: exploiting vibration noise to optimally detect edges in an image, *IEEE Trans. Pattern Analysis and Machine Intelligence*, vol. 25, no. 9, pp. 1051-1062.

Horsthemke, W. & Lefever, R. (2006). *Noise-induced Transitions: Applications to Physics, Chemistry and Biology*. New York: Springer.

Hu, G., Nicolis, G. & and Nicolis, C. (1990). Periodically forced Fokker-Planck equation and stochastic resonance, *Phys. Rev. A*, vol.42 issue 4, pp.2030-2041.

Jung, P., Behn, U., E. Pantazelou, E. & Moss, F. (1992). Collective response in globally coupled bistable systems, *Phys. Rev. A*, vol. 46, no. 4, pp. 1709-1712.

Kosko, B. & Mitaim, S. (2001). Robust stochastic resonance: Signal detection and adaptation in impulsive noise, *Phys. Rev. E*, vol. 64, 051110.

Kramers, H.A. (1940). Brownian motion in a field of force and the diffusion model of chemical reactions, *Physica*, vol.7, pp.284-304.

Lauterbur, P.C. & Liang, Z.-P. (2001). *Principles of Magnetic Resonance Imaging: A Signal Processing Perspective*. New York: IEEE Press.

Leng Y., Wang, T., Guo, Y., Xu, Y. & Fan, S. (2007). Engineering signal processing based on bistable stochastic resonance, *Mechanical Systems and Signal Processing,* vol.21, Issue 1, pp. 138-150.

Marks, R.J., Thompson, B., El-Sharkawi, M.A. Fox, W.L.J. & Miyamoto, R.T. (2002). Stochastic resonance of a threshold detector: image visualization and explanation, *IEEE International Symposium on Circuits and Systems,* vol. 4, pp.521-523.

McNamara, B. & Wiesenfeld, K. (1989). Theory of Stochastic Resonance, *Phys. Rev. A,* vol.39, pp. 4854-4869.

Morse, R.P. & Evans, E.F. (1996). Enhancement of vowel coding for cochlear implants by addition of noise, *Nature Medicine,* vol. 2, pp. 928-932.

Moss, F., Ward, L.M. & Sannita, W.G. (2004). Stochastic resonance and sensory information processing: a tutorial and review of application, *Clinical Neurophysiology,* vol. 115, no. 2, pp.267-281.

Nozaki, D., Mar D.J., Grigg, P. & Collins, J.J. (1999). Effects of colored noise on stochastic resonance in sensory neurons, *Phy. Rev. Lett.,* vol. 82, no. 11, pp. 2402-2405.

Park, K., Lai, Y. C., Liu, Z. & Nachman, A. (2004). Aperiodic stochastic resonance and phase synchronization, *Physics Letters A,* vol. 326 pp. 391-396.

Piana, M., Canfora, M. & Riani, M. (2000). Role of noise in image processing by the human perceptive system, *Phys. Rev. E,* vol.62, pp.1104-1109.

Rallabandi, V.P.S. & Roy, P.K. (2008). Stochastic resonance-based tomographic transform for image enhancement of brain lesions, *Jour. Comp. Asst. Tomography,* vol.28 issue 9, pp.966-974.

Rallabandi, V.P.S. & Roy, P.K. (2010). Magnetic resonance image enhancement using stochastic resonance in Fourier domain, *Magnetic Resonance Imaging,* vol.32 issue 6, pp.1361-1373.

Rousseau, D. & Chapeau-Blondeau, F. (2005). Stochastic resonance and improvement by noise in optimal detection strategies, *Digital Signal Processing,* vol. 15, pp. 19-32.

Rousseau, D., Anand, G.V. & Chapeau-Blondeau, F. (2006). Noise-enhanced nonlinear detector to improve signal detection in non-Gaussian noise, *Signal Processing,* vol. 86, Issue 11, pp. 3456-3465.

Simonotto, E., Riana, M., Seife, C., Roberts, M., Twitty, J. & Moss, F. (1997). Visual perception of stochastic resonance, *Phy. Rev. Lett.* , vol. 78, issue 6, pp. 1186-1189.

Stocks, N.G. (2001). Information transmission in parallel threshold arrays: Suprathreshold Stochastic Resonance, *Phys. Rev. E,* vol. 63, no. 4, 041114.

Sun, S. & Lei, B. (2008). On an aperiodic stochastic resonance signal processor and its application in digital watermarking, *Signal Processing* vol.88, Issue 8, pp.2085-2094.

Tan, J., Chen, X., Wang, J., Chen, H., Cao, H., Zi, Y. & He, Z. (2009). Study of frequency-shifted and re-scaling stochastic resonance and its application to fault diagnosis, *Mechanical Systems and Signal Processing,* vol.23, Issue 3, pp. 811-822.

Wang, Y. (2008). Nonlinear statistics to improve signal detection in generalized Gaussian Noise, *Digital Signal Processing, vol.* 18, Issue 3, pp. 444-449.

Wellens, T., Shatokhin, V. & Buchleitner, A. (2004). Stochastic resonance, *Reports on Progress in Physics,* vol. 67, pp. 45-105.

Xu, B., Duan, F. & Chapeau-Blondeau, F. (2004).Comparison of aperiodic stochastic resonance in a bistable system realized by adding noise and by tuning system parameters, *Phys. Rev. E,* vol. 69, 061110.

Ye, Q., Huang, H., He, X. & Zhang, C. (2003). A study on the parameters of bistable stochastic resonance systems and adaptive stochastic resonance systems, *Proc. 2003 IEEE Int. Conf. on Robotics, Int. Syst. and Signal Proc.* New York, pp. 484-488.

Zozor, S. & Amblard, P.O. (1999). Stochastic resonance in discrete time nonlinear AR (1) models, *IEEE Transactions on Signal Processing*, vol. 47, no. 1, pp. 108-122.

11

Thermodynamics of Amphiphilic Drug Imipramine Hydrochloride in Presence of Additives

Sayem Alam[1], Abhishek Mandal[2,3] and Asit Baran Mandal[2]
Council of Scientific and Industrial Research (CSIR)
- Central Leather Research Institute (CLRI)
[1]Industrial Chemistry Laboratory,
[2]Chemical Laboratory, Adyar, Chennai 600020,
[3]Present address: Department of Biomedical Engineering,
University of Saskatchewan, Saskatoon SK S7N 5A4,
[1,2]India
[3]Canada

1. Introduction

In aqueous environment, amphiphilic molecules (*viz.*, surfactants, drugs, polymers, etc.) or ions are frequently assemble at interfaces and self-associate in an attempt to sequester their apolar regions from contact with the aqueous phase (Attwood & Florence, 1983, Atherton & Barry, 1985, Attwood et al, 1989, Schreier et al, 2000, Attwood, 1995a,1995b, Mandal & Nair, 1988, 1991, Mandal et al, 1987, 1993, Geetha et al, 1993, 2003, Mandal, 1993, Mandal & Jayakumar, 1994, Geetha & Mandal, 1997a, 1997b, 2000, Rose & Mandal, 1996, Taboada et al, 2000, 2001, Junquera et al, 2001, Rodriguez et al, 2004, Misra et al, 2009, 2010, James et al, 2011, James & Mandal 2011, Mandal et al, 2010, Tiwary et al, 2011, Alam et al, 2007, Khan et al, 2009, 2010). A large number of drug molecules are amphiphilic and self-associate in aqueous solution to form small aggregates. These surface-active behavior among many diverse classes of drugs has been reported and attempts have been made to correlate surface activity and biological activity (Attwood & Florence, 1983, Atherton & Barry, 1985, Attwood et al, 1989, Schreier et al, 2000, Taboada et al, 2000, Junquera et al, 2001, Attwood, 1995, Geetha et al, 2003, Alam et al, 2007, 2008). The aggregation of the above drugs follows the same principles as of conventional surfactants (Schreier et al, 2000, Taboada et al, 2000, Junquera et al, 2001, Attwood, 1995, Geetha et al, 2003, Alam et al, 2007, 2008). The self-association of drug depends on the molecular structure of the drug, its concentration and the experimental conditions such as temperature, pH and salt concentration (Atherton & Barry, 1985, Taboada et al, 2000, Junquera et al, 2001, Attwood, 1995, Geetha et al, 2003, Alam et al, 2007, 2008). The "surfactant-like" behavior of these drugs is due to the presence of an almost planar tricyclic ring system and a short hydrocarbon chain carrying a terminal nitrogen atom (Taboada et al, 2000, Junquera et al, 2001).

The self-assembly and self-organization are natural and spontaneous processes, occurring mainly through non-covalent interactions such as, van der Waals, hydrogen-bonding, hydrophilic/hydrophobic, electrostatic, donor and acceptor, and metal-ligand coordination networks (Whitesides & Grzybowski, 2002). The interest in micelle solutions stems from their potential as functional molecular assemblies for use in many fields in pure and applied sciences, because they can be used as models for several biochemical and pharmacological systems and can solubilize water-insoluble substances (including certain medicines/drugs) in their hydrophobic cores (Barzykin et al, 1996).

The colloidal properties of amphiphilic drugs are largely determined by the nature of the aromatic ring system of their hydrophobic moieties, and such drugs are useful in probing the relationship between the molecular architecture and physicochemical properties (Attwood & Florence, 1983). In pharmacy, the interaction of small molecules with drugs is one of the most extensively studied. In this respect, many drugs, particularly those with local anesthetic, antidepressant, tranquillizer, and antibiotic actions, exert their activity by interaction with biological membranes, which can be considered as complex form of amphiphilic bilayers. Therefore, a full knowledge of the mechanism of the interactions of drugs with other foreign materials is required before the actual application in human body. This is due to the fact that drugs are always used in presence of a variety of additives (excipients).

Thermodynamic parameters of some amphiphilic drugs (viz. amitriptyline hydrochloride, imipramine hydrochloride, chlorpromazine hydrochloride, and promethazine hydrochloride) in presence of additives have recently been evaluated (Alam et al, 2010a, 2010b, 2010c, 2010d, 2010e). Micellar characteristics of various peptides, collagens, polymers in aqueous and non-aqueous media and their interactions with various surfactant micelles have widely been studied (Mandal et al, 1987, 1993, Mandal & Jayakumar, 1994, Jayakumar, et al, 1994, Geetha & Mandal, 1995, 1996, Jayakumar & Mandal, 1993, Ramya, et al, 2003, 2004, Khan et al, 2010a, 2010b) in light of aggregation, H-bonding, geometry, correlation times, conformation, hydrodynamic and thermodynamic studies.

It has been established from earlier studies on these drugs that aggregates of approximately 6–12 monomers are formed in water above the critical micelle concentration (cmc). The pK_a values of these drugs lie between 9.1-9.4 (Katzung, 2004), and depending upon the solution pH, the drug monomers may acquire cationic (i.e., protonated) or neutral (i.e., deprotonated) form (Kim, 2002).

It is well known that cmc of amphiphiles varies in presence of additives, because the interfacial and micellar properties of these compounds in solutions are governed by a delicate balance of hydrophobic and hydrophilic interactions. These characteristics can be modified in two ways: (i) through specific interactions with the amphiphile and (ii) by changing the nature of solvent (Ruiz & Sanchez, 1994). As drugs are used in combination with additives (e.g., surfactants), it is necessary to have a knowledge of additive effect on the cmc and their thermodynamics of amphiphilic drugs.

Clouding is a well-known phenomenon observed in non-ionic surfactants. The clouding phenomenon can be induced by changing the temperature of the solution. The temperature at which a clear, single phase becomes cloudy and phase-separates occur upon heating is known as the cloud point (CP) (Gu & Galera-Gomez, 1999). The mechanism of clouding in non-ionic surfactants, however, is not yet very clear, and continues to be a source of controversy among different research groups. However, the occurrence of CP in charged micelle (i.e., ionic surfactants) solutions is not usual except under special conditions, e.g., high salt concentration (Gomati et al, 1987, Kumar et al, 2000, 2001, 2002, 2003, Panizza et al, 1998), salt free aqueous solutions of certain surfactants with large headgroups (Kumar et al,

2001, Panizza et al, 1998) or large counterions (Kumar et al, 2001, 2003), and some mixed cationic and anionic surfactant solutions (Kim & Shah, 2002, 2003). The CP appearance in these systems is explained in terms of increased hydrophobic interactions, dehydration of hydrophilic group, and formation of large aggregates/clusters. Like ionic surfactants, some amphiphilic drugs undergo pH-, concentration-, and temperature- dependent phase separation (Kim & Shah, 2002, 2003, Kumar et al, 2006, Alam et al, 2006a, 2006b, 2007a, 2007b, 2007c, 2007d, 2008a, 2008b, 2008c, 2010a, 2010b, 2010c, Alam & Kabir-ud-Din, 2008a, 2008b). It was observed that their CP can vary with additives.

5-[3- (dimethylamino) propyl]- 10,11-dihydro -*5H*-dibenz[*b,f*]azepine hydrochloride (imipramine hydrochloride, IMP) is a tricyclic antdepressant amphiphilic drug with neuroleptic activity, showing a large capacity to interact with biological membranes and sometimes be used as a local anesthetic (Seeman, 1972). IMP possesses a rigid hydrophobic ring system and a hydrophilic amine portion, which becomes cationic at low pH values and neutral at high pH values (Scheme 1). Moreover, the pK_a value of this drug is 9.3 (Katzung, 2004). IMP is often regarded as a model drug for the investigation of interactions between drugs and biological or model membranes (Schreier el al, 2000). Amphiphilic antidepressant drugs aggregate in a micelle-like manner and the value of N_{agg} (aggregation number) being of the order of 6-15 (Attwood & Florence, 1983, Attwood, 1995, Schreier el al, 2000). As clouding is concentration, pH and temperature dependent, it is essential to have a knowledge of clouding behavior of the drug under varying conditions.

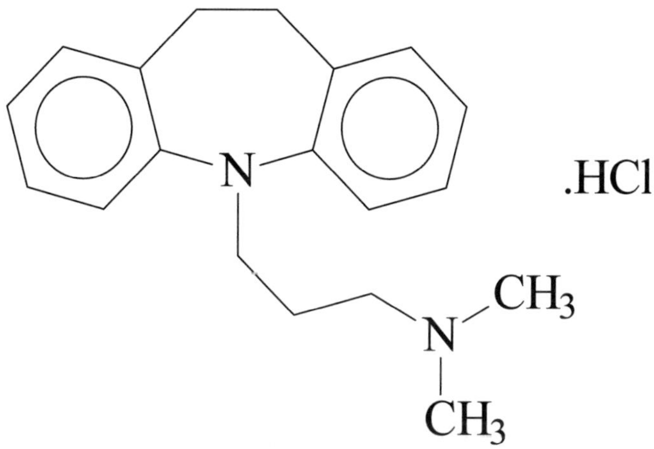

Scheme 1. The molecular structure of amphiphilic tricyclic antidepressant drug, 5-[3-(dimethylamino)propyl]-10,11-dihydro-*5H*-dibenz[*b,f*]azepine hydrochloride (imipramine hydrochloride, IMP) used in the present study.

In the present work, we report the micellization and clouding of an amphiphilic tricyclic antidepressant drug, IMP (see Scheme 1) in absence and presence of additives (KCl and TX-100). The work presented here is aimed at obtaining a better understanding of the role of the presence of additives in the thermodynamic quantities of micellization and clouding of the drug in absence and presence of additives. With this viewpoint surface tension, conductivity measurements and dye solubilization studies have been performed on aqueous solutions of IMP to determine the cmc of these drugs in presence of different additives. The surface properties (in water and in presence of varying mole fraction of TX-100) of IMP and the micellar and surface parameters *viz.*, cmc, Γ_{max} (maximum surface excess concentration at air/water interface) and A_{min} (minimum area per surfactant molecule at the air/water interface), interaction parameter, β^m, activity coefficeients (f_1, f_2) were evaluated. Using these data, we had evaluated Gibbs energies viz., Gibbs energies at air/water interface ($G_{min}^{(s)}$), the standard Gibbs energy of micellization ($\Delta_{mic}G^0$), the standard Gibbs free energy change of adsorption ($\Delta_{ads}G^0$), and the excess Gibbs energy change of micellization (ΔG_{ex}). We report the micellization and clouding of IMP in absence and presence of KCl. The thermodynamic parameters are evaluated (in micellization and at CP) in presence and absence of electrolyte (KCl). The results have relevance in drug delivery/model drug delivery.

2. Materials and methods

2.1 Materials

IMP (≥ 98 %, Sigma, USA), polyethylene glycol *t*-octylphenyl ether, TX-100 (≥ 99 %, Fluka, Switzerland), and KCl (≥ 99.9 %, Ranbaxy, India) were used as received. Doubly distilled and deionized water (sp. cond. = 1-2 $\mu S \cdot cm^{-1}$) was used as the solvent. Trisodium phosphate dodecahydrate (TSP), and sodium dihydrogen phosphate monohydrate (SDP) were of reagent grades obtained from Merck. 10 mM Sodium phosphate (SP) buffer solutions were used throughout as solvent. The pH of the IMP solutions was measured with an ELICO pH meter (model LI 120) using combined electrode.

2.2 Methods

2.2.1 Surface tension measurements

The cmc values of the drugs (with and without additives) in pure water were determined by measuring the surface tension (ST) of pure drug, as well as drug + additive (TX-100), solutions of various concentrations at ~ 300 K. The cmc values were obtained by plotting ST *vs* log [drug]. The ST values decrease continuously and then remain constant along a wide concentration range (see Figure 1). The point of break, when the constancy of ST begins, was taken as the cmc of the drug.

2.2.2 Conductivity measurements

GLOBAL conductivity meter (model DCM 900) and dip cell (cell constant 1.0 cm^{-1}) was employed to perform the conductivity measurements at different temperatures (viz., 293.15 , 303.15, 313.15 and 323.15 K). The stock solutions of IMP (with or without a fixed concentration of KCl) were prepared in double distilled water. The conductivity was measured by successive addition of concentrated solution in pure water (in case of without KCl) or in a fixed concentration of KCl solutions. A break in the specific conductivity versus drug concentration curve signals the onset of the micellization process (Figure 2).

2.2.3 Cloud point measurements

All CPs were obtained by placing Pyrex glass tubes (containing the drug solution) into a temperature controlled bath, the temperature was ramped at the rate of 0.1 K / min near the CP and onset of clouding was noted by visual inspection. The temperature, as the clouding commences, was taken as CP (Gu & Galera-Gomez, 1999, Kim & Shah, 2002, Kumar et al, 2006, Alam et al, 2006a, 2007a, 2008a, 2010a, Alam & Kabir-ud-Din, 2008a). The uncertainty in the measured CP was ± 0.5 K.

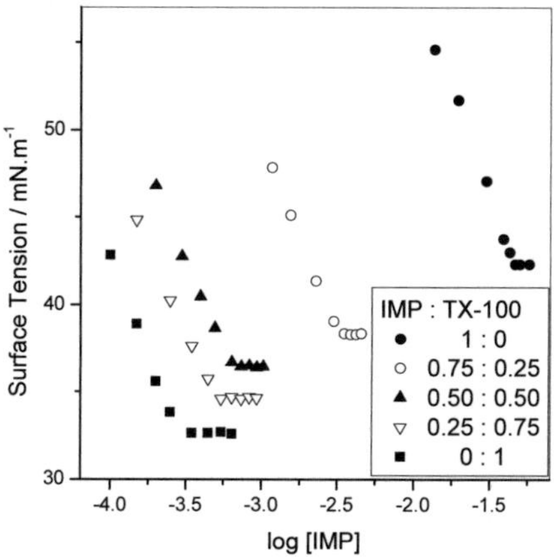

Fig. 1. Plots of surface tension *vs.* log [IMP].

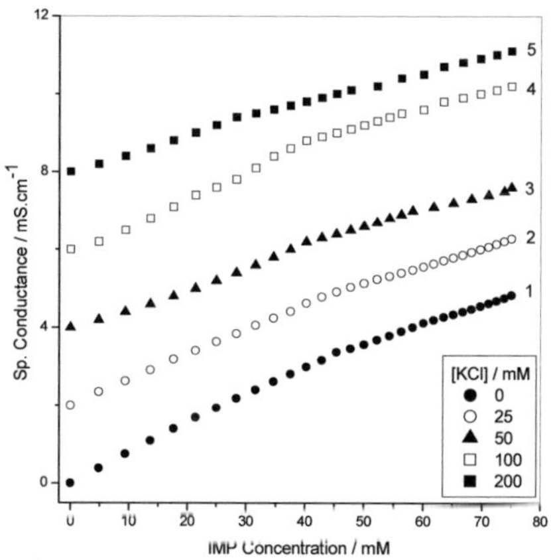

Fig. 2. Representative plots of specific conductance versus [IMP] in absence and presence of different fixed KCl concentrations at 303.15 K. The curves 2, 3, 4, and 5 have been shifted by 2, 4, 6, and 8 scale units (mS · cm^{-1}), respectively.

2.2.4 Dye solubilization measurements

Dye solubilization experiments for the aqueous drug solutions (with and without electrolyte) were performed at room temperature. The sample solutions with Sudan III dye (kept for 24 h) were filtered and then the spectra were recorded using a UV-visible Shimadzu spectrophotometer (model UV-1800).

3. Results and discussion

3.1 Micellization

3.1.1 Surface tension measurements

The value for pure drug has been found to be in good agreement with the literature value (Attwood & Florence, 1983), whereas the values decrease in the presence of additive (TX-100

– see Figure 3). The values of the surface pressure at the cmc (Π_{cmc}) were obtained by using the equation

$$\Pi_{cmc} = \gamma_0 - \gamma_{cmc} \qquad (1)$$

where γ_0, and γ_{cmc} (see Figure 1) are the surface tension of the solvent and the surface tension of the mixture at the cmc, respectively. With increasing the additives concentration, the values of Π_{cmc} increase, indicating that the efficiency increases (Table 1).

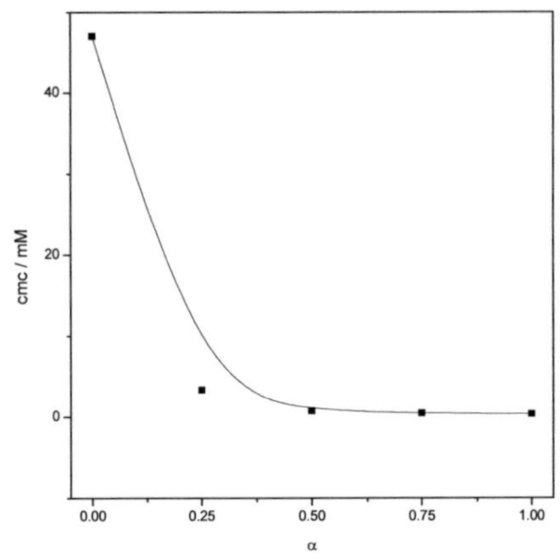

Fig. 3. Effect of additive (TX-100) on the cmc of amphiphilic drug IMP at 300.15 K.

It is well known that the air/solution interface of an amphiphile solution is well populated (Clint, 1992) by the adsorbed molecules. Accordingly, it has been shown that the concentration of the surfactant is always greater at the surface due to adsorption over and above the concentration of surfactant in the bulk. For calculation of Gibbs free energy changes, required different surface properties (e.g., the surface excess concentration, Γ_{max}, minimum area per surfactant molecule at the air/water interface, A_{min} etc.). The surface excess concentration is an effective measure of the Gibbs adsorption at liquid/air interface, which was calculated by applying equation (Chattoraj & Birdi, 1984)

$$\Gamma_{max} = -\frac{1}{2.303nRT}(d\gamma / d\log c)_T \qquad (2)$$

where γ, R, T and c are surface tension, gas costant, absolute temperature and concentration, respectively. The variable n is introduced to allow for the simultaneous adsorption of cations and anions. The expression used in the calculation of n was that proposed by Matejevic and Pethica (Matijevic & Pethica, 1958). $n = 1 + m/(m+m_s)$, where m_s is the concentration of the added electrolyte. Thus, n has a value of 2 in water and approaches 1 in the presence of excess inert electrolyte. The slope of the tangent at the given concentration of the γ *vs* log c plot was used to calculate Γ_{max}, and A_{min} was evaluated using the relation (Anand et al, 1991)

$$A_{min} = 10^{16} / N_A \Gamma_{max} \ (\text{Å}^2) \qquad (3)$$

where N_A is Avogadro number.

The data show the expected area decrease with increasing additive concentration. This is due to progressive charge shielding and closer packing of the drug ions in the surface. The low values of A_{min} suggest that the orientation of the surfactant molecule at the interface is almost perpendicular to the interface (Anand et al, 1991). The values of A_{min} for the drug are similar to those reported for other antidepressants (Taboada et al, 2001) and phenothiazines (Zografi & Zarenda, 1966).

X_{IPM}	cmc / mM	$10^{10}\bullet\Gamma_{max}$ / mol•m^{-2}	A_{min} / Å^2	Π_{cmc} / mN·m^{-1}
1	47.78	1.95	87.34	29.7
0.75	3.47	1.99	85.81	33.7
0.5	0.76	2.06	82.69	35.6
0.25	0.48	2.12	80.52	37.4
0	0.31	2.23	76.31	39.4

Table 1. Effect of additive concentrations on the cmc (determined by surface tension measurements), Γ_{max}, A_{min} and Π_{cmc} values of amphiphilic drug IMP in aqueous solutions at ~300 K.

Sugihara *et al* (Sugihara et al, 2003, 2004) have proposed a thermodynamic quantity for the evaluation of synergism in mixing, i.e., the free energy of the given air/water interface $G_{\rm min}^{(s)}$ which is defined as follows:

$$G_{\rm min}^{(s)} = A_{\rm min} \cdot \Pi_{cmc} \cdot N_{\rm A} \quad (4)$$

$G_{\rm min}^{(s)}$ regard as the work needed to make an interface per mole or the free energy change accompanied by the transition from the bulk phase to the surface phase of the solution components. In other words, the lower the values of $G_{\rm min}^{(s)}$, the more thermodynamically stable surface is found. The $G_{\rm min}^{(s)}$ values are decreased with increasing the additive concentration/mole fraction (Figure 4).

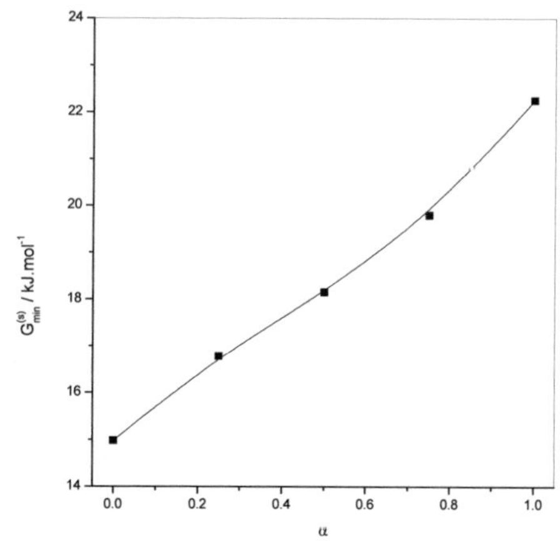

Fig. 4. Variation of Gibbs free energy at the air/water interface, $G_{\rm min}^{(s)}$ of the amphiphilic drug IMP at different concentration (mole fraction) of TX-100.

To quantify the effect of additives in the mixture on the micellization process, the standard Gibbs free energy change of micellization, $\Delta_{mic}G^0$, and the standard Gibbs energy of adsorption, $\Delta_{ads}G^0$, were calculated by using equations (5) and (6),

$$\Delta_{mic}G^0 = RT \ln \mathrm{cmc_m} \quad (5)$$

($\mathrm{cmc_m}$ is the cmc of the mixture of the two components at a given mole fraction)

$$\Delta_{ads}G^0 = \Delta_{mic}G^0 - \Pi_{\mathrm{cmc}} / \Gamma_{\mathrm{max}} \quad (6)$$

Figures 5 illustrates that $\Delta_{mic}G^0$ and $\Delta_{ads}G^0$ decrease with increasing the additive concentrations, respectively. The standard state for the adsorbed surfactant is a hypothetical monolayer at its minimum surface area per molecule, but at zero surface pressure. The last term in equation (6) expresses work involved in transferring the surfactant molecule from a monolayer at a zero surface pressure to the micelle. In all cases (in abcence and presence of additive), $\Delta_{mic}G^0$ values are negative and decreases with increasing additive concentration/mole fraction. This indicates that the micellization takes place more spontaneously in presence of additive (TX-100) (Figure 5). All the $\Delta_{ads}G^0$ values are negative, which implies that the adsorption of the surfactants at the air/mixture interface takes place spontaneously (see Figure 5).

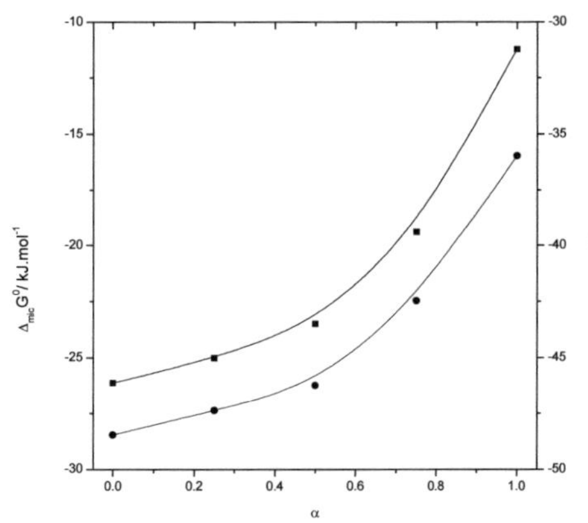

Fig. 5. Variation of the standard Gibbs energy change of micellization, $\Delta_{mic}G^0$ and the standard Gibbs free energy change of adsorption, $\Delta_{ads}G^0$ of the amphiphilic drug at different concentration/mole fraction of TX-100.

It has been reported that surfactants form mixed micelles with the drugs (Rodriguez et al, 2004, Alam et al, 2007). Mixed micelles are known to possess quite different physicochemical properties from those of pure micelles of the individual components. The micellar aggregation number and the association of counterions with micelles change dramatically with composition in mixed micelles. The degree of counterion association of an ionic micelle is about 0.7 for monovalent counterions. However, when an ionic surfactant is mixed with a non-ionic surfactant, the degree of the association falls to zero as the mole fraction of the non-ionic surfactant in the micelle increases (Meyer & Sepulveda, 1984, Jansson & Rymden,

1987). Most cmc's of binary mixtures fall between the cmc's of the two components but some are above (Sugihara et al, 1988) or below (Nguyen et al, 1986) this range. Our results for the cmc of drug in presence of TX-100 show the same behavior (Table 1). Addition of TX-100 assists in micelle formation of drug. TX-100 (by penetrating into the micelles) lowers the repulsive forces between the polar head groups of the drug (IMP).

Rodriguez et al (Rodriguez et al, 2004) who studied the effect of dodecyltrimethylammonium bromide concentration on the cmc of amitrityline hydrochloride in aqueous solution by conductivity and static fluorescence measurements explained their results on the basis of mixed micelle formation. Theoretical calculations predicted an apparent ideal but non-synergistic behavior of the mixed micelles. Our results do indicate mixed micelle formation (Table 1).

The nature and strength of the interactions between the two components (amphiphilic drug IMP and surfactant TX-100) can be determined by calculating the values of their β parameters (Rubingh, 1979).

The intermicellar interaction coefficient in the mixed micelles is calculated from:

$$\frac{[(x_1^{\mathrm{m}})^2 \cdot \ln(\mathrm{cmc} \cdot \alpha_1 / \mathrm{cmc}_1 \cdot x_1^{\mathrm{m}})]}{[(1-x_1^{\mathrm{m}})^2 \cdot \ln\{(\mathrm{cmc} \cdot (1-\alpha_1) / \mathrm{cmc}_2 \cdot (1-x_1^{\mathrm{m}})]} = 1 \tag{7}$$

and

$$\beta^{m} = \ln(cmc \cdot \alpha_1 \cdot x_1^{m}) / (1-x_1^{m})^2 \tag{8}$$

where x_1^m is the mole fraction of component 1 in the micelles and cmc_1 ; cmc_2 and cmc are the cmc's for component 1, component 2 and their mixture at mole fraction of component1, a_1, in the solution.

Equation (7) was solved iteratively for x_1^m, which was then substituted into equations (8) to calculate β^m values.

The activity coefficients f_1^m and f_2^m are related to β^m as

$$f_1^{m} = \exp\{\beta^{m} \cdot (1-x_1^{m})^2\} \tag{9}$$

$$f_2^{m} = \exp\{\beta^{m} \cdot (x_1^{m})^2\} \tag{10}$$

The evaluated parameters (x_1^m, f_1^m, f_2^m and β^m) are given in Table 2.

a_1	x_1^m	β^m	f_1^m	f_2^m
0.25	0.324	-3.14	0.043	0.486
0.50	0.447	-3.42	0.033	0.107
0.75	0.568	-3.5	0.03	0.002

Table 2. Micellar composition (x_1^m), interaction parameter (β^m), and activity coefficients (f_1^m, f_2^m) of binary mixtures of drug IMP and TX-00 at different mole fractions of IMP (a_1).

The composition of the adsorbed mixed monolayer of binary component systems in equilibrium with the singly dispersed components can be evaluated using Rosen's equations (Li et al, 2001, Zhou & Rosen, 2003). From analogy, using the derivation of Rubingh's equations for mixed micelles, the mole fraction of component 1, x_1^σ, in the mixed monolayer is related to α_1 as

$$\frac{[(x_1^\sigma)^2 \cdot \ln(\text{cmc} \cdot \alpha_1 / \text{cmc}_1 \cdot x_1^\sigma)]}{[(1-x_1^\sigma)^2 \cdot \ln\{(\text{cmc} \cdot (1-\alpha_1)/\text{cmc}_2 \cdot (1-x_1^\sigma)]} = 1 \tag{11}$$

and

$$\beta^\sigma = \ln(cmc \cdot \alpha_1 \cdot x_1^\sigma) / (1-x_1^\sigma)^2 \tag{12}$$

where cmc_1, cmc_2 and cmc are the molar concentrations of components 1, 2 and their mixture, at α_1, required to produce a given surface tension reduction (corresponds to $\gamma = 45$ mN•m^{-1}, determined from the plots of γ *vs* log [drug]), and β^σ is the interaction parameter for mixed monolayer formation at the aqueous solution/air interface.

Equation (11) was solved iteratively for x_1^σ, which was then substituted into equations (12) to calculate β values.

The activity coefficients f_1^σ and f_2^σ are related to β^σ as

$$f_1^\sigma = \exp\{\beta^\sigma \cdot (1-x_1^\sigma)^2\} \tag{13}$$

$$f_2^\sigma = \exp\{\beta^\sigma \cdot (x_1^\sigma)^2\} \tag{14}$$

The evaluated parameters (x_1^σ, f_1^σ, f_2^σ and β^σ) are given in Table 3.

α_1	x_1^σ	β^σ	f_1^σ	f_2^σ
0.25	0.394	-3.33	0.036	0.244
0.50	0.522	-3.58	0.028	0.014
0.75	0.633	-3.61	0.027	2.19E-05

Table 3. Monomer composition (x_1^σ), interaction parameter (β^σ), and activity coefficients (f_1^σ, f_2^σ) of binary mixtures of drug IMP and TX-00 at different mole fractions of IMP (α_1).

3.1.1.1 Significance of β

β indicates not only the degree of interaction between the two components but also accounts for the deviation from ideality. β assumes a value of zero for ideal mixing of two components. Positive β values means repulsion among mixed species. A negative β value implies an attractive interaction; the more negative its value, the greater the interaction. The β^m values are negative at all mole fractions of the mixed system (Tables 2 and 3), suggest that the interaction between the two components is more attractive in the mixed micelle than the self-interaction of the two components before mixing. As the mole fraction of additive (TX-100) increases, β^m values become more negative. This indicates an increase in the attractive interaction with the increase in additive concentration is also evident from the cmc values, which decrease with increasing additive concentration.

β^σ also follows similar trend (Tables 2 and 3). The mixtures of drugs/surfactants show stronger attractive interaction at the air/water interface. These interactions are stronger than in mixed micelles as evidenced by the fact that β^σ are more negative than β^m values. This is due to the steric factor, which is more important in micelle formation than in monolayer formation at a planar interface. Increased bulkiness in the hydrophobic group causes greater difficulty for incorporation into the curved mixed micelle compared to that of accommodating at the planar interface (Rosen et al, 1994).

The excess free energy change of micellization, ΔG_{ex}, calculated by the equation (15)

$$\Delta G_{ex} = [x_1^m \cdot \ln f_1 + (1 - x_1^m) \cdot \ln f_2]RT \qquad (15)$$

and shown in Figure 6. The values of ΔG_{ex} are negatives for all mole fraction/concentration of additives and the magnitude increases (ΔG_{ex} become more negative) with increasing the additives mole fractions/concentrations, indicating stability of the micelles (Figure 6).

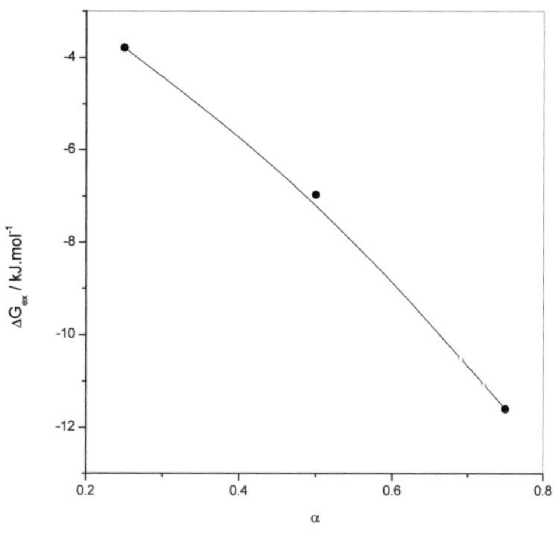

Fig. 6. Variation of the excess free energy change of micellization, ΔG_{ex} of the amphiphilic drug IMP at different concentration/mole fraction of TX-100.

3.1.2 Conductivity measurements

The cmc of IMP in absence and presence of fixed concentrations of KCl (25, 50, 100 and 200 mM) were determined by conductivivity method at different temperatures (293.15, 303.15, 313.15, and 323.15 K). Figure 2 shows the representative plots of specific conductivity vs. [IMP]. The cmc values of IMP are measured in absence as well as presence of a fixed concentration of KCl at different temperatures and listed in Table 4. The cmc values of IMP decrease with increasing the KCl concentration (see Figure 7), whereas the effect of temperature shows an opposite trend for all systems (i.e., increase with increasing temperature) (Figure 8).

The value of the cmc is dependent upon a variety of parameters including the nature of the hydrophilic and hydrophobic groups, additives present in the solution, and external influences such as temperature. The micellization takes place where the energy released as a result of association of hydrophobic part of the monomer is sufficient to overcome the electrostatic repulsion between the ionic head groups and decrease in entropy accompanying the aggregation. The cmc can also be influenced by the addition of a strong electrolyte into the solution. This serves to increase the degree of counterion binding, which has the effect of reducing head group repulsion between the ionic head groups, and thus decrease the cmc. This effect has been empirically quantified according to (Corrin & Harkins, 1947)

$$\log \mathrm{cmc} = -a \log C_t + b \tag{16}$$

where a and b are constants for a specific ionic head group ant C_t denotes the total conunterion concentration.

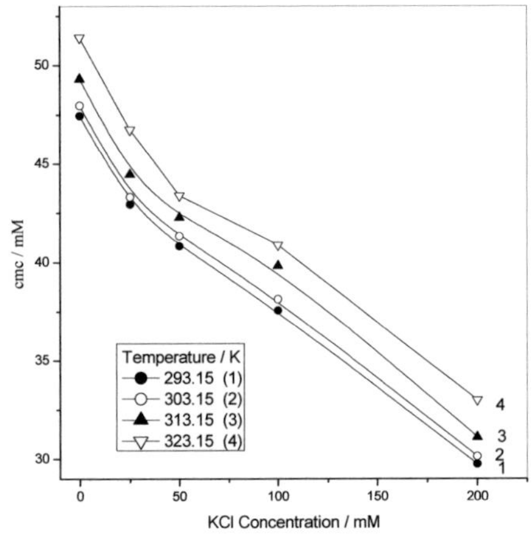

Fig. 7. Effect of KCl concentrations on the cmc of IMP solutions.

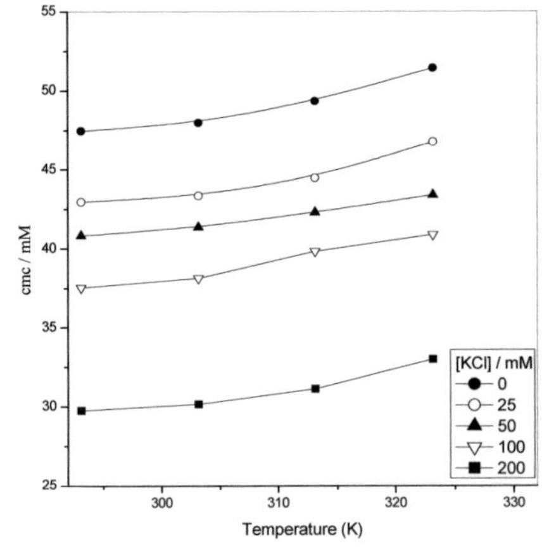

Fig. 8. Effect of temperature on the cmc of IMP solutions.

The degree of dissociation, x of the micelles was determined from the specific conductance vs. concentration of surfactants plot. Actually, x is the ratio of the post micellar slope to the premicellar slope of these plots. The counter ion association, y of the micelles is equal to $(1 - x)$. The results of cmc and y values obtained for IMP micelles in absence and presence of KCl at different temperatures are given in Table 4. It is found that the cmc of IMP in aqueous solution increased with increase in temperature, whereas the cmc of IMP decreased in the presence of additive (KCl) at all temperatures mentioned above (see Table 4). The increase in cmc and decrease in y values for IMP micelles in aqueous solution suggest that the micelle formation of IMP is hindered with the increase in temperature. However, the micelle formation of IMP is more facilitated in the presence of KCl even at higher temperatures showing lower cmc and higher y values (see Table 4).

[KCl] mM	cmc mM	y	ΔG_m^0 (kJ·mol⁻¹)	ΔH_m^0 (kJ·mol⁻¹)	ΔS_m^0 (kJ·K·mol⁻¹)
293.15 K					
0	47.45	0.3126	-29.07	-1.31	0.095
25	42.94	0.3231	-29.30	-1.05	0.096
50	40.82	0.3277	-29.42	-1.51	0.095
100	37.55	0.3682	-29.04	-1.76	0.093
200	29.74	0.3246	-30.77	-1.60	0.100
303.15 K					
0	47.97	0.3278	-29.74	-3.54	0.086
25	43.32	0.3169	-30.37	-3.34	0.089
50	41.34	0.3284	-30.36	-2.87	0.091
100	38.12	0.3377	-30.53	-5.54	0.082
200	30.14	0.3341	-31.58	-4.03	0.091
313.15 K					
0	49.32	0.3618	-29.98	-5.56	0.078
25	44.46	0.3571	-30.51	-6.72	0.076
50	42.28	0.3462	-30.93	-3.46	0.088
100	39.82	0.3520	-31.08	-3.53	0.088
200	31.11	0.3722	-31.74	-7.74	0.077
323.15 K					
0	51.42	0.4251	-29.57	-5.70	0.074
25	46.75	0.4343	-29.79	-6.82	0.071
50	43.38	0.4355	-30.09	-3.49	0.082
100	40.88	0.4268	-30.50	-3.59	0.083
200	32.98	0.4182	-31.58	-8.01	0.073

Table 4. The cmc and Various Thermodynamic Parameters for IMP Solutions in Absence and Presence of Different Fixed KCl Concentrations at Different Temperatures; Evaluated on the Basis of Conductivity Measurements.

3.1.2.1 Thermodynamics

In the van't Hoff method, the cmc of a surfactant is measured at different temperatures and the energetic parameters can be evaluated by the mass-action and pseudo-phase models (Attwood & Florence, 1983, Moroi, 1992, Moulik et al, 1996, Chaterjee et al, 2001, 2002, Dan et al, 2008, 2009). For calculating thermodynamic parameters, we have used the following equations:

$$\Delta G_m^0 = (2-\alpha)RT\ln \chi_{cmc} \tag{17}$$

$$\Delta H_m^0 = -(2-\alpha)RT^2\left(\frac{\partial \ln \chi_{cmc}}{\partial T}\right)_P \tag{18}$$

And

$$\Delta S_m^0 = \frac{\Delta H_m^0 - \Delta G_m^0}{T} \tag{19}$$

where ΔG_m^0, ΔH_m^0 and ΔS_m^0 are the standard Gibbs free energy, enthalpy and entropy of micellization, expressed per mole of monomer unit, respectively. The y, R, T and χ_{cmc} are the counterion association, universal gas constant, temperature in absolute scale and cmc in mole fraction unit, respectively. In the present case, all the ΔG_m^0 values are negative, which increase with increasing the electrolyte concentration (Table 4); this implies that the drug-electrolyte solutions are more stable. The values of ΔH_m^0 and ΔS_m^0 also agree with the low randomness and more stability (Table 4).

3.2 Clouding phenomena

3.2.1 Effect of KCl on the cloud point

The CP of the IMP solutions has been found highly sensitive to the solution pH (see Figure 9). The results show that the CP decreases as the value of pH increases (whether or not an electrolyte is present). In the pH range employed, this decrease in the CP is due to changes in the micellar surface charge. The ionization constant, pK_a, of IMP in free molecular state is 9.3 (Attwood & Florence, 1983, Katzung, 2004). The tricyclic part of IMP molecule (Scheme 1) is hydrophobic and the *t*-amine portion is hydrophilic. The protonation is highly dependent upon the solution pH. At low pH, the *t*-amine becomes protonated (i.e., cationic) and at high pH, the *t*-amine becomes deprotonated (i.e., neutral). The number of un-ionized (deprotonated) IMP molecules in micelles increases with the increase in solution pH. This, in turn, reduces both intra- as well as inter-micellar repulsions, leading to an increase in micellar aggregation and a decrease in CP (Schreier et al, 2000, Kim & Shah, 2002, Wajnberg et al, 1988, Mandal et al, 2010).

Figure 10 illustrates the variation of CP of 100 mM IMP solutions with KCl addition at different fixed pHs, prepared in 10 mM SP buffer. Here, the pH was varied from 6.5 to 6.8. It is seen that, as before (see Figure 9), CP decreases with increasing pH at all KCl concentrations (due to decrease in repulsions, as discussed above for Figure 3). The behavior of CP increases with increasing KCl concentration is found to follow a similar trend at all pH values. As discussed above, both charged and uncharged fractions of IMP molecules would be available for aggregate (so-called IMP micelle) formation. Thus, each micelle would bear a cationic charge. Increasing the amount of KCl would, therefore, cause the micellar size to increase progressively with the concomitant increase in CP (Kim & Shah, 2002).

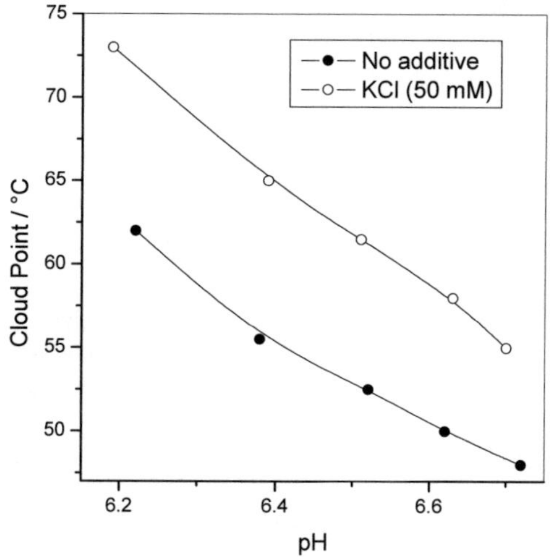

Fig. 9. Effect of pH on the CP of 100 mM IMP solution, prepared in 10 mM sodium phosphate buffer, containing no or a fixed KCl concentration (50 mM).

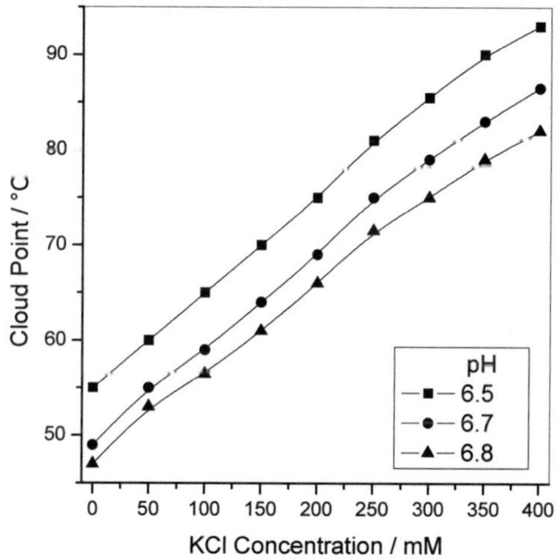

Fig. 10. Effect of KCl concentration on the CP of 100 mM IMP solution, prepared in 10 mM sodium phosphate buffer at different pHs.

Figure 11 displays the effect of KCl addition on the CP of IMP solutions of different fixed concentrations of the drug (100, 125 and 150 mM). At a constant KCl concentration, increase in drug concentration increases both the number and charge of micelles. This increases both inter- and intra-micellar repulsions, causing increase in CP.

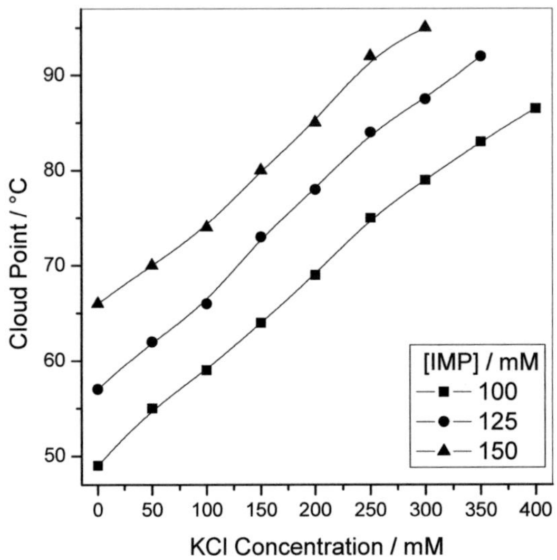

Fig. 11. Effect of KCl concentration on the CP of different fixed concentrations of IMP solution, prepared in 10 mM sodium phosphate buffer (pH = 6.7).

3.2.2 Thermodynamics at CP

As the clouding components above CP release their solvated water and separate out from the solution, the CP of an amphiphile can be considered as the limit of its solubility. Hence, the standard Gibbs energy of solubilization (ΔG_s^0) of the drug micelles can be evaluated from the relation

$$\Delta G_s^0 = -RT \ln \chi_s \tag{20}$$

where χ_s is the mole fraction concentration of additive at CP, R is gas constant and T is the clouding temperature in Kelvin scale.

The standard enthalpy and entropy of clouding, ΔH_s^0 and $T\Delta S_s^0$, respectively, can be calculated by

$$\Delta H_s^0 = \frac{\partial(\Delta G_s^0 / T)}{\partial(1/T)} \tag{21}$$

$$T\Delta S_s^0 = \Delta H_s^0 - \Delta G_s^0 \tag{22}$$

The energetic parameters were calculated using eqs. (20) to (22). The thermodynamic data of clouding for the drug IMP in the presence of KCl are given in Table 5. For IMP with and without KCl, the thermodynamic parameters, ΔG_s^0, ΔH_s^0 and $T\Delta S_s^0$ are found to be positive.

$\chi_{PMT} \cdot 10^3$	CP K	ΔG_s^0 kJ·mol^{-1}	ΔH_s^0 kJ·mol^{-1}	$T\Delta S_s^0$ kJ·K^{-1}·mol^{-1}
x = 0				
1.80	322.15	16.93	21.58	4.65
2.25	330.15	16.74		4.84
2.69	339.15	16.68		4.9
x = 50				
1.80	328.15	17.25	23.49	6.24
2.24	335.15	16.99		6.5
2.69	343.15	16.88		6.61
x = 100				
1.80	332.15	17.46	24.82	7.36
2.24	339.15	17.2		7.62
2.69	347.15	17.08		7.74
x = 150				
1.79	337.15	17.73	25.67	7.94
2.24	346.15	17.56		8.11
2.69	353.15	17.38		8.29
x = 200				
1.79	342.15	17.99	25.92	7.93
2.24	351.15	17.81		8.11
2.69	358.15	17.63		8.29
x = 250				
1.79	348.15	18.31	26.36	8.05
2.24	357.15	18.12		8.24
2.69	365.15	17.97		8.39
x = 300				
1.79	352.15	18.52	27.26	8.74
2.23	360.65	18.3		8.96
2.68	368.15	18.13		9.13
x = 350				
1.79	359.65	18.92	28.72	9.8
2.23	365.15	18.53		10.2

Table 5. Cloud Point (CP) and Energetic Parameters for Clouding of different fixed concentration (100, 125 and 150 mM) of IMP Prepared in 10 mM Sodium Phosphate Buffer Solutions (pH = 6.7) in Presence of x mM KCl.

3.3 Dye solubilization measurements

An important property of micelles that has particular significance in pharmacy is their ability to increase the solubility of sparingly soluble substances (Mitra et al, 2000, Kelarakis et al, 2004, Mata et al, 2004, 2005). A number of approaches have been taken to measure the solubilizing behavior of amphiphiles in which the solubilization of a water insoluble dye in the surfactant micelles was studied. The plots illustrated in Figure 12 clearly demonstrate that, in the presence of additives, micelle size increases due to the fact that more dye can solubilize in the aggregates.

The absorbance variations with KCl concentration in the absence as well as presence of different fixed concentrations of IMP are illustrated in Figure 12. The amount of solubilized dye depends on the state of aggregation. We see that the solubilizing power of the drugs markedly increases in the presence of additives. Figure 6 shows the visible spectra of Sudan III solubilized in 50 mM IMP in water containing different fixed amounts of the additive (KCl) concentrations. One can see that the absorbance increases on addition of KCl, increasing the concentration of KCl increases the absorbance. Addition of KCl raises the aggregation number of ionic micelles due to electrostatic effects (Evans & Wennerstrom, 1999). The absorbance increase with increasing concentration of KCl suggests that the micellar growth is substantial with KCl addition.

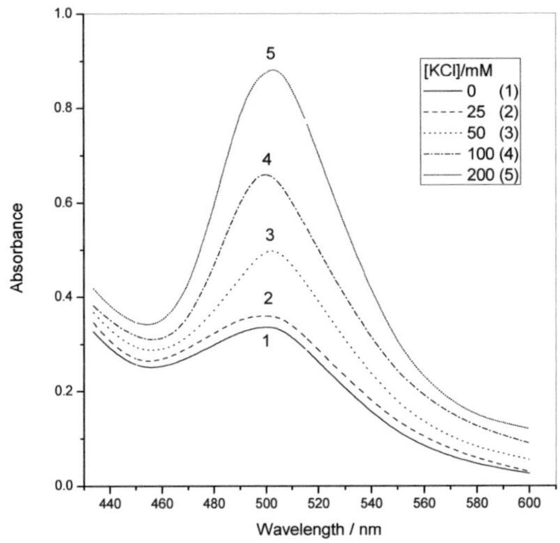

Fig. 12. Visible spectra of Sudan III solubilized in the PMT (50 mM) containing no or a fixed concentration of KCl.

4. Conclusion

We have studied the thermodynamics of a tricyclic antidepressant drug imipramine hydrochloride (IMP). The mixed micelles of IMP and non-ionic surfactant polyethylene glycol *t*-octylphenyl ether (TX-100) has been investigated using surface tension measurements and evaluated Gibbs energies (at air/water interface ($G^{(s)}_{min}$), the standard Gibbs energy change of micellization ($\Delta_{mic}G^0$), the standard Gibbs energy change of adsorption ($\Delta_{ads}G^0$), the excess free energy change of micellization (ΔG_{ex})). The micellization at different fixed temperatures (*viz.*, 293.15, 303.15, 313.15 and 323.15 K), and clouding behavior of IMP in absence and presence of KCl. The *critical micelle concentration* (*cmc*) of IMP is measured by conductivity method and the values decrease with increasing the KCl concentration, whereas with increasing temperature the *cmc* values increase. The thermodynamic parameters *viz.*, standard Gibbs energy (ΔG^0_m), standard enthalpy (ΔH^0_m), and standard entropy (ΔS^0_m) of micellization of IMP are evaluated, which indicate more stability of the IMP solution in presence of KCl. IMP undergoes concentration-, pH-, and

temperature-dependent phase separation, also known as "*clouding*", which is a well known phenomenon with non-ionic surfactants. The temperature at which phase separation occurs is called '*cloud point*' (CP). Studies on the CP of IMP have been made to see the effect of KCl. Strong dependence on the concentration of the KCl has been observed. A pH increase in the presence as well as in the absence of electrolyte decreased the CP. Drug molecules become neutral at high pH and therefore, head group repulsion decreases which lead to CP decrease. Effect of KCl at different fixed drug concentrations showed that at all electrolyte concentrations the CP value was higher for higher drug concentrations. However, variation of pH produced opposite effect: CP at all KCl concentrations decreased with increasing pH. The results are interpreted in terms of micellar growth. Furthermore, the thermodynamic parameters are evaluated at CP.
The surface properties, Gibbs energies of an amphiphilic drug IMP in water are evaluated in absence and presence of additive (TX-100), and the micellization and clouding behavior of IMP in absence and presence of KCl have studied and the results obtained are as:

i. With TX-100, increase in Γ_{max} and decrease in cmc/A_{min} are due to the formation of mixed micelles with the drug.
ii. The drug/surfactant systems show an increase in synergism with the increase in surfactant concentration.
iii. Rosen's approach reveals increased synergism in the mixed monolayers in comparision to in the mixed micelles.
iv. In all cases (in presence and absence of additive) the G^{s}_{min} values decrease with increasing the additives concentrations, indicating thermodynamically stable surface.
v. The $\Delta_{mic}G^0$ values are negative and decreases with increasing the additive concentration indicate that the micelle formation takes place spontaneously.
vi. The negative $\Delta_{ads}G^0$ values indicate that the adsorption of the surfactant at the air/solution interface takes place spontaneously.
vii. The values ΔG_{ex} are negative for all mole fractions of additives indicating the stability of the micelles.
viii. Knowledge of self-aggregation and clouding behavior of amphiphilic drugs and effect of additives on clouding will allow the better designing of effective therapeutic agents.
ix. The critical micelle concentration (cmc) of IMP decreases with increasing KCl concentration, whereas with increasing temperature the cmc values increases.
x. The thermodynamic parameters are evaluated, which indicate more stability of the IMP solution in presence of KCl.
xi. The IMP also shows phase-separation. The cloud point (CP) of IMP decreases with increase in pH of the drug molecules because of deprotonation.
xii. The CP values increase with increasing KCl and IMP concentrations leading to micellar growth.

5. Acknowledgment

Md. Sayem Alam is grateful to Prof. Kabir-ud-Din, Aligarh Muslim University, Aligarh and Dr. Sanjeev Kumar, M. S. University, for their constant encouragement. The support of the University of Saskatchewan, Canada to Abhishek Mandal in the form of research grand during his Ph. D. Program is gratefully acknowledged.

6. References

Alam, Md. S.; Kabir-ud-Din; Mandal, A. B. Evaluation of Thermodynamic Parameters of Amphiphilic Tricyclic Antidepressant Drug Imipramine Hydrochloride—Additive Systems at the Cloud Point. *Colloids Surf. B*, 2010, *76*, 577-584.

Alam, Md. S.; Kabir-ud-Din; Mandal, A. B. Thermodynamics at the Cloud Point of Phenothiazine Drug Chlorpromazine Hydrochloride—Additive Systems. *J. Chem. Eng. Data*, 2010, *55*, 1693-1699.

Alam, Md. S.; Kabir-ud-Din; Mandal, A. B. Amphiphilic Drug Promethazine Hydrochloride—Additive Systems: Evaluation of Thermodynamic Parameters at Cloud Point. *J. Chem. Eng. Data*, 2010, *55*, 1893-1896.

Alam, Md. S.; Kabir-ud-Din; Mandal, A. B. Thermodynamics of Some Amphiphilic Drugs in Presence of Additives. *J. Chem. Eng. Data*, 2010, *55*, 2630–2635.

Alam, Md. S.; Kabir-ud-Din; Mandal, A. B. *J. Dispersion Sci. Technol.*, 2010 (in press).

Alam, Md. S.; Ghosh, G.; Kabir-ud-Din Light Scattering Studies of Amphiphilic Drugs Promethazine Hydrochloride and Imipramine Hydrochloride in Aqueous Electrolyte Solutions. *J. Phys. Chem. B*, 2008, *112*, 12962-12967.

Alam, Md. S.; Mandal, A.; Mandal, A. B. Effect of KCl on the Micellization and Clouding Phenomenon of Amphiphilic Phenothiazine Drug Promethazine Hydrochloride: Some Thermodynamic Properties. *J. Chem. Eng. Data* 2011, *56*, 1540–1546.

Alam, Md. S.; Kumar, S.; Naqvi, A. Z.; Kabir-ud-Din Study of the Cloud Point of an Amphiphilic Antidepressant Drug: Influence of Surfactants, Polymers and Non-electrolytes. *Colloids Surf. A*, 2006, *287*, 197–202.

Alam, Md. S.; Kumar,S.; Naqvi, A. Z.; Kabir-ud-Din Effect of Electrolytes on the Cloud Point of Chlorpromazine Hydrochloride Solutions. *Colloids Surf. B*, 2006, *53*, 60–63.

Alam, Md. S.; Naqvi, A. Z.; Kabir-ud-Din Influence of Electrolytes/Non-electrolytes on the Cloud Point Phenomenon of the Aqueous Promethazine Hydrochloride Drug Solution. *J. Colloid Interface Sci.*, 2007, *306*, 161-165.

Alam, Md. S.; Naqvi, A. Z.; Kabir-ud-Din Tuning of Cloud Point of Promethazine Hydrochloride with Surfactants and Polymers. *J. Surf. Detergents*, 2007, *10*, 35-40.

Alam, Md. S.; Naqvi, A. Z.; Kabir-ud-Din Role of Surfactants in Clouding Phenomenon of Imipramine Hydrochloride. *Colloids Surf. B*, 2007, *57*, 204-208.

Alam, Md. S., Naqvi, A. Z.; Kabir-ud-Din Influence of Organic Additives on the Clouding Phenomena of Promethazine Hydrochloride Solutions. *Colloid Polym. Sci.*, 2007, *285*, 1573–1579.

Alam, Md. S.; Naqvi, A. Z.; Kabir-ud-Din Study of the Cloud Point of the Phenothiazine Drug Chlorpromazine Hydrochloride: Effect of Surfactants and Polymers. *J. Dispersion Sci. Technol.*, 2008, *29*, 274–279.

Alam, Md. S.; Naqvi, A. Z.; Kabir-ud-Din Cloud Point Phenomenon in Amphiphilic Drug Promethazine Hydrochloride + Electrolyte Systems. *J. Dispersion Sci. Technol.*, 2008, *29*, 783-786.

Alam, Md. S.; Kabir-ud-Din Cloud Point and Dye Solubilization Studies on the Micellar Growth of Amphiphilic Drug Chlorpromazine Hydrochloride: Influence of Electrolytes. *Acta Phys. –Chim. Sin.*, 2008, *24*, 411–415.

Alam, Md. S.; Naqvi, A. Z.; Kabir-ud-Din Influence of Additives on the Clouding Phenomenon of Chlorpromazine Hydrochloride Solutions. *Colloids Surf. B*, 2008, *63*, 122–128.

Alam, Md. S.; Kabir-ud-Din Investigation of the Role of Electrolytes and Non-electrolytes on the Cloud Point and Dye Solubilization in Antidepressant Drug Imipramine Hydrochloride Solutions. *Colloids Surf. B*, 2008, *65*, 74-79.

Alam, Md. S.; Naqvi, A. Z.; Kabir-ud-Din Phase Separation Study of Imipramine Hydrochloride-Additive Systems. *J. Dispersion Sci. Technol.*, 2010, *31*, 449–455.

Anand, K.; Yadav, O. P.; Singh, P. P. Studies on the Surface and Thermodynamic Properties of Some Surfactants in Aqueous and Water+1,4-Dioxane Solutions *Colloids Surf.* 1991, *55*, 345-358.

Atherton, A. D.; Barry, B. W. Micellar Properties of Phenothiazine Drugs: A Laser Light Scattering Study. *J. Colloid. Interface. Sci.*, 1985, *106*, 479- 489.

Attwood, D. The Mode of Association of Amphiphilic Drugs in Aqueous Solution. *Adv. Colloid Interface Sci.*, 1995, *55*, 271-303.

Attwood, D. and Florence, A.T. *Surfactant Systems: Their Chemistry, Pharmacy and Biology*, Chapman and Hall, New York, 1983.

Attwood, D.; Mosquera, V.; Villar, V. P. Thermodynamic Properties of Amphiphilic Drugs in Aqueous Solution. *J. Chem. Soc., Faraday Trans. 1*, 1989, *85*, 3011-3017.

Barzykin, A. V.; Tachiya, M. Reaction Kinetics in Micro- Disperse Systems. *Heterog. Chem. Rev*. 1996, *3*, 105- 167.

Chaterjee, A.; Moulik, S.P.; Sanyal, S.K.; Mishra, B.K.; Puri, P.M. Thermodynamics of Micelle Formation of Ionic Surfactants: A Critical Assessment for Sodium Dodecyl Sulfate, Cetyl Pyridinium Chloride and Dioctyl Sulfosuccinate (Na Salt) by Microcalorimetric, Conductometric, and Tensiometric Measurements. *J. Phys. Chem. B*, 2001, *105*, 12823–12831.

Chatterjee, A.; Maiti, S.; Sanyal, S. K.; Moulik, S. P. Micellization and Related Behaviors of *N*-Cetyl-*N*-ethanolyl-*N*,*N*-dimethyl and *N*-Cetyl-*N*,*N*-diethanolyl-*N*-methyl Ammonium Bromide. *Langmuir*, 2002, *18*, 2998-3004.

Chattoraj, D. K.; Birdi, K. S. *Adsorption and the Gibbs Surface Excess*, Plenum: New York, 1984.

Clint, J. H. *Surfactant Aggregation*, Blackie: London, 1992.

Corrin, M. L.; Harkins, W. D. *J. Am. Chem. Soc.*, 1947, *69*, 683-688.

Dan, A.; Ghosh, S.; Moulik, S. P. The Solution Behavior of Poly(vinylpyrrolidone): Its Clouding in Salt Solution, Solvation by Water and Isopropanol, and Interaction with Sodium Dodecyl Sulfate. *J. Phys. Chem. B* 2008, *112*, 3617-3624.

Dan, A.; Ghosh, S.; Moulik, S. P. Physicochemistry of the Interaction between Inulin and Alkyltrimethylammonium Bromides in Aqueous Medium and the Formed Coacervates. *J. Phys. Chem. B*, 2009, *113*, 8505-8513.

Evans, D. F.; Wennerstrom, H. The Colloidal Domain: Where Physics, Chemistry and Biology Meet, 2nd ed., Wiley-VCH, New York, 1999.

Geetha, B.; Mandal, A. B.; Ramasami, T. Synthesis, Characterization and Micelle Formation in Aqueous Solution of Methoxy Polyethylene Glycol Macromonomer, Homopolymer and Graft Copolymer. *Macromolecules*, 1993, *26*, 4083-4088.

Geetha, B.; Gaspar, M.; Mandal, A. B. Surface Characterizations of Comblike Copolymers from Hexadecyl Acrylamide and Acrylic Acid at the Air/Water Interface. *Langmuir*, 2003, *19*, 9051-9057.

Geetha, B.; Mandal, A. B. Molecular Dynamics of Methoxy Polyoxyethylene Macromonomer Micelles in Absence and Presence of SDS Micelles Using ^{1}H NMR Spin-Lattice Relaxation Time Measurements. *Chem. Phys. Letts.*, 1997, *266*, 443-450.

Geetha, B.; Mandal, A. B. Determination of the Critical Micelle Concentration of the Methoxy Polyethylene Glycol Based Macromonomer, and Partition Coefficient of a New Electrochemical Probe Using Cyclic Voltammetric Technique. *Langmuir,* 1997, *13,* 2410-2414.

Geetha, B.; Mandal, A. B. 2-Dimensional Surface Properties of w-Methoxy Poly(ethyleneglycol) Macromonomer in Absence and Presence of Stearic Acid and Its Interaction with SDS As a Function of Different Mole Compositions at Various Temperatures at Air-Water Interface: Thermodynamic studies Using Langmuir Film Balance (Monolayer) Technique. *Langmuir,* 2000, *16,* 3957-3963.

Geetha, B.; Mandal, A. B. Self-Diffusion Studies on Methoxy Polyethyleneglycol Macromonomer Micelles By Using Cyclic Voltammetric and Fourier Transform Pulsed Gradient Spin Echo Nuclear Magnetic Resonance Techniques. *Langmuir,* 1995, *11,* 1464-1467.

Geetha, B.; Mandal, A. B. The Shape, Size, Aggregation, Hydration, Correlation Times and Thermodynamic Studies on Macromonomer Micelles in Aqueous Solutions. *J. Chem. Phys.,* 1996, *105,* 9649-9656.

Gomati, R.; Appell, J.; Bassereau, P.; Marignan, J.; Porte, G. Influence of the Nature of the Counterion and of Hexanol on the Phase Behavior of the Dilute Ternary Systems: Cetylpyridinium Bromide or Chloride-Hexanol-Brine. *J. Phys. Chem.,* 1987, *91,* 6203–6210.

Gu, T.; Galera-Gomez, P. A. The Effect of Different Alcohols and Other Polar Organic Additives on the Cloud Point of Triton X-100 in Water. *Colloids Surf. A,* 1999,*147,* 365-370.

James, J; Ramalechume, C.; Mandal, A. B. Two-Dimensional Surface Properties of PEO-PPO-PEO Triblock Copolymer Film at the Air/Water Interface in the Absence and Presence of Tyr-Phe Dipeptide, Val-Tyr-Val Tripeptide, SDS and Stearic Acid. *Colloid Surf. B,* 2011, *82,* 345-353.

James, J.; Mandal, A. B. Micelle Formation of Tyr-Phe Dipeptide and Val-Tyr-Val Tripeptide in Aqueous solution and their Influence on the Aggregation of SDS and PEO–PPO–PEO Copolymer Micelles. *Colloid Surf. B,* 2011, *84,* 172-180.

Jansson, M.; Rymden, R. Counterion Binding in Aqueous Solutions of Mixed Micellar Aggregates from Self-Diffusion Measurements. *J. Colloid Interface Sci.* 1987, *119,* 185 -193.

Jayakumar, R.; Mandal, A. B.; Manoharan, P. T. Aggregation, Hydrogen-Bonding and Thermodynamic Studies on Boc-Val-Val-Ile-OMe Tripeptide Micelles in Chloroform. *J. Chem. Soc. Faraday Trans.,* 1994, *90,* 2725-2730.

Jayakumar, R.; Mandal, A. B.; Manoharan, P. T. The Micelle Formation of Boc-Val-Val-Ile-OCH3 Tripeptide in $CHCl_3$ and Its Conformational Analysis. *J. Chem. Soc. Chem. Commun.,* 1993, 853-855.

Junquera, E.; Romero, J. C.; Aicart, E. Behavior of Tricyclic Antidepressants in Aqueous Solution: Self-Aggregation and Association with β–Cyclodextrin. *Langmuir,* 2001, *17,* 1826-1832.

Kabir-ud-Din; Rub, M. A.; Alam, Md. S. Micellization and Clouding Phenomenon of Phenothiazine Drug Promethazine Hydrochloride: Effect of NaCl and Urea Addition. *J. Dispersion Sci. Technol.,* 2010, *31,* 1182–1187.

Katzung, B. G. *Basic and Clinical Pharmacology,* 9th ed., McGraw Hill: New York, 2004.

Kelarakis, A.; Mai, S. –M.; Havredaki, V.; Brett, A.; Booth, C. Thermodynamics of Micellization of Tapered Statistical Copolymers of Ethylene Oxide and Propylene Oxide in Water. *J. Colloid Interface Sci.* 2004, *275*, 439-444.

Khan, I. A.; Mohammad, R.; Alam, Md. S.; Kabir-ud-Din Effect of Alkylamine Chain Length on the Critical Micelle Concentration of Cationic Gemini Butanediyl–α, ω-bis(dimethylcetylammonium bromide) Surfactant. *J. Dispersion Sci. Technol.*, 2009, *30*, 1486–1493.

Khan, I. A.; Mohammad, R.; Alam, Md. S.; Kabir-ud-Din Mixed micellization of cationic gemini surfactants with primary linear alkylamines. *J. Surf. Detergents,* 2010, 13, 179–188.

Khan, I. A.; Mohammad, R.; Alam, Md. S.; Kabir-ud-Din Surface Properties and Mixed Micellization of Cationic Gemini Surfactants Ethyleaneamines. *J. Chem. Eng. Data,* 2010, *55*, 370–380.

Khan, I. A.; Mohammad, R.; Alam, Md. S.; Kabir-ud-Din The Interaction of Cationic Gemini Surfactant 1, 4-Butanediyl- α, ω-bis(dimethylcetylammonium bromide) with Primary Linear Alkanols. *J. Dispersion Sci. Technol.*, 2010, *31*, 129–137.

Kim, E. J.; Shah, D. O. Cloud Point Phenomenon in Amphiphilic Drug Solutions. *Langmuir* 2002, *18*, 10105-10108.

Kim, E. J.; Shah, D. O. A Cloud Point Study on the Micellar Growth of an Amphiphilic Drug in the Presence of Alcohol and Ionic Surfactant. *J. Phys. Chem. B*, 2003, *107*, 8689–8693.

Ko, J-S.; Oh, S-W.; Kim, Y-S.; Nakashima, N.; Nagadome, S.; Sugihara, G. Adsorption and Micelle Formation of Mixed Surfactant Systems in Water. IV. Three Combinations of SDS with MEGA-8,-9 and -10. *J. Oleo Sci.* 2004, *53*, 109-126.

Kumar, S.; Sharma, D.; Kabir-ud-Din Cloud Point Phenomenon in Anionic Surfactant + Quaternary Bromide Systems and its Variation with Additives, *Langmuir*, 2000, *16*, 6821-6824.

Kumar, S.; Sharma, D.; Khan, Z. A.; Kabir-ud-Din Occurrence of Cloud Points in Sodium Dodecyl Sulfate–Tetra-*n*-Butylammonium Bromide System, *Langmuir*, 2001, *17*, 5813-5816.

Kumar, S.; Sharma, D.; Khan, Z. A.; Kabir-ud-Din Salt-induced Cloud Point in Anionic Surfactant Solutions: Role of the Headgroup and Additives. *Langmuir*, 2002, *18*, 4205-4209.

Kumar, S.; Sharma, D.; Kabir-ud-Din Temperature-[Salt] Compensation for Clouding in Ionic Micellar Systems Containing Sodium Dodecyl Sulfate and Symmetrical Quaternary Bromides. *Langmuir*, 2003, *19*, 3539-3541.

Kumar, S.; Alam, Md. S.; Parveen, N.; Kabir-ud-Din Influence of Additives on the Clouding Behavior of Amphiphilic Drug Solutions. *Colloid Polym. Sci.*, 2006, *284*, 1459–1463.

Li, F.; Rosen, M. J.; Sulthana, S. B. Surface Properties of Cationic Gemini Surfactants and Their Interaction with Alkylglucoside or -Maltoside Surfactants. *Langmuir* 2001, *17*, 1037-1042.

Mandal, A. B.; Ramesh, D. V.; Dhar, S. C. Physicochemical studies of Micelle formation on sepia cartilage collagen solutions (pepsin extracted) in acetate buffer and its interaction with ionic and nonionic micelles : Hydrodynamic and thermodynamic studies, *FEBS, Eur. J. Biochem.*, 1987, *169*, 617-628.

Mandal, A. B.; Nair, B. C. U.; Ramaswamy, D. Determination of the CMC of Various Surfactants and Partition Coefficient of an Electrochemical Probe Using Cyclic Voltammetry. *Langmuir*, 1988, *4*, 736-740.

Mandal, A. B.; Nair, B. C. U. Cyclic Voltammetric Technique for the Determination of the Critical Micelle Concentration of Surfactants, Self-diffusion Coefficient of the Micelles and Partition Coefficient of an Electrochemical Probe. *J. Phys. Chem.*, 1991, *95*, 9008-9013.

Mandal, A. B.; Jayakumar, R. A New Micelle-Forming Peptide. *J. Chem. Soc. Chem. Commun.*, 1993, 237-238.

Mandal, A. B. Self-diffusion Studies on Various Micelles Using Ferrocene as Electrochemical Probe. *Langmuir*, 1993, *9*, 1932-1933.

Mandal, A. B.; Jayakumar, R. Aggregation, Hydrogen-Bonding and Thermodynamics on Tetrapeptide Micelles. *J. Chem. Soc. Faraday Trans.*, 1994, *90*, 161-165.

Mandal, A. B.; Wang, L.; Brown, K.; Verrall, R. E. Conductance and NMR Studies of Cetyltrimethylammonium Bromide and Chloride Micelles in the Presence of Several Additives. *J. Colloid Interface Sci.*, 1993, *161*, 292-298.

Mandal, A. B.; Alam, Md. S.; Mandal, A.; Baskar, G. Prospect, Retrospect and Resurgence of Micelles. *J. Surface Sci. Technol.*, 2010, *26*, 321–352.

Mandal, A.; Krishnan, R.S.G.; Thennarasu. S.; Panigrahi, S.; Mandal, A. B. Two-Dimensional Surface Properties of an Antimicrobial Hydantoin at the Air-Water Interface: An Experimental and Theoretical Study. *Colloids Surf. B*, 2010, *79*, 136-141.

Mata, J.; Varade, D.; Bahadur, P. Aggregation Behavior of Quaternary Salt Based Cationic Surfactants. *Thermochim. Acta* 2005, *428*, 147-155.

Mata, J.; Varade, D.; Ghosh, G.; Bahadur, P. Effect of Tetrabutylammonium Bromide on the Micelles of Sodium Dodecyl Sulfate. *Colloids Surf. A* 2004, *245*, 69-73.

Matijevic, E.; Pethica, B. A. The Properties of Ionized Monolayers. Part 1.—Sodium Dodecyl Sulphate at the Air/Water Interface. *Trans. Faraday Soc.* 1958, *54*, 1382-1389.

Meyer, M.; Sepulveda, L. Counterion Association in Mixed Micelles of Cationic and Nonionic Detergents. *J. Colloid Interface Sci.* 1984, *99*, 536-542.

Misra, P. K.; Mishra, H. P.; Das, U.; Mandal, A.B. Organization of Amphiphiles Part-X: Studies on the Interaction Between Polyoxyethylated (30) Octylphenol and Cetyltrimethylammonium Bromide in Aqueous Solution. *J.Colloid Interface Sci.*, 2009, *333*,590-598.

Misra, P. K.; Panigrahi, S.; Dash, U.; Mandal, A. B. Organization of amphiphiles, Part XI: Physico-Chemical Aspects of Mixed Micellization involving normal Conventional Surfactant and a Non-Ionic Gemini Surfactant. *J.Colloid Interface Sci.*, 2010,*345*, 392-401.

Mitra, S.; Dungan, S. R. Micellar Properties of Quillaja Saponin. 2. Effect of Solubilized Cholesterol on Solution Properties. *Colloids Surf. B* 2000, *17*, 117-133.

Moroi, Y. Micelles, Theoretical and Applied Aspects, Plenum Press, New York, 1992.

Moulik, S. P.; Haque, Md. E.; Jana, P. K.; Das, A. R. Micellar Properties of Cationic Surfactants in Pure and Mixed States. *J. Phys. Chem.*, 1996, *100*, 701–708.

Nguyen, C. N..; Rathman, J. F.; Scamehorn, J. F. Thermodynamics of Mixed Micelle Formation. *J. Colloid Interface Sci.* 1986, *112*, 438-446.

Panizza, P.; Cristobal, G.; Curely, J. Phase Separation in Solution of Worm-like Micelles: A Dilute $n \to 0$ Spin-Vector Model. *J. Phys.: Condens. Matter*, 1998, *10*, 11659–11678.

Ramya, S.; Thennarasu, S.; Mandal, A. B. Self-assembling Characteristics of a Hydantoin Drug. *Chem. Phys.*, 2003, *291*, 195.

Ramya, S.; Thennarasu, S.; Mandal, A. B. Self-assembling Characteristics of a New Class of Non-ionic Gemini Surfactant *viz.* bis-Amide. *J. Phys. Chem. B.*, 2004, *108*, 8806-8816.

Rose, C.; Mandal, A. B. Interaction of SDS and Urea with Cat Fish Collagen: Hydrodynamic, Thermodynamic, Conformation and Geometric Studies. *Int. J. Biological Macromolecules*, 1996, *18*, 41-53.

Rodriguez, A.; Junquera, E.; del Burgo, P.; Aicart, E. Conductometric and Spectrofluorimetric Characterization of the Mixed Micelles Constituted by Dodecyltrimethylammonium Bromide and a Tricyclic Antidepressant Drug in Aqueous Solution. *J. Colloid Interface Sci.* 2004, *269*, 476-483.

Rosen, M. J.; Gao, T; Nakasiyi, Y; Masuyama, A. Synergism in Binary Mixtures of Surfactants. 12. Mixtures Containing Surfactants with Two Hydrophilic and Two or Three Hydrophobic Groups. *Colloids Surf. A* 1994, *88*, 1-11.

Rubingh, D. N.; in *Solution Chemistry Chemistry of Surfactants*, Ed. Mittal, K. L., Vol. 1, Plenum: New York, 1979.

Ruiz, C. C.; Sanchez, F. G. Effect of Urea on Aggregation Behavior of Triton X-100 Micellar Solutions: A Photophysical Study. *J. Colloid Interface Sci.* 1994, *165*, 110-115.

Schreier, S.; Malheiros, S. V. P.; de Paula, E. Surface Active Drugs: Self-association and Interaction with Membranes and Surfactants. Physicochemical and Biological Aspects. *Biochim. Biophys. Acta*, 2000, *1508*, 210-234.

Seeman, P. The Membrane Actions of Anesthetics and Tranquilizers. *Pharmacol. Rev.*, 1972, *24*, 583- 655.

Sugihara, G.;. Yamamoto, M.; Wada, Y.; Murata, Y.; Ikawa, Y. Mixed Micelle Formation in Aqueous Solutions of Nonyl-*N*-Methylglucamine with Sodium Perfluorooctanoate at Different Pressures. *J. Solution Chem.* 1988, *17*, 225-235.

Sugihara, G.; Miyazono, A.; Nagadome, S.; Oida, T.; Hayashi, Y.; Ko, J-S. Adsorption and Micelle Formation of Mixed Surfactant Systems in Water. II: A Combination of Cationic Gemini-type Surfactant with MEGA-10. *J. Oleo Sci.* 2003, *52*, 449-461.

Taboada, P.; Attwood, D.; Ruso, J. M., Garcia, M.; Mosquera, V. Static and Dynamic Light Scattering Study on the Association of Some Antidepressants in Aqueous Electrolyte Solutions. *Phys. Chem. Chem. Phys.* 2000, *2*, 5175-5179.

Taboada, P.; Ruso, J. M.; Garcia, M.; Mosquera, V. Surface Properties of Some Amphiphilic Antidepressant Drugs. *Colloids Surf. A* 2001, *179*, 125-128.

Taboada, P.; Ruso, J. M.; Garcia, M.; Mosquera, V. Surface Properties of Some Amphiphilic Antidepressant Drugs. *Colloids Surf. A* 2001, *179*, 125-128.

Tiwary, L. K.; Mandal, A.; Alam, Md. S.; Thennarasu, S.; Mandal, A. B. Thermodynamics Studies on Tyrosine-Hydantoin Drug–Cetyltrimethylammonium Bromide Mixed Micellar System. Colloids Surf. B, 2011, *82*, 126–133.

Wajnberg, E.; Tabak, M.; Nussenzveig, P. A.; Lopes, C. M.; Louro, S. R. W. pH-Dependent Phase Transition of Chlorpromazine Micellar Solutions in the Physiological Range. *Biochim. Biophys. Acta*, 1988, *944*, 185-190.

Whitesides, G. M.; Grzybowski, B. Self-Assembly at All Scales. *Science*, 2002, *295*, 2418- 2421.

Zografi, G.; Zarenda, I. The Surface Activity of Phenothiazine Derivatives at the Air-Solution Interface. *Biochem. Pharmacol.* 1966, *15*, 591-598.

Zhou, Q; Rosen, M. J. Molecular Interactions of Surfactants in Mixed Monolayers at the Air/Aqueous Solution Interface and in Mixed Micelles in Aqueous Media: The Regular Solution Approach. *Langmuir* 2003, *19*, 4555-4562.

12

Nonequilibrium Thermodynamics of Ising Magnets

Rıza Erdem[1] and Gül Gülpınar[2]
[1]*Department of Physics, Akdeniz University, Antalya,*
[2]*Department of Physics, Dokuz Eylül University, İzmir,*
Turkey

1. Introduction

Real magnets and Ising models have provided a rich and productive field for the interaction between theory and experiment over the past 86 years (Ising, 1925). In order to identify the real magnets with a simple microscopic Hamiltonian, one needs to understand the behaviour of individual magnetic ions in crystalline environment (Wolf, 2000). Spin-1/2 Ising model and its variants such as Blume-Capel, Blume-Emery-Griffiths and mixed spin models were regarded as theoretical simplifications, designed to model the essential aspects of cooperative systems without detailed correspondence to specific materials. The similarities and differences between theoretical Ising models and a number of real magnetic materials were widely reviewed by many authors. The early experiments were focused on identifying Ising-like materials and characterizing the parameters of the microscopic Hamiltonian. Various approximate calculations were then compared with thermodynamic mesurements. Although both the theoretical and experimental studies concerning Ising-like systems have concentrated on static properties, very little has been said about its dynamic characteristics.

Lyakhimets (Lyakhimets, 1992) has used a phenomenological description to study the magnetic dissipation in crystalline magnets with induced magnetic anisotropy. In his study, the components of the second-order tensor which describes the induced anisotropy of the magnet were taken as thermodynamic variables and the nonequilibrium linear Onsager thermodynamics was formulated for the system. Such an approach reflects all symmetry characteristics of the relaxation problem. The relaxation parameters and their angular denpendencies were formulated for spin waves and moving domain walls with the help of the dissipation function. The implications of nonequilibrium thermodynamics were also considered for magnetic insulators, including paramagnets, uniform and nonuniform ferromagnets (Saslow & Rivkin, 2008). Their work was concentrated on two topics in the damping of insulating ferromagnets, both studied with the methods of irreversible thermodynamics: (a) damping in uniform ferromagnets, where two forms of phenomenological damping were commonly employed, (b) damping in non-uniform insulating ferromagnets, which become relavent for non-monodomain nanomagnets. Using the essential idea behind nonequilibrium thermodynamics, the long time dynamics of these systems close to equilibrium was well defined by a set of linear kinetic equations for the magnetization of insulating paramagnets (and for ferromagnets). The dissipative properties of these equations were characterized by a matrix of rate coefficients in the linear relationship of fluxes to appropriate thermodynamic forces.

Investigation of the relaxation dynamics of magnetic order in Ising magnets under the effect of oscillating fields is now an active research area in which one can threat the sound propagation as well as magnetic relaxation. In most classes of magnets, a very important role is played by the order parameter relaxation time and it is crucial parameter determining the sound dynamics as well as dynamic susceptibility. As a phenomenological theory, nonequilibrium thermodynamics deals with approach of systems toward steady states and examines relaxation phenomena during the approach to equilibrium. The theory also encompasses detailed studies of the stability of systems far from equilibrium, including oscillating systems. In this context, the notion of nonequilibrium phase transitions is gaining importance as a unifying theoretical concept.

In this article, we will focus on a general theory of Ising magnets based on nonequilibrium thermodynamic. The basics of nonequilibrium thermodynamics is reviewed and the time-reversal signature of thermodynamic variables with their sources and fluxes are discussed in Section 2. Section 3 then considers Ising spin models describing statics of ferromagnetic and antiferromagnetic orders in magnets. Section 4 contains a detailed description of the kinetic model based on coupled linear equations of motion for the order parameter(s). The effect of the relaxation process on critical dynamics of sound propagation and dynamic response magnetization is investigated in Section 5. Comparison with experiments is made and reasons for formulating a phenomenological theory of relaxation problem are given in Section 6. Finally, the open questions and future prospects in this field are outlined.

2. Basics of nonequilibrium thermodynamics

Nonequilibrium thermodynamics (NT), a scientific discipline of 20 th century, was invented in an effort to rationalize the behavior of irreversible processes. The NT is a vast field of scientific endeavour with roots in physics and chemistry. It was developed in the wake of the great success of certain symmetry relations, known as Onsager reciprocal relations in the phenomenological laws. These symmetry relations between irreversible phenomena have found a wide field of application in all branches of the physical science and engineering, and more recently in a number of interdisciplinary fields, including environmental research and, most notably, the biological sciences. Above applications can be classified according to their tensorial character. First one has scalar phenomena. These include chemical reactions and structural relaxation phenomena. Onsager relations are of help in this case, in solving the set of ordinary differential equations which describe the simultaneous relaxation of a great number of variables. Second group of phenomena is formed by vectorial processes, such as diffusion, heat conduction and their cross effects (e.g. thermal diffusion). Viscous phenomena and theory of sound propagation have been consistently developed within the framework of nonequilibrium thermodynamics.

Before introducing the notion of nonequilibrium thermodynamics we shall first summarize briefly the linear and nonlinear laws between thermodynamic fluxes and forces. A key concept when describing an irreversible process is the macroscopic state parameter of an adiabatically isolated system. These parameters are denoted by A_i . At equilibrium the state parameters have values A_i^0, while an arbitrary state which is near or far from the equilibrium may be specified by the deviations α_i from the equilibrium state:

$$\alpha_i = A_i - A_i^0 \,. \tag{1}$$

It is known empirically that the irreversible flows, time derivatives of deviations ($J_i = \dot{\alpha}_i$), are linear functions of the thermodynamic forces (X_i)

$$J_i = \sum_j L_{ij} X_j \,, \tag{2}$$

where the quantities L_{ij} are called the phenomenological coefficients and the Eqs. (2) are referred to as the phenomenological equations. The coefficients L_{ij} obey either Onsager's reciprocal relations $L_{ij} = L_{ji}$ or Casimir's one $L_{ij} = -L_{ji}$. These relations, also known as Onsager-Casimir reciprocal relations (Onsager, 1931; Casimir, 1945; De Groot, 1963), express an important consequence of microscopic time-reversal invariance for the relaxation of macroscopic quantities in the linear regime close to thermodynamic equilibrium. The proof of these relations involves the assumption that the correlation functions for the thermal fluctuations of macroscopic quantities decay according to the macroscopic relaxation equation.

It is well known that the entropy of an isolated system reaches its maximum value at equilibrium: so that any fluctuation of the thermodynamic parameters results with a decrease in the entropy. In response to such a fluctuation, entropy-producing irreversible process spontaneously drive the system back to equilibrium. Consequently, the state of equilibrium is stable to any perturbation that reduces the entropy. In contrast, one can state that if the fluctuations are groving, the system is not in equilibrium. The fluctuations in temperature, volume, magnetization, kuadrupole moment, etc. are quantified by their magnitude such as δT, δV, δM and δQ the entropy of a magnetic system is a function of these parameters in general one can expand the entropy as power series in terms of these parameters:

$$S = S_{eq} + \delta S + \frac{1}{2}\delta^2 S + O(\delta^3 S) \,, \tag{3}$$

In this expansion, the second term represents the *first-order* terms containing δT, δV, δM, δQ, etc., the third term indicates the *second-order* terms containing $(\delta T)^2$, $(\delta V)^2$, $(\delta M)^2$, $(\delta Q)^2$, etc., and so on. On the other hand, since the entropy is maximum, the *first-order* terms vanishes wheares the leading contribution to the increment of the entroıpy originates from the *second-order* term $\delta^2 S$ (Kondepudi & Prigogine, 2005).

The thermodynamic forces in Eqs. (2) are the intensive variables conjugate to the variables α_i:

$$X_i = \left.\frac{\partial S}{\partial \alpha_i}\right|_{\alpha_i} \,, \tag{4}$$

where S is the entropy of the system described by the fundamental relation $S = S(\alpha_1, ..., \alpha_n)$. The Eqs. (2) could be thought of as arising from a Taylor-series expansion of the fluxes in terms of the forces. Such a Taylor series will only exist if the flux is an analytic function of the forces at $X=0$:

$$J_i(X) = J_i(0) + \sum_j \frac{\partial J_i}{\partial X_j}\bigg|_{X=0} X_j + \sum_{j,k} \frac{1}{2!}\frac{\partial^2 J_i}{\partial X_j \partial X_k}\bigg|_{X=0} X_i X_j + O(X^3)\,. \tag{5}$$

Clearly the first term in Eq. (5) is zero as the fluxes vanish when the thermodynamic forces are zero. The term which is linear in the forces is evidently derivable, at least formally, from the equilibrium properties of the system as the functional derivative of the fluxes with respect to the forces computed at equilibrium, $X=0$. The quadratic term is related to what are known as the nonlinear contributions to the linear theory of irreversible thermodynamics. In general, Eq. (5) may be written as nonlinear functions of the forces in the expanded form

$$J_i(X) = \sum_j L_{ij} X_j + \sum_{j,k} M_{ijk} X_j X_k + \sum_{j,k,l} N_{ijkl} X_j X_k X_l + \ldots, \tag{6}$$

where the coefficients defined by

$$L_{ij} = \frac{\partial J_i}{\partial X_j}\bigg|_{X=0},\; M_{ijk} = \frac{1}{2}\frac{\partial^2 J_i}{\partial X_j \partial X_k}\bigg|_{X=0},\; N_{ijkl} = \frac{1}{6}\frac{\partial^3 J_i}{\partial X_j \partial X_k \partial X_l}\bigg|_{X=0}. \tag{7}$$

Here the coefficients L_{ij} are the cross coefficients which are scalar in character. The second order coefficients M_{ijk} are vectorial. The third order coefficients N_{ijkl} are again scalar. Within the linear range, there is a lot of experimental evidence of Onsager relation.

In the nonlinear thermodynamic theory, a nonlinear generalization of Onsager's reciprocal relations was obtained using statistical methods (Hurley & Garrod, 1982). Later, the same generalization was also proved with pure macroscopic methods (Verhas, 1983). The proof of the generalization is based on mathematical facts. None of these generalizations are of general validity. The principle of macroscopic reversibility proposed by Meixner gives a good insight to the structure of the Onsager-Casimir reciprocal relations and says that the entropy production density in invariant under time inversion if it is quadratic function of independent variables. Demanding its validity to higher order leads to conflict only with the rules of the chemical reactions (Meixner, 1972).

3. Ising model and equilibrium properties based on the mean field approximation

In this section, we consider the Ising model on a regular lattice where each interior site has the same number of nearest-neighbour sites. This is called the coordination number of the lattice and will be denoted by z. We assume that, in the thermodynamic limit, boundary sites can be disregarded and that, with N sites, the number of nearest-neighbour site pairs is $Nz/2$. The standard Hamiltonian for the the simplest Ising model is given by

$$\mathrm{H} = -J\sum_{\langle ij\rangle} s_i s_j - h\sum_i s_i\,,$$

with

$$s_i = \pm 1 , \tag{8}$$

where h is the external magnetic field at the site i and the summation $\sum_{\langle ij \rangle}$ is performed for nearest-neighbour sites. J is the exchange interaction between neighbouring sites $\langle ij \rangle$. Two distinctive cases corresponding to different signs of intersite interaction is considered, i.e., $J < 0$ (ferromagnetic coupling) and $J > 0$ (antiferromagnetic coupling). On the other hand, Eq. (8) may be extended by allowing values $s = 0, \pm 1, \pm 2, ..., \pm S$ for the variables. It is then possible to consider higher order interactions such as $K\sum_{\langle ij \rangle} s_i^2 s_j^2$ or a chemical potential such as $\Delta \sum_i s_i^2$. These generalizations are regarded as extensions of the Blume-Emery-Griffiths model (BEG) (Blume *et al.*, 1971). Recently, there have been many theoretical studies of mixed spin Ising systems. These are of interest because they have less translational symmetry than their single-spin counterparts since they consist of two interpenetrating inequivalent sublattices. The latter property is very important to study a certain type of ferrimagnetism, namely molecular-based magnetic materials which are of current interest (Kaneyoshi & Nakamura, 1998).

For sake of the brevity, here we will focus on the equilibrium properties of the $S = 1/2$ case which is described by the Hamiltonian given in Eq. (8). The Gibbs free energy depends on the three extensive variables ψ, N, and V. Here ψ, N, and V are magnetization per site, the total number of Ising spins and the volume of the lattice, respectively. Using the definition of the entropy the configurational Gibbs free energy in the Curie-Weiss approximation G ($G = E - TS - h\psi$) is obtained

$$G(V(a),\psi,h,T) = G_0(V,T) - \frac{1}{2}NJz\psi^2 + NkT\left[\left(\frac{1+\psi}{2}\right)\ln\left(\frac{1+\psi}{2}\right) + \left(\frac{1-\psi}{2}\right)\ln\left(\frac{1-\psi}{2}\right)\right] - h\psi , \tag{9}$$

where a, k, T are the lattice constant, the Boltzmann factor, the absolute temperature, respectively. $G_0(V,T)$ is the lattice free energy which is independent of spin configuration. One can see that G is an even function of ψ. Thus the second derivative of G with respect to ψ is

$$\frac{\partial^2 G}{\partial \psi^2} = -NJz + \frac{NkT}{1-\psi^2} , \tag{10}$$

and we define the critical temperature T_C by

$$T_C = Jz . \tag{11}$$

From Eq. (10), it is seen that the G vs. ψ curve is convex downwards for all ψ in the range $(-1, +1)$ for $T > T_C$, as shown in Figure 1. At $T = T_C$ the curvature changes sign to becomes convex upwards for $T < T_C$. The magnetic field h is conjugate to magnetization density ψ and from the fundamental relations of the thermodynamics one can write the following expression

$$h = \frac{\partial G}{\partial \psi} = -NJz\psi + \frac{1}{2}NkT\ln\frac{1+\psi}{1-\psi}. \tag{12}$$

To find the magnetization we solve Eq. (12) for ψ and obtain the equation of state (or self-consistent equation):

$$\psi = \tanh(\beta(Jz\psi + h)). \tag{13}$$

Now, using the definition

$$\chi(T) = \left.\frac{\partial \psi}{\partial h}\right|_{h\to 0}, \tag{14}$$

one obtains the following expression for the susceptibility

$$\chi(T) = \frac{1-\tanh^2(z\psi/T + h)}{T\left\{1 - \frac{(1-\tanh^2(z\psi/T + h))z}{T}\right\}}. \tag{15}$$

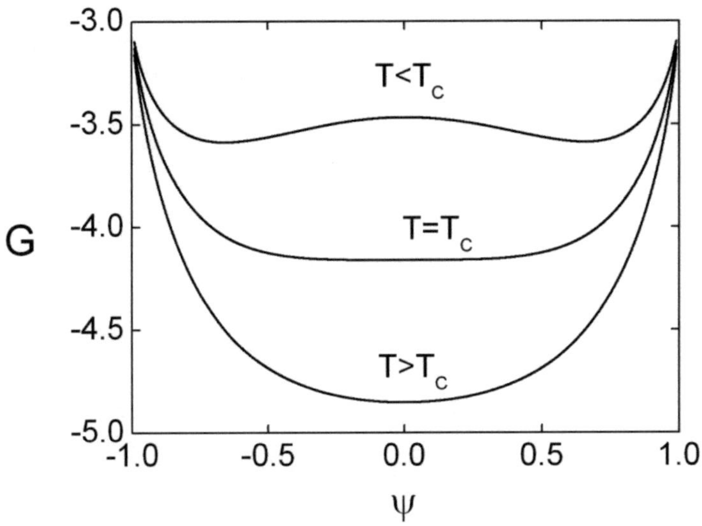

Fig. 1. Free energy-magnetization isotherms ($z = 6$)

Among the physical systems which undergo phase transitions, the most interesting class is the ferromagnet-paramagnet transtions in simple magnets. The free energy in such systems is nonanalytical function of its arguments. This is a manifestation of very strong fluctuations of quantity called order parameter. Phase transformations in ferromagnets are the continuous phase transitions which show no latent heat, seen in Figure 2. On the other hand, many physical quantities such as specific heat and static susceptibility diverge to infinity or tend to zero when approaching the critical temperature T_C. The behaviour of the static susceptibility of an Ising ferromagnet on a simple cubic lattice ($z = 6$) in the neighborhood of the critical point is shown in Figure 3. One can see that the static susceptibility diverges at the critical point on both sides of the critical region (Lavis & Bell, 1998).

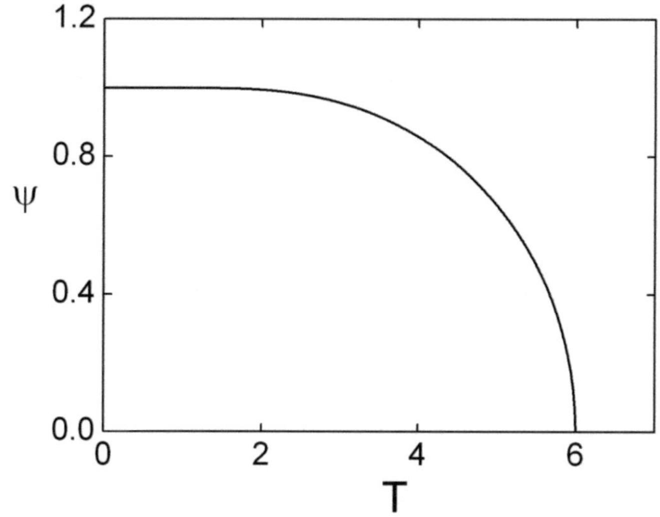

Fig. 2. The spontaneous magnetization plotted against temperature ($z = 6$)

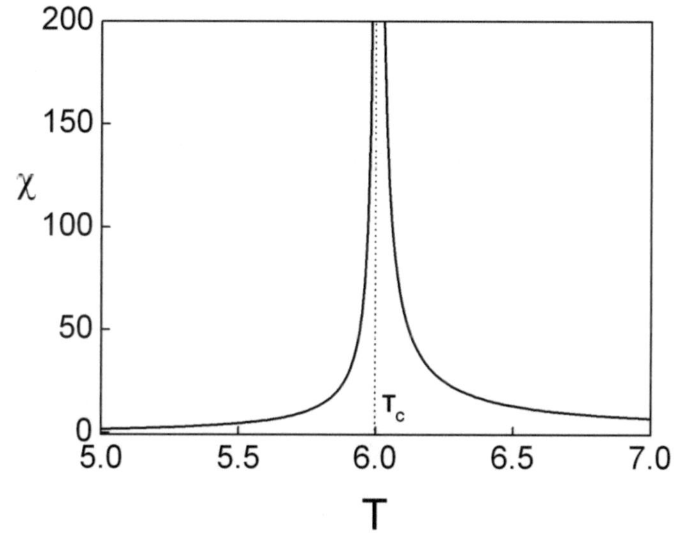

Fig. 3. The temperature dependence of the static susceptibility for a cubic lattice ($z = 6$)

4. Thermodynamic description of the kinetic model

In this section, a molecular-field approximation for the magnetic Gibbs free-energy production is used and a generalized force and a current are defined within the irreversible thermodynamics. Then the kinetic equation for the magnetization is obtained within linear response theory. Finally, the temperature dependence of the relaxation time in the neighborhood of the phase-transition points is derived by solving the kinetic equation of the magnetization. For a simple kinetic model of Ising magnets, we first define the time-dependent long-range order parameter $\psi(t)$ (or magnetization), describing the ferromagnetic ordering, as the thermodynamic variable. In the nonequilibrium theory of the Ising system, the relaxation towards equilibrium is described the equation

$$\dot{\psi} = -\frac{\psi - \overline{\psi}}{\tau}, \tag{16}$$

where τ is the relaxation time characterizing the rate at which the magnetization ψ approaches the equilibrium ($\overline{\psi}$). Eq. (16) is the simplest equation of irreversible thermodynamics (De Groot & Mazur, 1962) and can also be written in the form

$$\dot{\psi} = LX, \tag{17}$$

where L is the rate constant (or kinetic coefficient) and X is the thermodynamic force which causes the current $\dot{\psi}$. In Eq. (17) X is found from the derivative of mean-field Gibbs energy production (ΔG) with respect to deviation of magnetization from the equilibrium:

$$X = \frac{d(\Delta G)}{d(\psi - \overline{\psi})}, \tag{18}$$

with

$$\Delta G = \frac{1}{2}\Big[A(\psi - \overline{\psi})^2 + 2B(\psi - \overline{\psi})(h - \overline{h}) + C(h - \overline{h})^2 + D(\psi - \overline{\psi})(a - \overline{a})$$

$$+E(h - \overline{h})(a - \overline{a}) + F(a - \overline{a})^2 + G'(h - \overline{h})\Big], \tag{19}$$

In Eq. (19), the coefficients are called as Gibbs production coefficients:

$$A = \left(\frac{\partial^2 G}{\partial \psi^2}\right)_{eq} = -\frac{N(-Jz + Jz\overline{\psi}^2 + kT)}{-1 + \overline{\psi}^2}, \tag{20}$$

$$B = \left(\frac{\partial^2 G}{\partial \psi \partial h}\right)_{eq} = -1, \tag{21}$$

$$C = \left(\frac{\partial^2 G}{\partial h^2}\right)_{eq} = 0, \tag{22}$$

$$D = \left(\frac{\partial^2 G}{\partial \psi \partial a}\right)_{eq} = -Nz\overline{\psi}\left(\frac{\partial J}{\partial a}\right)_{eq}, \tag{23}$$

$$E = \left(\frac{\partial^2 G}{\partial h \partial a} \right)_{eq} = 0, \tag{24}$$

$$F = \left(\frac{\partial^2 G}{\partial a^2} \right)_{eq} = \left(\frac{\partial^2 G_0}{\partial a^2} \right)_{eq} - \frac{1}{2} N z \overline{\psi}^2 \left(\frac{\partial^2 J}{\partial a^2} \right)_{eq}, \quad G' = \left(\frac{\partial G}{\partial h} \right)_{eq} = -\overline{\psi}. \tag{25}$$

The rate (or kinetic) equation is obtained using Eqs. (18)-(25) in the relaxation equation (Eq. (17)):

$$\dot{\psi} = LA(\psi - \overline{\psi}) + LB(h - \overline{h}) + DL(a - \overline{a}). \tag{26}$$

In order to find the relaxation time (τ) for the single relaxation process, one considers the rate equation when there is no external stimulation, i.e., $h = \overline{h}$, $a = \overline{a}$. Eq. (26) then becomes

$$\dot{\psi} = LA(\psi - \overline{\psi}). \tag{27}$$

Assuming a solution of the form $\psi - \overline{\psi} \approx \exp(-t/\tau)$ for Eq. (27), one obtains

$$\frac{1}{\tau} = -LA. \tag{28}$$

Using Eq. (20) yields

$$\tau = -\frac{1 - \overline{\psi}^2}{NL(-Jz + Jz\overline{\psi}^2 + kT)}. \tag{29}$$

The behaviour of the relaxation time near the phase-transition points can be derived analytically from the critical exponents. It is a well-known fact that various thermodynamic functions represents singular behavior as one approaches the critical point. Therefore, it is convenient to introduce an expansion parameter, which is a measure of the distance from the critical point ($\varepsilon = T - T_c$). Here T_c is the critical temperature given by Eq. (11). In the neighborhood of the transition point the relaxation time of the Ising model can be written in the form,

$$\tau(\varepsilon) = -\frac{1 - (\overline{\psi}(\varepsilon))^2}{NL(-Jz + Jz(\overline{\psi}(\varepsilon))^2 + k(\varepsilon + T_C))}. \tag{30}$$

In the vicinity of the second-order transition the magnetization vanishes at T_c as

$$\overline{\psi}(\varepsilon) = (\varepsilon)^{1/2}. \tag{31}$$

The critical exponent for the function $\tau(\varepsilon)$ is defined as

$$\lambda = \lim_{\varepsilon \to 0} \frac{\ln \tau(\varepsilon)}{\ln \varepsilon}. \tag{32}$$

This description is valid for all values of λ, where the negative value corresponds to the divergence of the variable $\tau(\varepsilon)$ as ε goes to zero, positive value corresponding to relaxation time that approaches zero, and the zero value corresponding to logarithmic divergence, jump singularity or a cusp (the relaxation time is finite at the critical point but one of its derivative diverges (Reichl, 1998). On the other hand, in order to distinguish a cusp from a logarithmic divergence, another type of critical exponent, $\lambda^{'}$, is introduced. To find the exponent $\lambda^{'}$ that describes the singular parts of τ with a cusplike singularity, we first find the smallest integer m for which the derivative $\tau^{(m)} = \left(\partial^m \tau / \partial \varepsilon^m\right)$ diverge as $\varepsilon \to 0$:

$$\lambda^{'} = m + \lim_{\varepsilon \to 0} \frac{\ln \tau(\varepsilon)}{\ln \varepsilon}. \quad (33)$$

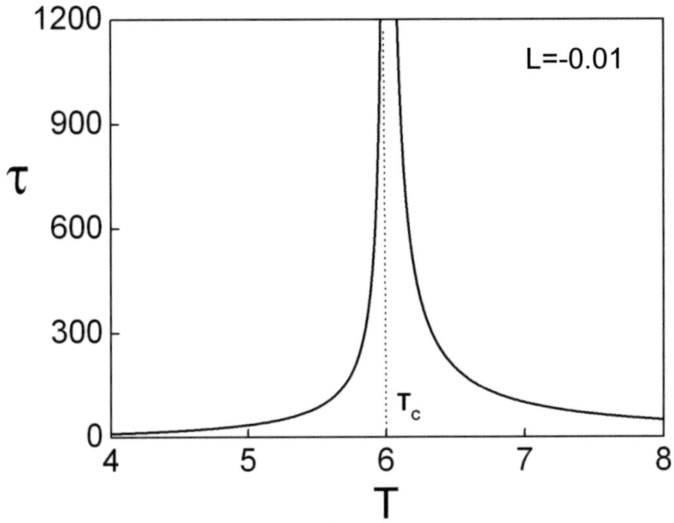

Fig. 4. Relaxation time vs temperature in the neighbourhood of critical point

The behavior of the relaxation time τ as a function of temperature is given in Figure 4. One can see from Figure 4 that τ grows rapidly with increasing temperature and diverges as the temperature approaches the second-order phase-transition point. In accordance with this behavior, the critical exponent of τ is found to be $\lambda = -1.0$. On the other hand, the scaling form of the relaxation time reads $\tau \approx \xi^{\,z} \approx |T - T_c|^{\nu z}$, where ξ, ν and z are the correlation length, critical exponent for ξ and dynamical critical exponent, respectively (Ray *et al.*, 1989). According to mean-field calculations, the dynamic critical exponent of the Ising model is $z = 2$ at the critical point. In addition to studies on Blume-Capel model which undergoes first-order phase transitions and represents rich variety of phase diagrams has revealed the fact that the dynamical critical exponent is also $z = 2$ at the critical endpoint, and double critical endpoints as well as tricritical point, whereas $z = 0$ for first-order critical transition points (Gulpinar & İyikanat, 2011). We should note that the analysis used in this article is identical to Landau-Ginzburg kinetic theory of phase transitions of a spatially homogenous system. As is discussed extensively by Landau and Lifshitz (Landau & Lifshitz, 1981), in the case of spatially inhomogeneous medium where $\psi - \bar{\psi} = \partial\psi(t,r)$, the Landau-Ginzburg kinetic theory of critical phenomena reveals the fact that the relaxation time becomes finite for $T = T_c$ for components with $q \neq 0$. Here q is the Fourier transform of the spatial variable r. On the other hand, the renormalization-group formalism has proved to be very useful in calculating not only the static behavior but also the dynamic scaling. By making use of this method, Halperin *et al.* (Halperin *et al.*, 1974) found the critical-point singularity of the linear dynamic response of various models. The linear response theory, however, describes the reaction of a system to an infinitesimal external disturbance, while in experiments and computer simulations it is often much easier to deal with nonlinear-response situations, since it is much easier to investigate the response of the system to finite changes in the thermodynamic variables. A natural question is whether the critical-point singularity of the linear and nonlinear responses is the same. The answer is yes for ergodic systems, which reach equilibrium independently of the initial conditions (Racz, 1976). The assumption that the initial and intermediate stages of the relaxation do not affect the divergence of the relaxation time (motivated by the observation that the critical fluctuations appear only very close to equilibrium) led to the expectation that in ergodic systems τ^{nl} and τ^{l} diverge with same critical exponent. This view seemed to be supported by Monte Carlo calculations (Stoll et al., 1973) and high-temperature series expansion of the two-dimensional one-spin flip kinetic Ising model. Later, Koch *et al.* (Koch *et al., 1996)* presented field-theoretic arguments by making use of the Langevin equation for the one-component field $\phi(r,r)$ as well as numerical studies of finite-size effects on the exponential relaxation times τ_1 and τ_2 of the order parameter and the square of the order parameter near the critical point of three-dimensional Ising-like systems.

For the ferromagnetic interaction, a short range order parameter as well as the long range order is introduced (Tanaka et al., 1962; Barry, 1966) while there are two long range sublattice magnetic orders and a short range order in the Ising antiferromagnets (Barry & Harrington, 1971). Similarly the number of thermodynamic variables (order parameters) also increases when the higher order interactions are considered (Erdem & Keskin, 2001; Gülpınar et al., 2007; Canko & Keskin, 2010). For a general formulation of Ising spin kinetics with a multiple number of spin orderings (ψ_i), the Gibbs free energy production is written as

$$\Delta G = \frac{1}{2}\left\{ \sum_{i,j=1}^{n} \beta_{ij}(\psi_i - \bar{\psi}_i)(\psi_j - \bar{\psi}_j) + 2\sum_{i=1}^{n}\left[\sum_{k=1}^{m}\left[\gamma_{ik}(\psi_i - \bar{\psi}_i)(h_k - \bar{h}_k) \right.\right.\right.$$
$$\left.\left.\left. + \phi_k(h_k - \bar{h}_k)^2 + \eta_k(h_k - \bar{h}_k)(a - \bar{a}) \right] + \mu_i(\psi_i - \bar{\psi}_i)(a - \bar{a}) + \sigma(a - \bar{a})^2 \right]\right\}, \tag{34}$$

where the coefficients are defined as

$$\beta_{ij} = \left(\frac{\partial^2 G}{\partial\psi_i \partial\psi_j} \right)_{eq}, \qquad \gamma_{ik} = \left(\frac{\partial^2 G}{\partial\psi_i \partial h_k} \right)_{eq}, \qquad \phi_k = \left(\frac{\partial^2 G}{\partial h_k^2} \right)_{eq},$$

$$\eta_k = \left(\frac{\partial^2 G}{\partial h_k \partial a}\right)_{eq}, \qquad \mu_i = \left(\frac{\partial^2 G}{\partial \psi_i \partial a}\right)_{eq}, \qquad \sigma = \left(\frac{\partial^2 G}{\partial a^2}\right)_{eq}. \tag{35}$$

Then a set of linear rate equations may be written in terms of a matrix of phenomenological coefficients which satisfy the Onsager relation (Onsager, 1931):

$$\begin{bmatrix} \dot{\psi}_i \\ \cdot \\ \cdot \\ \cdot \\ \dot{\psi}_n \end{bmatrix} = \begin{bmatrix} L_{i1} & \dots & L_{in} \\ \cdot & \cdot & \cdot \\ \cdot & \cdot & \cdot \\ \cdot & \cdot & \cdot \\ L_{n1} & \dots & L_{nn} \end{bmatrix} \begin{bmatrix} X_i \\ \cdot \\ \cdot \\ \cdot \\ X_n \end{bmatrix}, \tag{36}$$

where the generalized forces are

$$X_j = \frac{\partial(\Delta G)}{\partial(\psi_j - \overline{\psi}_j)} = \sum_{i,j=1}^{n} \beta_{ij}(\psi_i - \overline{\psi}_i) + \sum_{i=1}^{n}\left[\mu_i(a - \overline{a}) + \sum_{k=1}^{m} \gamma_{ik}(h_k - \overline{h}_k)\right]. \tag{37}$$

The matrix equation given by Eq. (36) can be written in component form using Eq. (37), namely a set of n coupled, linear inhomogenous first-order rate equations. Embedding this relation into Eq. (36) one obtains the following matrix equation for the fluxes:

$$\hat{\dot{\psi}} = \hat{L}\hat{\beta}\partial\hat{\psi} + \hat{L}\hat{\gamma}\hat{h} + \hat{L}a\hat{\mu}, \tag{38}$$

where the matrixes are defined by

$$\hat{\beta} = \begin{bmatrix} \beta_{i1} & \dots & \beta_{in} \\ \cdot & \cdot & \cdot \\ \cdot & \cdot & \cdot \\ \cdot & \cdot & \cdot \\ \beta_{n1} & \dots & \beta_{nn} \end{bmatrix}, \qquad \hat{\gamma} = \begin{bmatrix} \gamma_{i1} & \dots & \gamma_{im} \\ \cdot & \cdot & \cdot \\ \cdot & \cdot & \cdot \\ \cdot & \cdot & \cdot \\ \gamma_{n1} & \dots & \gamma_{nm} \end{bmatrix}, \qquad \hat{L} = \begin{bmatrix} L_{i1} & \dots & L_{in} \\ \cdot & \cdot & \cdot \\ \cdot & \cdot & \cdot \\ \cdot & \cdot & \cdot \\ L_{n1} & \dots & L_{nn} \end{bmatrix},$$

$$\hat{\dot{\psi}} = \begin{bmatrix} \dot{\psi}_1 \\ \cdot \\ \cdot \\ \cdot \\ \dot{\psi}_n \end{bmatrix}, \qquad \partial\hat{\psi} = \begin{bmatrix} \psi_1 - \overline{\psi}_1 \\ \cdot \\ \cdot \\ \cdot \\ \psi_n - \overline{\psi}_n \end{bmatrix}, \qquad \hat{\mu} = \begin{bmatrix} \mu_1 \\ \cdot \\ \cdot \\ \cdot \\ \mu_n \end{bmatrix}, \qquad \hat{h} = \begin{bmatrix} h_1 - \overline{h}_1 \\ \cdot \\ \cdot \\ \cdot \\ h_m - \overline{h}_m \end{bmatrix} \tag{39}$$

Since the phenomenological coefficients L_{ij} in matrix $\hat{L}$ obey one of the reciprocal relations $L_{ij} = \pm L_{ji}$ according to microscopic time-reversal invariance of relaxing macroscopic quantities $\psi_i(t)$, the matrix may be symmetric or antisymmetric. In order to obtain the relaxation times, one considers the corresponding inhomogenous equations (Eq. (38)) resulting when the external fields are equal to their equilibrium values, i.e., $h_k = \overline{h}_k$ for $k = 1,...,m$ and $a = \overline{a}$. In the neighbourhood of the equilibrium states, solutions of the form

$\psi_i - \overline{\psi}_i \approx \exp(-t/\tau_i)$ are assumed for the linearized kinetic equations and approaches of the order parameters $\psi_i(t)$ to their equilibrium values are described by a set of characteristic times, also called relaxation times τ_i. To find each time (τ_i) one must solve the secular equation. Critical exponents (λ_i and $\lambda_i^{'}$, $i=1,...,n$) for the functions $\tau_i(\varepsilon)$ are also calculated using Eqs. (32) and (33) to see the divergences, jumps, cusps etc. for the relaxation times $\tau_i(\varepsilon)$ at the transition points.

5. Critical behaviours of sound propagation and dynamic magnetic response

In this section, we will discuss the effect of the relaxation process on critical dynamics of sound propagation and dynamic response magnetization for the Ising magnets with single order parameter (ψ). Firstly we study the case in which the lattice is under the effect of a sound wave. Then the sound velocity and sound attenuation coefficient of the system are derived using the phenomenological formulation based on the method of thermodynamics of irreversible processes. The behaviors of these quantities near the phase transition temperatures are analyzed. Secondly, we consider case where the spin system is stimulated by a small uniform external magnetic field oscillating at an angular frequency. We examine the temperature variations of the non-equilibrium susceptibility of the system near the critical point. For this aim, we have made use of the free energy production and the kinetic equation describing the time dependency of the magnetization which are obtained in the previous section. In order to obtain dynamic magnetic response of the Ising system, the stationary solution of the kinetic equation in the existence of sinusoidal external magnetic field is performed. In addition, the static and dynamical mean field critical exponents are calculated in order to formulate the critical behavior of the magnetic response of a magnetic system.

In order to obtain the critical sound propagation of an Ising system we focus on the case in which the lattice is stimulated by the sound wave of frequency ω for the case $h=\overline{h}$. In the steady state, all quantities will oscillate with the same frequency ω and one can find a steady solution of the kinetic equation given by Eq. (26) with an oscillating external force $a-\overline{a}=a_1 e^{i\omega t}$. Assuming the form of solution $\psi(t)-\overline{\psi}=\psi_1 e^{i\omega t}$ and introducing this expression into Eq. (26), one obtains the following inhomogenous equation for ψ_1

$$i\omega \psi_1 e^{i\omega t} = LA\psi_1 e^{i\omega t} + LDa_1 e^{i\omega t} \tag{40}$$

Solving Eq. (40) for ψ_1 / a_1 gives

$$\frac{\psi_1}{a_1} = \frac{LD}{i\omega - LA} = \frac{\tau LD}{1+i\omega\tau}. \tag{41}$$

The response in the pressure $(p-\overline{p})$ is obtained by differentiating the minimum work with respect to $(V-\overline{V})$ and using Eqs. (9) and (19)

$$p-\overline{p} = \frac{\partial \Delta G}{\partial (V-\overline{V})} = -\frac{\overline{a}}{3\overline{V}}\frac{\partial \Delta G}{\partial (a-\overline{a})}, \tag{42}$$

then

$$p-\bar{p}=-\frac{\bar{a}}{3\bar{V}}\left[D(\psi-\bar{\psi})+F(a-\bar{a})\right] \quad . \tag{43}$$

Finally, the derivative of the pressure with respect to volume gives

$$\left(\frac{\partial p}{\partial V}\right)_{sound}=\left(\frac{\bar{a}}{3\bar{V}}\right)^2\left[F+D\frac{\psi_1}{a_1}\right] \quad . \tag{44}$$

Here F and D are given by Eqs. (23) and (25). Introducing the relation (41) and the density $\rho=M/\bar{V}$ into Eq. (44) one obtains

$$\left(\frac{\partial p}{\partial \rho}\right)_{sound}=\frac{\bar{a}^2}{9M}\left[\left(\frac{\partial^2 G_0}{\partial a^2}\right)_{eq}-\frac{1}{2}Nz\bar{\psi}^2\left(\frac{\partial^2 J}{\partial a^2}\right)_{eq}+\frac{\tau LD^2}{1+i\omega\tau}\right]. \tag{45}$$

From the real and imaginary parts of Eq. (45) one obtains the velocity of sound and attenuation coefficient for a single relaxational process as

$$c(\omega,T)=\mathrm{Re}\left[c_{sound}\right]=c_0\left[1-\frac{Nz\bar{a}^2\bar{\psi}^2}{18Mc_0^2}\left(\frac{\partial^2 J}{\partial a^2}\right)+LD^2\frac{\tau}{1+\omega^2\tau^2}\right], \tag{46}$$

$$\alpha(\omega,T)=\mathrm{Im}\left[-\frac{\omega}{c_{sound}}\right]=LD^2\frac{\omega^2\tau^2}{1+\omega^2\tau^2}, \tag{47}$$

where c_0 is the velocity of sound at very high frequencies or at very high temperatures and $c_{sound}=(\partial p/\partial \rho)^{1/2}$ is the a complex expression for sound velocity. We perform some calculations for the frequency and temperature dependencies of $c(\omega,T)$ and $\alpha(\omega,T)$. Figures (5) and (6) show these dependencies. From the linear coupling of a sound wave with the order parameter fluctuations $(\psi-\bar{\psi})$ in the Ising system, the dispersion which is relative sound velocity change displays a frequency-dependent velocity or dispersion minimum (Figure 5) while the attenuation exhibits a frequency-dependent broad peak (Figure 6) in the ordered phase. Calculations of $c(T)$ and $\alpha(T)$ for the simple Ising spin system reveals the same features as in real magnets, i.e. the shifts of the velocity minima and attenuation maxima to lower temperatures with increasing frequency are seen. The velocity minima at each frequency occur at temperatures lower than the corresponding attenuation maxima observed for the same parameters used. The notions of minimum in sound velocity and maximum in attenuation go back to Landau and Khalatnikov (Landau & Khalatnikov, 1954; Landau & Khalatnikov, 1965) who study a more general question of energy dissipation mechanism due to order parameter relaxation. Their idea was based on the slow relaxation of the order parameter. During this relaxation it allows internal irreversible processes to be switch on so as to restore local equilibrium; this increases the entropy and involves energy dissipation in the system. In the critical region, behaviours of both quantities are verified analytically from definition of critical exponents given in Eq. (32) for the functions $c(\varepsilon)$ and

$\alpha(\varepsilon)$. It is found that the dispersion just below the critical temperature is expressed as $c(\varepsilon) \propto \varepsilon^0$ while the attenuation goes to zero as $\alpha(\varepsilon) \propto \varepsilon$.
In the presence of many thermodynamic variables for more complex Ising-type magnets, there exist more than one relaxational process with relaxational times (τ_i). Contribution of these processes to the sound propagation were treated in more recent works using the above technique in the general phenomenological formulation given in the previous section. Dispersion relation and attenuation coefficient for the sound waves of frequency ω were derived for sevaral models with an Ising-type Hamiltonian (Keskin & Erdem, 2003; Erdem & Keskin, 2003; Gulpinar, 2008; Albayrak & Cengiz, 2011). In these works, various mechanisms of the sound propagation in Ising-type magnets were given and origin of the critical attenuation with its exponent was discussed.

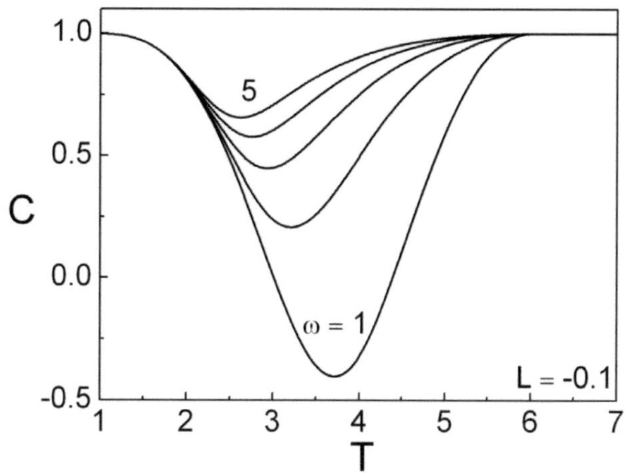

Fig. 5. Sound dispersion $c(T)$ at different frequencies ω for $L=10$

Similarly, theoretical investigation of dynamic magnetic response of the Ising systems has been the subject of interest for quite a long time. In 1966, Barry has studied spin–1/2 Ising ferromagnet by a method combining statistical theory of phase transitions and irreversible thermodynamics (Barry, 1966). Using the same method, Barry and Harrington has focused on the theory of relaxation phenomena in an Ising antiferromagnet and obtained the temperature and frequency dependencies of the magnetic dispersion and absorption factor in the neighborhood of the Neel transition temperature (Barry & Harrington, 1971). Erdem investigated dynamic magnetic response of the spin–1 Ising system with dipolar and quadrupolar orders (Erdem, 2008). In this study, expressions for the real and imaginary parts of the complex susceptibility were found using the same phenomenological approach proposed by Barry. Erdem has also obtained the frequency dependence of the complex susceptibility for the same system (Erdem, 2009). In Ising spin systems mentioned above, there exist two or three relaxing quantities which cause two or three relaxation contributions to the dynamic magnetic susceptibility. Therefore, as in the sound dynamics case, a general formulation (section 4) is followed for the derivation of susceptibility expressions. In the following, we use, for simplicity, the theory of relaxation with a single characteristic time to obtain an explicit form of complex susceptibility.

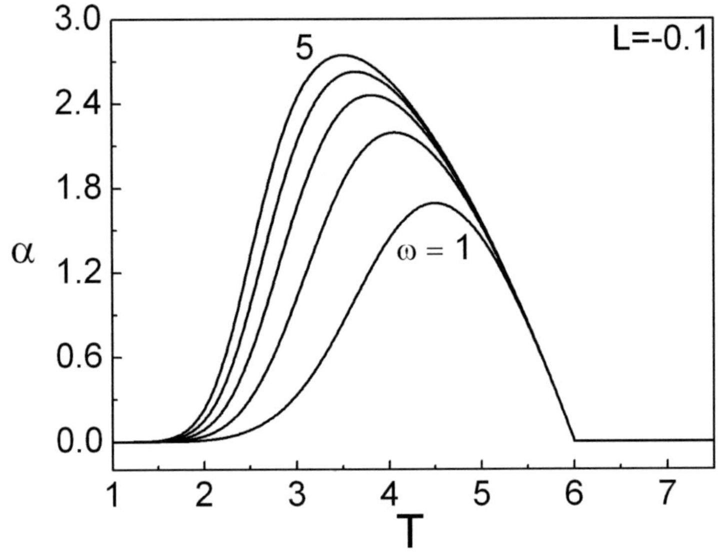

Fig. 6. Sound attenuation $\alpha(T)$ at different frequencies ω for $L = 10$

If the spin system descibed by Eq. (8) is stimulated by a time dependent magnetic field $h(t) = h_1 e^{i\omega t}$ oscillating at an angular frequency ω, the order parameter of the system will oscillate near the equilibrium state at this same angular frequency at the stationary state:

$$\psi(t) - \overline{\psi} = \psi_1 e^{i\omega t}, \tag{48}$$

If this equation is substituted into the kinetic equation Eq. (17) we find following form:

$$i\omega\psi_1 e^{i\omega t} = LA\psi_1 e^{i\omega t} + LBh_1 e^{i\omega t}. \tag{49}$$

Solving Eq. (49) for ψ_1 / h_1 gives

$$\frac{\psi_1}{h_1} = \frac{LB}{i\omega - LA} \tag{50}$$

Eq. (50) is needed to calculate the complex initial susceptibility $\chi(\omega)$. The Ising system induced magnetization (total induced magnetic moment per unit volume) is given by

$$\psi(t) - \psi_\infty = \mathrm{Re}\left(\psi_1 e^{i\omega t}\right), \tag{51}$$

where ψ_∞ is the magnetization induced by a magnetic field oscillating at ω. Also, by definition, the expression for $\chi(\omega)$ may be written

$$\psi(t)-\psi_\infty=\mathrm{Re}\left[\chi(\omega)h_1e^{i\omega t}\right], \tag{52}$$

where $\chi(\omega)=\chi^{'}(\omega)-i\chi^{''}(\omega)$ is the complex susceptibility whose real and imaginary parts are called as magnetic dispersion and absorption factors respectively. Comparing Eqs. (38) to Eq. (40) one may write

$$\chi(\omega)=\frac{\psi_1}{h_1}. \tag{53}$$

Finally the magnetic dispersion and absorbtion factors become

$$\chi^{'}(\omega)=\frac{AL^2}{A^2L^2+\omega^2}=LB\frac{\tau}{1+\omega^2\tau^2}. \tag{54}$$

$$\chi^{''}(\omega)=\frac{L\omega}{A^2L^2+\omega^2}=LB\frac{\tau^2\omega}{1+\omega^2\tau^2}. \tag{55}$$

In Figures 7 and 8 we illustrate the temperature variations of the magnetic dispersion and absorption factor in the low frequency limit $\omega\tau<<1$. These plots illustrate that both $\chi^{'}(\omega)$ and $\chi^{''}(\omega)$ increase rapidly with temperature and tend to infinity near the phase transtion temperature. The divergence of $\chi^{'}(\omega)$ does not depend on the frequency while the divergence of $\chi^{''}(\omega)$ depends on ω and gets pushed away from the critical point as ω increases. When compared with the static limit ($\omega\to0$) mentioned in section 3, a good agreement is achieved. Above critical behaviours of both components for the regime $\omega\tau<<1$ may be verified by calculating the critical exponents for the functions $\chi^{'}(\varepsilon)$ and $\chi^{''}(\varepsilon)$ using Eq. (32). Results of calculation indicates that $\chi^{'}(\varepsilon)$ and $\chi^{''}(\varepsilon)$ behave as ε^{-1} and ε^{-2}, respectively.

Finally the high frequency behavior ($\omega\tau>>1$) of the magnetic dispersion and absorption factor are given in Figures 9 and 10. The real part $\chi^{'}(\omega)$ has two frequency-dependent local maxima in the ordered and disordered phase regions. When the frequency increases, the maximum observed in the ferromagnetic region decreases and shifts to lower temperatures. The peak observed in the paramagnetic region also decreases but shifts to higher temperatures. On the other hand, the imaginary part $\chi^{''}(\omega)$ shows frequency-dependent maxima at the ferromagnetic-paramagnetic phase transtion point. Again, from Eq. (32), one can show that the real part converges to zero ($\chi^{'}(\varepsilon)\propto\varepsilon$) and the imaginary part displays a peak at the transition ($\chi^{''}(\omega)\propto\varepsilon^{0}$) as $\varepsilon\to0$.

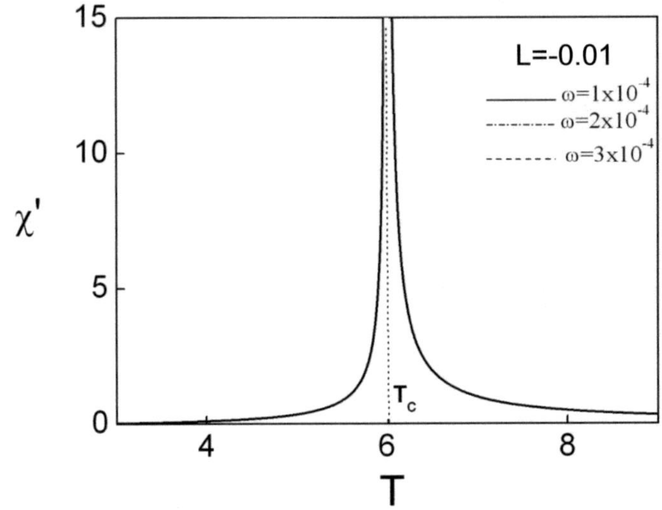

Fig. 7. Magnetic dispersion $\chi^{'}(\omega)$ vs temperature for the low frequency limit ($\omega\tau << 1$ in the neighbourhood of critical point

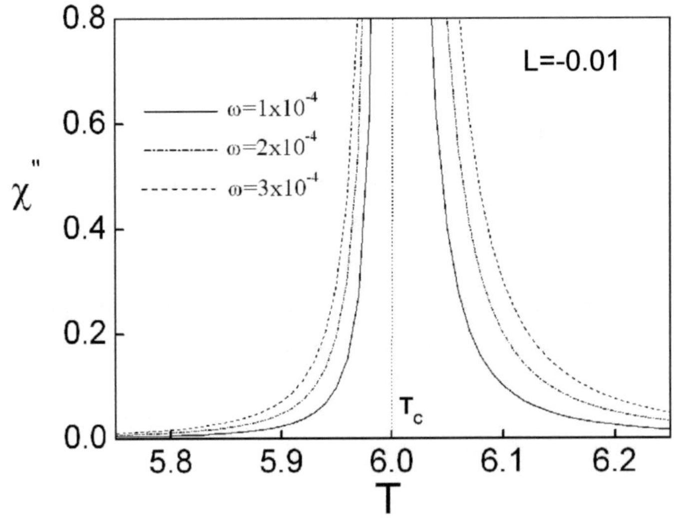

Fig. 8. Same as Figure 7 but for the magnetic absorption factor $\chi^{''}(\omega)$

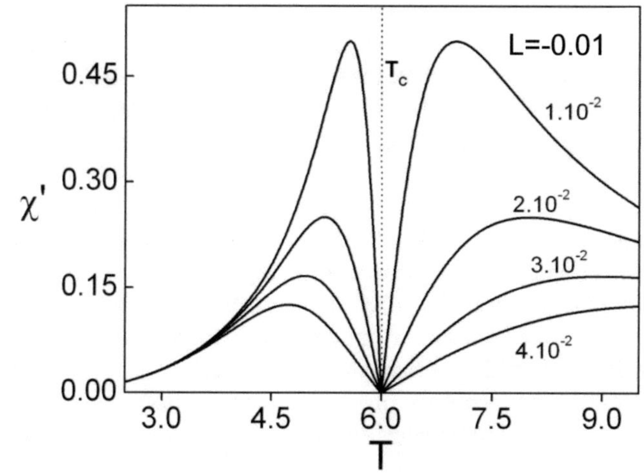

Fig. 9. Magnetic dispersion $\chi^{'}(\omega)$ vs temperature for the high frequency limit ($\omega\tau >> 1$) in the neighbourhood of critical point

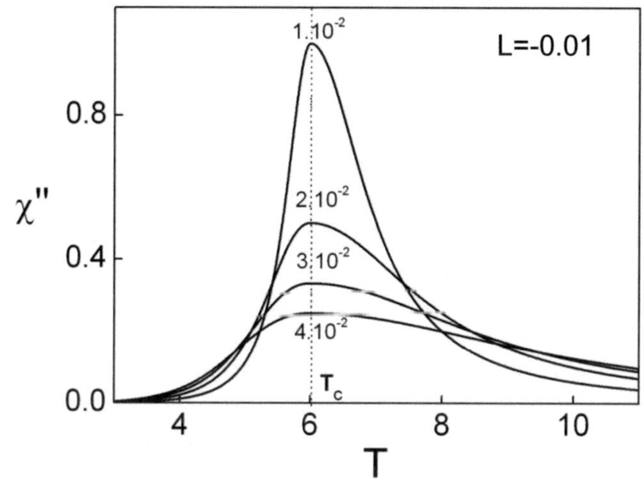

Fig. 10. Same as Figure 9 but for the magnetic absorption factor $\chi^{''}(\omega)$

6. Comparison of theory with experiments

The diverging behavior of the relaxation time and corresponding slowing down of the dynamics of a system in the neighborhood of phase transitions has been a subject of experimental research for quite a long time. In 1958, Chase (Chase, 1958) reported that liquid helium exhibits a temperature dependence of the relaxation time consistent with the scaling relation $(T - T_c)^{-1}$. Later Naya and Sakai (Naya & Sakai, 1976) presented an analysis of the critical dynamics of the polyorientational phase transition, which is an extension of the statistical equilibrium theory in random phase approximation. In addition, Schuller and

Gray (Schuller & Gray, 1976) have shown that the relaxation time of the superconducting order parameter diverges close to the transition temperature, in accordance with the theoretical prediction of several authors (Lucas & Stephen, 1967; Schmid & Schon, 1975). Recently, Sperkach et al. (Sperkach et al, 2001) measured the temperature dependence of acoustical relaxation times in the vicinity of a nematic-isotropic phase-transition point in 5CB liquid crsystal. Comparing Figures 2(a) and Fig. 5 of their work one can observe the similarity between the temperature-dependent behavior of the low-frequency relaxation time of the 5CB liquid crystal and the Blume-Capel model with random single-ion anisotropy (Gulpinar & İyikanat, 2011). Moreover, very recently, Ahart et al. (Ahart et al., 2009) reported that a critical slowing down of the central peak. These results indicate that the relaxation time of the order parameter for an Ising magnet diverges near the critical point, which corresponds to a familiar critical slowing down.

It is well known fact that measurements of sound propagation are considered useful in investigating the dynamics of magnetic phase transitions and therefore many experimental and theoretical studies have been carried out. Various aspects of ultrasonic attenuation in magnetic insulators (Lüthi & Pollina, 1969; Moran & Lüthi, 1971) and in magnetic metals (Lüthi et al., 1970; Maekawa & Tachiki, 1978) have been studied. In these works, the transtion temperature was associated with the experimentally determined peaks whose maximum shift towards the lower temperatures as the sound frequency increases. Similarly, acoustic studies, especially those of dispersion, have also been made on several magnetic systems such as transition metals (Golding & Barmatz, 1969), ferromagnetic insulators (Bennett, 1969) and antiferromagnetic semiconductors (Walter, 1967). It was found that the critical changes in sound velocity show a uniform behaviour for all substances studied, namely, a frequency-independent and weak temperature-dependent effect. It was also found that, in the ordered phase, the minima of the sound velocity shifted to lower temperatures with increasing frequency (Moran & Lüthi, 1971).

Dynamic response of a spin system to a time-varying magnetic field is an important subject to probe all magnetic systems. It is also called AC or dynamic suceptibility for the magnetization. The dynamic susceptibility is commonly used to determine the electrical properties of superconductors (Kılıç et al, 2004) and magnetic properties of some spin systems such as spin glasses (Körtzler & Eiselt, 1979), cobal-based alloys (Durin et al., 1991), molecule-based magnets (Girtu, 2002), magnetic fluids (Fannin et al., 2005) and nanoparticles (Van Raap et al., 2005). The dynamic magnetic response of these materials and the development of methods for its modification are important for their potential applications. For example, cores made of cobalt-based alloys in low signal detectors of gravitational physics contribute as a noise source with a spectral density proportional to the ac susceptibility of the alloy. The knowledgement of dynamic susceptibility for nanocomposite particles is very important for the design of magneto-optical devices.

7. Conclusion

In this chapter, we have discussed a simple kinetic formulation of Ising magnets based on nonequilibrium thermodynamics. We start with the simplest relaxation equation of the irreversible thermodynamics with a characteristic time and mention a general formulation based on the research results in the literature for some well known dynamic problems with more than one relaxational processes. Recent theoretical findings provide a more precise description for the experimental acoustic studies and magnetic relaxation measurements in real magnets.

The kinetic formulation with single relaxation process and its generalization for more coupled irrevesible phenomena strongly depend on a statistical equilibrium description of free energy and its properties near the phase transition. The effective field theories of equilibrium statistical mechanics, such as the molecular mean-field approximation is used as this century-old description of free energy. However, because of its limitations, such as neglecting fluctuation correlations near the critical point and low temperature quantum excitatitions, these theories are invaluable tools in studies of magnetic phase transitions. To improve the methodology and results of mean-field analysis of order parameter relaxation, the equilibrium free energy should be obtained using more a reliable theory including correlations. This was recently given on the Bethe lattice using some recursion relations. The first major application of Bethe-type free energy for the relaxation process was on dipolar and quadrupolar interactions to study sound attenuation problem (Albayrak & Cengiz, 2011).

Bethe lattice treatment of phenomenological relaxation problem mentioned above has also some limitations. It predicts a transition temperature higher than that of a bravais lattice. Also, predicting the critical exponents is not reliable. Therefore, one must consider the relaxation problem on the real lattices using more reliable equilibrium theories to get a much clear relaxation picture. In particular, renormalization group theory of relaxational sound dynamics and dynamic response would be of importance in future.

8. Acknowledgements

We thank to M. Ağartıoğlu for his help in the preparation of the figures. This work was suppoted by by the Scientific and Technological Research Council of Turkey (TUBITAK), Grant No. 109T721

9. References

Ahart, M., Hushur, A., Bing, Y., Ye, Z. G., Hemley. R. J. & Kojima, S. (2009), Critical slowing down of relaxation dynamics near the Curie temperature in the relaxor $Pb(Sc_{0.5}Nb_{0.5})O_3$, *Applied Physics Letters*, 94, pp. 142906.

Albayrak, E. & Cengiz, T. (2011). Sound attenuation for the spin-1 Ising model on the Bethe lattice, *Journal of the Physical Society of Japan*, 80, pp. 054004.

Barry, J. H. (1966). Magnetic relaxation near a second-order phase-transition point, *Journal of Chemical Physics*, 45 (11), pp. 4172.

Barry, J. H., & Harrington, D. A. (1971). Theory of relaxation phenomena in Ising antiferromagnets, *Physical Review B*, 4 (9), pp. 3068.

Bennet, H. S. (1969). Frequency shifts of acoustic phonons in Heisenberg paramagnets. III, *Physical Review*, 185 (2), pp. 801.

Blume, M., Emery, V. J., & Griffiths R. B. (1971). Ising model for the λ transition and phase separation in $He^3 - He^4$ mixtures, *Physical Review A*, 4 (3), pp. 1070.

Canko, O., & Keskin, M. (2010). Relaxation theory of spin-3/2 Ising system near phase transition temperatures, *Chinese Physics B*, 19 (8), pp. 080516.

Casimir, H. B. G. (1945). On Onsager's principle of microscopic reversibility, *Reviews of Modern Physics*, 17 (2–3), pp. 343.

Chase, C. E. (1958). Propagation of ordinary sound in liquid helium near the λ point, *Physics of Fluids*, 1, pp. 193.

De Groot S. R. & Mazur, P. (1962). Non-equilibrium Thermodynamics, North-Holland Publishing Company, Amsterdam.

De Groot S. R. (1963). On the development of nonequilibrium thermodynamics, *Journal of Mathematical Physics*, 4 (2), pp. 147.

Durin, G., Bonaldi, M., Cerdonio, M., Tommasini, R. & Vitale, S. (1991). Magnetic viscosity of Co-based amorphous alloys between 0.02 and 4.2 K, *Journal of Magnetism and Magnetic Materials*, 101 (1–3), pp. 89.

Erdem, R., & Keskin, M. (2001). Dynamics of a spin–1 Ising system in the neighbourhood of the equilibrium states, *Physical Review E*, 64, pp. 026102.

Erdem. R. & Keskin, M. (2003). Sound dispersion in a spin–1 Ising system near the second-order phase transition point, *Physics Letters A*, 310, pp. 74.

Erdem, R., (2008). Magnetic relaxation in a spin–1 Ising model near the second-order phase-transition point, *Journal of Magnetism and Magnetic Materials*, 320, pp. 2273.

Erdem, R., (2009). Frequency dependence of the complex susceptibility for a spin–1 Ising model, *Journal of Magnetism and Magnetic Materials*, 321, pp. 2592.

Fannin, P. C., Marin, C. N., Malaescu., I. & Giannitsis, A. T. (2005). Microwave absorption of composite magnetic fluids, *Journal of Magnetism and Magnetic Materials*, 289, pp. 78.

Girtu, M. A. (2002). The dynamic susceptibility of a quasi-one-dimensional *Mn* Porphyrin-based hybrid magnet: Cole-cole analysis, *Journal of Optoelectronics and Advaced Materials*, 4 (1), pp. 85.

Golding, B. & Barmatz, M. (1969). Ultrasonic propagation near the magnetic critical point of Nickel, *Physical Review Letters*, 23 (5), pp. 223.

Gulpinar, G., Demirhan, D., & Buyukkilic, F. (2007). Relaxation dynamics of Iron-group dihalides, *Physical Review E*, 75, pp. 021104.

Gulpinar, G. (2008). Critical behaviour of sound attenuation in metamagnetic Ising system, *Physics Letters A*, 372 (2), pp. 98.

Gulpinar, G., & Iyikanat, F. (2011). Dynamics of the Blume-Capel model with quenched diluted single-ion anisotropy in the neighbourhood of the equilibrium states, *Physical Review E*, 83, pp. 041101.

Halperin, B. I., Hohenberg, P. C., & Ma, S. (1974). Renormalization-group methods for critical dynamics: I. recursion relations and effects of energy conservation, *Physical Review B*, 10, pp. 139.

Hurley, J. & Garrod, C. (1982). Generalization of Onsager reciprocity theorem, *Physical Review Letters*, 48 (23), pp. 1575.

Ising, I. (1925). Beitrag zur theorie des ferromanetismus, *Zeitschrift für Physik*, 31, pp. 253.

Kaneyoshi, T. & Nakamura, Y. (1998). A theoretical investigation for low-dimensional molecular-based magnetic materials, *Journal of Physics: Condesed Matter*, 10, pp. 3003.

Keskin, M. & Erdem, R. (2003). Critical behaviours of the sound attenuation in a spin–1 Ising model, *Journal of Chemical Physics*, 118 (13), pp. 5947.

Kılıç, A., Okur, S., Güçlü, N., Kölemen, U., Uzun, O., Özyüzer, L. & Gencer, A. (2004). Structural and low-field characterization of superconducting MgB_2, *Physica C*, 415 (1–2), pp. 51.

Koch, W., Dohm, V., & Stauffer, D. (1996). Order-parameter relaxation times of finite three-dimensional Ising-like systems, *Physical Review Letters*, 77 (9), 1789.

Kondepudi, D. & Prigogine, I. (2005). Modern Thermodynamics From Heat Engines to Dissipative Structures, John Wiley & Sons, USA.

Kötzler, J., & Eiselt, G. (1979). Observation of spin cluster freezing in dilute $(Eu_xSr_{1-x})S$ by low-frequency magnetic absorption, *J. Phys. C: Solid State Physics*, 12, pp. L469.

Landau, L. D. & Khalatnikov, I. M. (1954), Dokl. Akad. Nauk SSSR, 96, pp. 469.

Landau, L. D. & Khalatnikov, I. M. (1965), English Translation: in Collected Papers of L. D. Landau, edited by D. Ter Haar, Pergamon, London. pp. 626.

Landau, L. D., & E. M. Lifshitz, E. M. (1981). Physical Kinetics, Course of Theoretical Physics, Vol. 10, Pergamon Press, New York.

Lavis, D. A., & Bell, G. M. (1998). Statistical Mechanics of Lattice Systems I, Springer, London.

Lucas, G. & Stephen M. J. (1967). Relaxing of the superconducting order parameter, *Physical Review*, 154 (2), pp. 349.

Lüthi, B. & Pollina, R. J. (1969). Sound propagation near the magnetic phase transition point. *Physical Review Letters, 22* (14), pp. 717.

Lüthi, B., Moran, T. J. & Pollina, R. J. (1970). Sound propagation near the magnetic phase transition, *Journal of Physics and Chemistry of Solids, 31* (8), pp. 1741.

Lyakhimets, S. N. (1992). Phenomenological description of magnetic relaxation in magnetic crystals with induced anisotropy, *Physica Status Solidi (b)*, 174, pp. 247.

Maekawa, S. & Tachiki, M. (1978). Ultrasonic attenuation and exchange striction in rare-earth metals, *Physical Review B*, 18 (7), pp. 3736.

Meixner, J. (1972). The fundamental inequality in thermodynamics, *Physica*, 59, pp. 305.

Moran, T. J. & Lüthi, B. (1971). High-frequency sound propagation near magnetic phase transition, *Physical Review B*, 4, pp. 122.

Naya, S. & Sakai, Y. (1976). Critical dynamics of polyorientational phase transition -linear relaxation theory-, *Journal of the Physical Society of Japan*, 40, pp. 319.

Onsager, L. (1931). Reciprocal relations in irreversible processes I, *Physical Review*, 37, pp. 405.

Racz, Z. (1976). Nonlinear Relaxation Near the Critical Point: Molecular-field and Scaling Theory, *Physical Review B*, 13, pp. 263.

Ray, T. S., Tamayo, P., & Klein, W. (1989). Mean-field study of the Swendsen-Wang dynamics, *Physical Review A*, 39, pp. 5949.

Reichl, L. E. (1998). A Modern Course in Statistical Physics, John-Wiley & Sons, New York.

Saslow, W. M. & Rivkin, K. (2008). Irreversible thermodynamics of non-uniform insulating ferromagnets, *Journal of Magnetism and Magnetic Materials*, 320, pp. 2622.

Schmid, A. & Schon, G. (1975). Linearized kinetic equations and relaxation processes of a superconductor near T_c, *Journal of Low Temperature Physics*, 20, pp. 207.

Schuller, I. & Gray, K. E. (1976). Experimental observation of the relaxation time of the order parameter in superconductors, *Physical Review Letters*, 36, pp. 429.

Sperkach, Y. V., Sperkach, V. S., Aliokhin, O., Strybulevych, A. L. & Masuko, M. (2001). Temperature dependence of acoustical relaxation times involving the vicinity of N-

I phase transition point in 5CB liquid crystal, *Molecular Crstals and Liquid Crstals*, 366, pp. 183.

Stoll, E., Binder. K., & Schneider, T. (1973). Monte-Carlo investigation of dynamic critical phenomena in the two-dimensional kinetic Ising model, *Physical Review B*, 8 (7), pp. 3266.

Tanaka, T., Meijer, P. H. E. & Barry, J. H. (1962). Theory of relaxation phenomena near the second-order phase-transition point, *Journal of Chemical Physics*, 37 (7), pp. 1397.

Van Raap, M. B. F., Sanches, F. H., Torres, C. E. R., Casas, L., Roig, A. & Molins, E. (2005). Detailed dynamic magnetic behaviour of nanocomposite iron oxide aerogels, *Journal of Physics: Condensed Matter*, 17 (41), pp. 6519.

Verhas, J. (1983). An extension of the govering principle of dissipative processes to nonlinear constitutive-equations, *Annalen Der Physik*, 40 (4–5), pp. 189.

Walther, K. (1967). Ultrasonic relaxation at the Neel temperature and nuclear acoustic resonance in MnTe, *Solid State Communications*, 5 (5), pp. 399.

Wolf, W. P. (2000). The Ising model and real magnetic materials, *Brazilian Journal of Physics*, 30 (4), pp. 794.

13

The Thermodynamics of Defect Formation in Self-Assembled Systems

Colm T. O'Mahony, Richard A. Farrell, Tandra Goshal, Justin D. Holmes and Michael A. Morris
The Tyndall National Institute and University College Cork and CRANN, Trinity College Dublin Ireland

1. Introduction

The self-assembly of matter into highly ordered structures at the nanometer scale or beyond, is a topic that has attracted significant research over the past two decades. The term self-assembly is used to describe spontaneous processes where nanoscale entities pack into regular arrangements in order to attain a minimum free energy through minimisation of repulsive and maximisation of attractive molecular interactions (Whitesides & Grzybowski, 2002). Note that in most self-assembly processes, the formation of regular arrangements are enthalpy driven but in certain circumstances, entropy driven processes can produce ordered arrangements as discussed briefly below. These pattern formed in a self-assembly process have potential importance, as they may provide techniques to generate nanostructured surfaces by a simple, cost-effective method as opposed to current, expensive lithographical processes. The formation of these 'self-assembled' structures is thermodynamically driven and derived from the lower free energy of the structured, assembled system compared to that of the random structure. The lower free energy is usually a result of weaker intermolecular forces between, assembling or organising, moieties and is essentially enthalpic in nature. Pattern formation is a thermodynamic comprise between pattern generation, rate of pattern formation and the degree of disorder (Whitesides & Mathis, 1991).

Disorder can arise in self assembled systems in two ways; intrinsic and extrinsic sources. Thermodynamically intrinsic defect formation is defined by the entropy of a system and the free energy of defect formation. Disorder can also occur from kinetically derived extrinsic defects arising from mass-transport imitations in pattern formation. Extrinsic defects can also arise from contamination, moiety size irregularities, substrate effects, mechanical damage, etc. Understanding defect formation, the resultant density and control of defect concentrations is of critical importance is the possible application of these materials. The term self-assembly has also been applied to processes not involving individual entities but also has been used to describe processes such as phase separation within a single component (as in the example described here, block copolymer (BCP) microphase separation, and this is discussed in depth below). Phase separation can probably be more correctly described by the related term self-organisation. The difference between self-assembly and self-organisation can be difficult to differentiate (Misteli, 2001) and this is given consideration below. Very often in chemistry self-assembly and self-organisation are used interchangeably and we will continue this practice in this article. As more explicitly stated below, self-assembly is generally reserved for systems that are driven to equilibrium via physical interactions between entities (a free energy minimum). Self-organisation refers to a dynamic process where the assembled or organised structure is in a steady state. In block copolymer self-assembly, the interactions between different blocks results in a free energy minimum for specific ordered arrangements of the chains. These structures have a nano-dimension (i.e. the same scale as the polymer blocks) because complete phase separation cannot occur because of the bonds between blocks. For BCP systems, a well-defined order-disorder transition is present defined by a temperature where the chains are too mobile to form discrete ordered arrangements. As described below, these systems reach a true equilibrium and the term self-assembly can be properly used as outlined further below.

1.1 Self-assembly

Self-assembly is an equilibrium process that represents a balance between repulsive and attractive forces between entities. These forces are manifest as a minimum in potential energy with distance apart and are discussed further below. This provides a useful framework for understanding and modelling the microphase separation of BCPs. The thermodynamics of the self-assembly process can be represented by a simple Gibbs Free Energy equation:

$$\Delta G_{SA} = \Delta H_{SA} - T\Delta S_{SA} \tag{1}$$

where self-assembly is a spontaneous process if ΔG_{SA} is negative. ΔH_{SA} is the enthalpy change of the process and is largely determined by the potential energy/intermolecular forces between the assembling entities. ΔS_{SA} is the change in entropy associated with the formation of the ordered or hierarchical arrangement. Since the organisation is generally (but not always) accompanied by an entropy decrease, for self-assembly to be spontaneous the enthalpy term must be negative and in excess of the entropy term. The equation shows that the self-assembly process will become progressively less likely as the magnitude of $T\Delta S_{SA}$ approaches the magnitude of ΔH_{SA} and above a critical temperature, spontaneous self-assembly will not occur. Note, that in many examples of self-assembly it may be more useful to think about self-assembly bringing about a reduction of the Helmoltz free energy since reactions are carried out in closed reaction cells.

Self-assembly is a classic example of thermodynamics. In thermodynamic terms, self-assembly is due to the minimization of free energy in a closed system and the result is an equilibrium state (Jones, 2004). In systems which are enthalpy driven and entropy appears to decrease, there is no contravention of the Second Law of Thermodynamics when the whole of the system rather than the aggregation moieties is considered. In a self-assembly process, individual lower thermodynamically stable species are used to generate more thermodynamically stable aggregates at a higher hierarchical level. The reaction proceeds via temporal changes within the system such as diffusion and other forms of mass transport which allow the aggregated self-assembled structure or a precursor there of to be 'sampled'. The lifetime (τ) of the aggregated structure is given by and equation of the form such as:

$$\tau = \tau_o \left(\exp\left(\Delta G_{SA}/RT\right)\right) \tag{2}$$

where τ_o represents the average time between collisions or vibrations or similar events that may lead to dissociation of the aggregate. This is an important equation because it suggests that although these self-assembled arrangements may be experimentally observed microscopically, spectroscopically and structurally, they may be in constant flux and changing during the observation. It is, of course, an absolute pre-requisite that the system is in some sort of flux because of the nature of the equilibrium process. The individual components (molecules, particles, polymer chains etc.) must 'sample' the organised structure for the self-assembly process to occur. Kinetic energy is thus an important parameter in a self-assembling system. If it is too low, the rate of formation of the organised structure will be to low to be practical. If it is too high, the organised structure will not form. Understanding the self-assembly is a fundamental challenge in modern science. It is at the very heart of nature and is responsible for all life and the thermodynamics of this process have been considered in depth not only scientifically but form a more philosophical view (Bensaude-Vincent, 2009).

It should also be noted that not all self-assembly processes are examples of simple thermodynamics. Self-assembly can result in metastable states whose form is dependent on an external source of energy. This is often described as dynamic self-assembly. Temperature, magnetic fields, chemical reactions, light etc. can be used to select or modify the metastable states that are formed (Grzybowski et al., 2009). The term dynamic self-assembly was coined by Whitesides and Grzybowski (Whitesides and Grzybowski, 2002). However, the definition they used is more applicable to the use of the term self-organisation. Self-assembly, certainly in the field of nanotechnology, has been used interchangeably with self-organisation but it is becoming clear that an exact differentiation between these is necessary at least in the biological field Halley and Winkler, 2008). Halley and Winkler outline the main distinctions between these two processes that can result in the formation of well-organised patterns from smaller entities. Firstly, whilst self-assembly is a true equilibrium process, self-organisation is not and requires an external (to the system) energy source. It is this definition that defines the term dynamic self-assembly as a true self-organisation phenomenon. It should be noted that practically the two processes can be separated experimentally because the organisation decays one the energy source is terminated. Secondly, self-assembly produces a well-defined, stable structure determined by the components of the system and the interactions between them whilst self-organisation is more variable with the structures being less stable and prone to change (Gerhart and Kirschner, 1997). Finally, self-assembly can be associated with a very limited number of components but in self-organisation there is a much higher number of components below which the organisation cannot occur (Nicholas and Prigogine, 1995). Whilst, the definition described here seems clear cut, this remains an area of considerable debate and arguments for the foreseeable future will continue.

It was suggested that above that generally, self-assembly is an enthalpy driven process. This is certainly true for work described in the area of nanotechnology on which this article focuses. However, in biology as well as a few examples in materials science, self-assembly is entropy not enthalpy driven. This may seem counter-intuitive but is well known and documented for systems such as micelles (Capone et al., 2009), some liquid-crystal molecules and colloidal particles (Adams et al., 1998) as well as various biological systems such as viruses (Frared et al., 1989). The reason the entropy increase favours more organised structures is because these structures allow more degrees of freedom within the system. E.g. increased rotation when rod-like structures are arranged parallel to one another rather than in an entanglement or increased degrees of freedom or water molecules in cell-like structures formed by micelles. In certain cases, self-assembly can be driven by both entropy and enthalpy (Thomas et al., 2001)

Because self-assembly is an equilibrium process, the assembled components are in equilibrium with the individual components. The assembly (in practical circumstances there will be very many assemblies formed) is governed by the normal processes of nucleation and growth. Small assemblies are formed because of their increased lifetime as the attractive interactions between the components lower the Gibbs free energy, equation 2. As the assembly grows the Gibbs free energy continues to decrease until the assembly becomes stable for long periods. The necessity for this to be an equilibrium process is defined by the organisation of the structure which requires non-ideal arrangements to be sampled before the lowest energy configuration is found. If the process is to occur around room temperature the nature of the forces between the components must necessarily by quite small i.e. of the order of kT. Thus, the forces tend to be intermolecular in type rather than ionic or covalent which would 'lock' the assembly into non-equilibrium structures. The types of forces seen in most self-assembly processes are van der Waals, hydrogen bonds, weak polar forces, metal chelation, etc (Lindoy and Atkinson, 2001).

As regular structural arrangements are frequently observed (and, in fact, why many self-assembling processes find potential application in modern technologies) in self-assembly (although the formation of ordered structures is not necessary in a self-assembling system) it is clear that there must be a balance of attractive and repulsive forces between entities or else an equilibrium distance would not exist between the particles. The repulsive forces can be electron cloud - electron cloud overlap, electrostatic or derive from the differences in cohesive energies between the assembling components. Since, for practical reasons, the assembly is generally not at the atomic or small molecule scale (for practical reasons outlined below) it is generally necessary that both attractive and repulsive forces are long range interactions (as distinct from short range chemical bonds) if the separation distance between features is to be in the nanometre range. This can be illustrated using very simple consideration of the intermolecular forces between the entities. If we assume that the attractive intermolecular forces can be modelled as an attractive potential between similar point charges (Q), the potential energy (V_{att}) follows a 1/r dependence and can be written as:

$$V_{att} = -Q^2/4\pi\varepsilon_o r \quad (3)$$

where r is the separation of the entities. The repulsive charge can be modelled as $V_{rep} \alpha 1/r^n$. Assuming a charge of around 1.37×10^{-19} C and a repulsive constant = 10000 $(10^{-9})^n$ kJ mol^{-1} the variation in the total potential-distance curves ($V_{tot} = V_{att} + V_{rep}$) as a function of n in the repulsion term can be plotted, figure 1(A). The curves describe a classic potential energy well with a minimum V_{tot} at an equilibrium separation distance between the entities. It is worth noting since RT ~ 2.5 kJ mol^{-1}, that the minimum value of V_{tot} must be significantly greater than this to provide a driving force for assembly that compensates for an entropy decrease (equation 1). The minimum value of V_{tot} can be approximately associated with ΔH_{SA} assuming that there is no volume or temperature change during self-assembly. The attractive term is long range in nature and the width of the potential energy well that is formed is defined, in this case, the repulsive term. At n = 4 there is a well-defined potential energy minimum. This is important because it will precisely define an equilibrium distance between entities that is necessary if structural regularity is going to be high. The effect of increasing n, *i.e.*, increasing the short-range nature of the repulsive forces, is to reduce the value of the potential energy minimum, increase the width of the potential energy well and move the minimum to greater distances. The increasing shallowness of the well is a major problem in terms of generating patterns of high structural regularity because it ensures a variation in spacing between entities (or features in phase-separated systems outlined below) can exist with little energy cost.

The effect of increasing the dependence of potential energy with distance, *i.e.*, increasing the short-range nature of the attractive potential, also has a dramatic effect on the potential energy curve. This is modelled using a V_{att} that follows a $1/r^n$ dependence whilst using a constant repulsive term that varies as $1/r^6$. Illustrative data are shown in Figure 1B. Increasing the value of n reduces the value of the potential energy minimum, increases the width of the potential energy well and moves the minimum to greater distances as described for the repulsive forces above. However, as n increases the decrease in the value of the potential energy minimum is very considerable such that changing n from 1 to 3 reduces the potential energy minimum by a factor of around ~x 150. Although these are quite simple calculations they do illustrate some important concepts in self-assembly on the mesoscale. Firstly, if the entities are to be relatively large distances apart, the repulsive and attractive forces between the entities will need to be relatively high or the potential energy will not provide an effective driving force at room temperature. Secondly, as the spacing between entities or features increases, variations in the separation distance within the self-assembled structure will increase dramatically and lead to poor structural regularity. Finally, for self-assembly to be effective, there needs to be a delicate balance of the intermolecular forces and because of this self-assembly with high structural regularity is not common-place and will require careful molecular or particle design coupled to optimisation of the process.

Finally, it should be stressed that self-assembly is a spontaneous chemical process where entities or components within a mixture arrange themselves in a structured manner and these processes take place in normal chemistry environments *e.g.*, solution mediated. Normally the self-organisation is borne from an initially disordered system. Importantly the equilibrium low-energy arrangement is reached from positional fluctuations as a result of thermal effects. Thus, the effective interaction potential between the entities or components cannot exceed thermal energy by too great a factor or else it will not be possible to minimise positional errors in the in the arrangement. Alternatively, there has to be enough difference between thermal energy and the interaction potential energy to maintain order within the pattern.

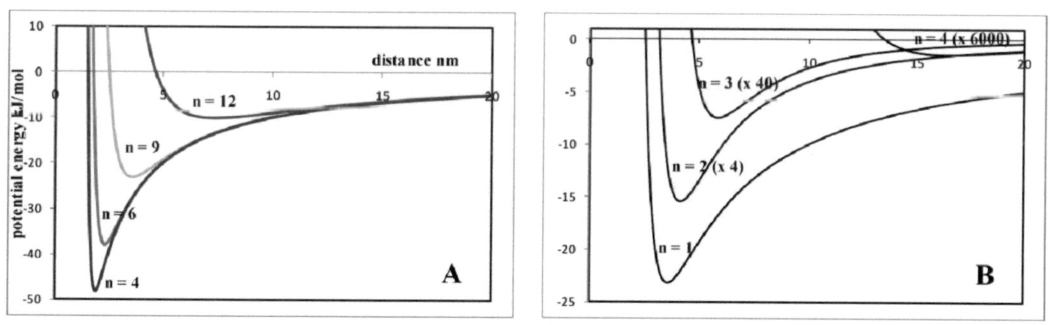

Fig. 1. Potential energy against distance curves. A – result of increasing the short range nature of the repulsive forces between entities in a self-assembly process. Note the increasing width of potential energy well. B – result of increasing the short range nature of the attractive forces between entities in a self assembly process. Note the dramatic decrease in the depth of the well.

1.2 The need for low defect concentrations in self-assembled systems

The self-assembly of BCP (block copolymer) systems can be more properly described as microphase separation and is becoming a subject of research for potential commercial development. Few self-assembling systems can approach the structural regularity of BCP systems, with perhaps only mesoporous silicates (Rice et al., 2007), track-etched polycarbonate membrane (Fang and Leddy, 1995) and porous anodic alumina (Petkov et al., 2007) rivalling the polymer systems in the feature size range of interest and having been shown to offer opportunities for controlled alignment. It can be seen from the arguments made above that the high regularity of the systems is because the intermolecular forces that drive the self-assembly are such that highly periodic structures are favoured and ordering can be attained at practical temperatures. One of the advantages of the BCP self-assembly methodology compared to the other forms mentioned (in the other cases the final structure is not reversible on application of temperature because the structure has been templated via components that have been subsequently removed or through techniques where chemical bonds are either made or destroyed) is that the film can be 'annealed' after their formation to improve the regularity of the self-assembled structures and this approach is described further below. Other self-assembled systems such as nanoparticle superlattices (Pileni, 2001) also produce highly regular and sometimes complex structures but transferring them to a macroscopic scale is difficult. In general, nanoparticle self-assembly development has been limited because the synthesis of size mono- dispersed particles is challenging for all but a few systems and thin films of these tend to lack thermal and mechanical robustness. Compared to other techniques, BCP systems offer a combination of experimental advantages; thin films can be formed from simple solutions, the resultant films are robust, the feature size is highly controllable using polymer engineering and the films are readily processed (*e.g.*, in pattern transfer where the polymer pattern is transferred to the surface by selective etch processes (Borah et al., 2011 and Farrell et al., 2010). Authors have demonstrated many applications for microphase separated BCP thin films. BCP micelle systems have found commercial use in applications such as drug delivery but these are not the focus of the work described here and the reader is directed to some excellent reviews (Lodge, 2003) Applications for BCP films in the general area of materials science include solid state battery electrolytes (Soo et al., 1999) and membrane separation technologies (Ulbricht, 2006). Park and co-workers have provided an extensive review of technologies that might be developed using BCP thin films and these are largely in the area of development of strategies to develop nanoscale electronics, magnetics and photonics. These ICT focussed technologies include low dielectric materials for electrical insulation and reduction of crosstalk, high density magnetic storage media and photonic band gap crystals. By far the most researched area for use of these materials is as potential alternatives to conventional mask-based photolithography for fabrication of nanoelectronic circuitry. Photo-lithography has been the cornerstone of the electronics industry since the advent of the first silicon devices (Pease & Chou, 2008). The photolithographic process has been continually developed to allow the size of devices to be decreased and the density of devices constantly increased so that individual transistor sizes have shrank from cm type sizes to around 50 nm. The trend in resolution enhancement was, for many years, achieved by reducing the dimensions of the mask patterns whilst simultaneously decreasing the wavelength of the radiation (light) (Bloomstein et al., 2006). Currently, techniques such as immersion technologies whereby a liquid (usually water) is placed directly between the final lens and photoresist surface resulting in a resolution enhancement defined by the refractive index of the liquid have allowed device engineers to pattern transfer feature sizes (65 or 45 nm generation) that are actually less than the wavelength of light used (193 nm). Device performance is ultimately limited by the density of transistors on the chip and it is clear that patterning requirements will continue to the 32 nm node and beyond. Although photolithography can potentially be used to create sub-10 nm device structures for high volume manufacturing processes, it will necessitate the use of deep UV (13 nm) and x-ray sources and these are associated with high costs and materials implications for the masks and resists (ITRS roadmap, 2005).

For these reasons, self-assembly may have importance for transistor manufacture beyond the 22 nm node. The advantages of self-assembly over conventional and non-lithographic methods include: (i) the reduction of source costs, (ii) elimination of masks and photoresists, (iii) non-existence of proximity affects, (iv) the possibility of developing 3D patterning techniques, (v) absence of diffraction restrictions to resolution and (vi) can be used to pattern materials with precision placement techniques by availing of templating (*i.e.*, deposition of materials within the structure, known as graphoepitaxy) or a chemical pattern (alternating surface chemistries). The microphase separation of block copolymers is emerging as the most promising method of assembling highly ordered nanopatterns at dimensionalities and regularity approaching the future device dimension requirements. These requirements are extremely challenging for self-assembly and lithography alike and include sub-nm line edge roughness and sub-4 nm positioning (of a feature expressed from the overlay registry requirements) accuracy for the 16 nm technology node. The potential application of BCPs in this area has been extensively reported and reviewed (Jeong et al., 2008). These reviews also detail the methods by which the polymer nanopatterns can be processed into active components (*i.e.*, nanowires, nanodots of semiconducting, magnetic or conducting materials). If BCPs are to ever contribute to the development of devices with these types of dimensions then control and minimisation of defects is essential. In the remainder of this chapter we will explore the thermodynamics of defect formation in BCP thin films.

2. Block copolymer systems

In order to understand how defects form in BCP thin films, it is first important to understand the energetic of BCP self -assembly. BCPs have become increasingly more important materials as routine design and synthesis of these polymers has become practical. BCPs were first developed to essentially tune the properties of the macromolecule between that of the two blocks individually. The advantage of using a single macromolecule rather than a blend is that the macroscopic phase separation in mixtures cannot occur. However, the chemical mismatch does lead to microphase separation as described below. Industrial synthesis of BCPs was first demonstrated in the 1950s by scientists at BASF and ICI around the development of triblock systems of poly(ethylene oxide) and poly(propylene oxide) as a new class of synthetic surfactants. Amongst many applications these found widespread use not only as surfactants but also as anti-foaming agents, cosmetic materials and drug release materials (D'Errico, 2006). More recently they have found use as versatile 'templating agents' for the generation of ordered nanoporous silicates allowing precise control of pore diameters. Spandex was the first BCP to be widely known because of its use in textiles (spandex is an anagram of expands) and was invented by the DuPont chemist J. Shivers. It became apparent that the possibility of forming macromolecules with blocks of differing chemical properties could yield materials where the interaction of the different blocks would ordain important physical properties. Many aspects of BCPs have been reviewed in depth (Hamely, 2004). This article will be restricted to discussion of the formation of nanopatterns of these materials in thin film form on substrates. The nanopatterns are essentially the result of the self-organisation *via* microphase separation of the BCP at the surface and not *via* micelle formation and related phenomena of the BCP in solution. The term microphase separation is used because it is the chemical dissimilarity of the blocks which drive the different blocks apart but complete phase separation is not possible because of the covalent bonds linking the blocks. These bonds act as a restoring force and result in a series of ordered patterns as discussed below. Lyotropic phases will not be discussed at length here, however, solvent effects cannot be completely ignored because it is convenient and practical (particularly for the thin films discussed here) that the polymers are solvent cast onto the substrate surface by techniques such as dip- and spin-coating. Further, a technique known as solvent-annealing or solvent-swelling is becoming common place as a means of attaining high degrees of structural regularity. This ordering is a result of the increased mobility within the macromolecule block network related to the decrease in the glass transition temperature as a result of solvent molecule inclusion (Kim & Libera, 1998).

2.1 The thermodynamics of microphase separation in block copolymers

The thermodynamics of microphase separation in BCPs has been reviewed several times following the original work of Bates (Bates, 1991). The theory will not be detailed in depth here except to show how it relates to intermolecular forces through the solubility parameter and how the thermodynamics of defect formation in these systems can be properly understood. Most of the understanding of microphase separation of BCPs is centred on a term known as the interaction parameter χ. Assuming a simple di-block copolymer made up of sub-units A and B, the χ value resulting from the interactions between block A and block B can be written as:

$$\chi = z\Delta w/kT \tag{4}$$

and χ is the exchange energy per molecule normalised by the thermal energy kT and is dimensionless. The number of neighbours surrounding one block is z. Δw is the exchange energy which is the difference in energy between the interaction between block A and block B and the average of the self interactions between block A-block A and block B-block B. That is, Δw is the energy cost of taking a block of A from surrounding A blocks and placing in a B block environment and doing the same for a B block (from a B environment to an A environment). The interaction parameter can be related directly to the molar enthalpy change of mixing, ΔH_m, by:

$$\Delta H_m = f_A f_B \chi RT \tag{5}$$

where f_A and f_B are the volume fractions of the blocks. By conventional solution theory and assuming no volume change on mixing, it can be shown that:

$$\chi = V_m(\delta_A - \delta_B)^2/RT \tag{6}$$

where δ_A and δ_B are the solvent parameters (see below) of the two blocks. Therefore:

$$\Delta H_m = f_A f_B V_m(\delta_A - \delta_B)^2 \quad (7)$$

This is important because it shows that any block copolymer system where the blocks have different solubility parameters (*i.e.*, different strengths and forms of intermolecular interactions) will have a positive enthalpy of mixing and will, thus, have a tendency to microphase separate and self-assemble provided the entropy change (which in the case of BCP self-assembly always decreases as discussed above) associated with the process is not too large as to overcome the enthalpy contribution. Flory-Huggins theory has been the basis for modelling the behaviour of block copolymers since their invention and remains the most used model to date providing a robust basis for the prediction of morphology seen in BCP microphase separated systems. Using this formulism the configurational entropy of phase separation is assumed as the only major contribution to energy such that the entropy associated with microphase separation $\Delta S_m = k\ln\Omega$ where Ω is the number of possible ways of arranging the system. *Via* Stirling's approximation the entropy change can be written as can be written as:

$$\Delta S_m/RT = (1/N_A)\ln f_A + (1/N_B)\ln f_B \quad (8)$$

where N_A, N_B are the degrees of polymerisation of each block such that $f_A = N_A/(N_A + N_B)$. Since the entropy decreases in the system on mixing and using equation 5:

$$\Delta G_m/RT = f_A f_B \chi + (1/N_A)\ln f_A + (1/N_B)\ln f_B \quad (9)$$

This Equation specifically relates to the mixing process and not phase separation. The implication is that the free energy of mixing is always likely to be positive bearing in mind the definition of χ given in equations 4 and 6. The driving force for self-assembly is the minimisation of the free energy of mixing by the regular patterns formed by microphase separation. For illustrative purposes consider the formation of a regular, microphase separated, lamellar phase consisting of alternating stripes of blocks from a AB block copolymer with $f_A = f_B$. The lamellar structure is a common motif in phase separation because it is achieved with lowest mass transport limitations. This is particularly important considering that phase separation is limited by the covalent bonding between blocks and all theories suggest this is the lowest energy structure for BCPs. As can be seen from equation 4, ordered self-assembly/microphase separation will occur provided that $\Delta G_{SA} = G_{mix} - G_{PS}$ is negative. The free energy change in forming the lamellar structure ($\Delta G_{SA,L}$) can be described by modelling G_{mix} as a sum of AB contacts and G_{PS} as a Hookian term describing the balance of repulsive enthalpic and attractive/restorative entropic forces (as detailed above) plus an interfacial term. In this way (Matsen & Bates, 1996) the G_{PS} can be written:

$$G_{PS} = 1.19(\chi_{AB}N)^{1/3} \quad (10)$$

and the equilibrium spacing between stripes in the lamellar structure (L) as:

$$L = 1.03a(\chi AB)1/^6N^{2/3} \quad (11)$$

Since the sum of simple contacts in the mixed system allows G_{mix} to be estimated as $(\chi_{AB}N)/4$ it is possible to write that for microphase separation to occur G_{mix} must be greater or equal to G_{PS} and the minimum condition is:

$$1.19(\chi_{AB}N)^{1/3} = (\chi_{AB}N)/4 \tag{12}$$

Thus, for microphase separation $\chi_{AB}N$ must be greater than 10.4. Since χ_{AB} is a measure of the chemical dissimilarity between the units (mers) in the blocks $\chi_{AB}N$ represents the total dissimilarity over the whole macromolecule. Using equation 12 the minimum value of $(\chi_{AB}N)_{min}$ to bring about phase separation is about 10.4. This very simple approach provides a value for $(\chi_{AB}N)_{min}$ which is very similar to much more complex theories developed by Leibler using self-consistent field theory (Leibler, 1980). A summary of recent theoretical developments in block copolymer phase separation has been provided by Grason (Grason, 2006).

3. Origin of defects

In a self-assembled structure there are likely to be reasonable concentrations of defects. This is suggested in equation 1, $\Delta G_{SA} = \Delta H_{SA} - T\Delta S_{SA}$, because in most cases the thermodynamic driving force for self-assembly is provided by weak intermolecular interactions and is usually of the same order of magnitude as the entropy term. Practically, for any self-assembling system to reach the minimum free energy configuration there must be enough thermal energy to allow the mass transport of the self-assembling moieties. In these circumstances, obtaining defect free self-assembly over macroscopic areas is improbable. A self-assembled nanopatterned surface is likely to show a number of distinct irregularities or defects and these can take many forms as outlined below. The origins of these defects are manifold but each defect comes with an energy cost because it disrupts the arrangement of the polymer blocks which provides a free energy minimum within the film.

3.1 Equilibrium defects

As above, the thermodynamically defined concentration of defects originates from a balance of configurational entropy and the energy cost associated with the defect. These defects are 'statistical' in nature and while individual defects may have limited lifetimes a population of them will always exist at a concentration defined by conditions. The thermodynamics of defects in fully equilibrated systems is well understood but care must be taken to separate the free energy defining self-assembly and pattern formation from the free energy of defect formation so that the role of intermolecular forces can be well understood. For defect formation the free energy of single defect formation is given by:

$$\Delta G_{DF} = \Delta H_{DF} - T\Delta S_{DF} \tag{13}$$

The enthalpy term, ΔH_{DF}, does not necessarily reflect the intermolecular forces between blocks - it is the energy cost associated with disrupting the pattern and may be thought of as a region where optimum arrangement does not occur and the reduction of enthalpy associated with ideal self-assembly is not realised. For example, a system of hexagonally packed cylinders may exhibit defect regions of lamellar structure. The enthalpy of defect formation is partially related to the enthalpic difference between the two structural arrangements and this might be much less than the intermolecular forces between blocks. If the difference in enthalpy of two different arrangements is small a relatively high equilibrium concentration of defects might be expected compared to one where the enthalpy difference is large. The entropy difference now reflects the order change between the prefect and defective structural arrangement. Note here that the enthalpy cost of creating a defect is not only determined by the respective differences in structural arrangement but also by strain energy (due to tensile and compressive forces that are associated with defect insertion) as well as any interfacial effects arising from intermolecular interactions in the areas around the defect . If ΔG_{DF} is negative there will be a finite number of defects in the system and the concentration will be given by:

$$N/N_0 = \exp(-\Delta E_{act}/RT) \tag{14}$$

N is the number of defects in a matrix of N_0 self-assembled moieties or features and ΔE_{act} is the activation energy of defect formation. The activation energy ΔE_{act} should not be confused with ΔH_{DF}. The activation energy represents energy difference between the initial ideally arranged state and a transition state towards the defective structure. Equation 13 can be used to estimate the defect concentration by use of the Boltzmann formula to estimate entropy S

$$S = k\ln W \tag{15}$$

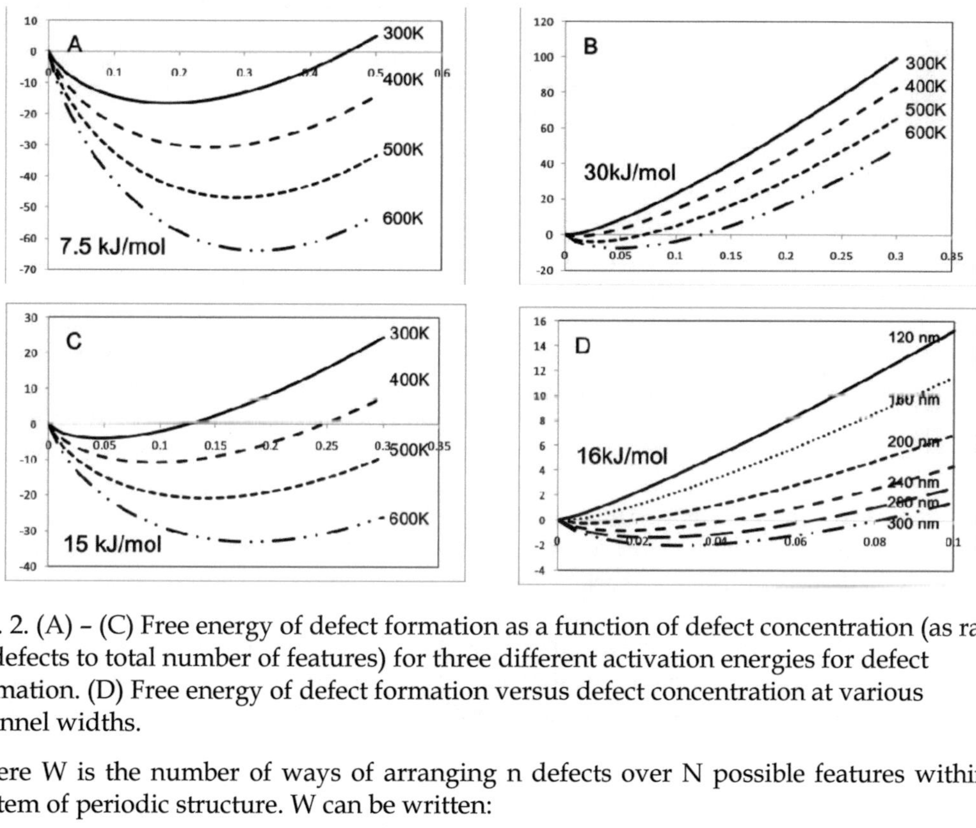

Fig. 2. (A) - (C) Free energy of defect formation as a function of defect concentration (as ratio of defects to total number of features) for three different activation energies for defect formation. (D) Free energy of defect formation versus defect concentration at various channel widths.

where W is the number of ways of arranging n defects over N possible features within a system of periodic structure. W can be written:

$$W = N!/((N-n)!n!) \tag{16}$$

and using Stirling's approximation we can write that :

$$\ln N! \sim N \ln N - N \tag{17}$$

so that:

$$T\Delta S_{DF} = kT[N \ln N - n \ln n - (N - n) \ln (N-n)] \tag{18}$$

and:

$$\Delta G_{DF(n)} = n\Delta H_{DF} - kT[N \ln N - n \ln n - (N - n) \ln (N-n)] \tag{19}$$

where $\Delta G_{DF(n)}$ represents the free energy cost of forming n defects in the system. This formulism allows a plot of free energy against n/N, the defect ratio, as shown in Figure 3. Three plots (Figures 3A-C) are shown for three different defect formation energies typical of what might be expected for a BCP system *e.g.*, Hammond (Hammond et al., 2008) has measured the defect formation energy in a PS-*b*-P2VP system at around 30 kJ mol^{-1}. Although these are simple calculations they illustrate the salient features of equilibrium defect formation.

At low defect concentrations defect formation is entropy driven until a critical concentration of defects allows the activation energy term to compensate for entropy. There is usually an equilibrium defect density indicated at the minimum free energy. As might be expected, as the activation energy for defect formation increases this equilibrium defect density. At high activation energy values (*e.g.*, around 30 kJ mol^{-1}) and low temperature (300 K) there is no thermodynamic driving force for defect formation and the data suggests that in BCP systems it should be possible to form highly regular structures.

3.2 Non-equilibrium defects

Practically, there are few examples of defect free microphase separation of BCP thin film systems even in cases where the number of equilibrium defects is vanishingly small. As these films are normally prepared by non-equilibrium methods such as spin- or dip-coating the microphase separated structure evolves by either thermal or solvent annealing and defects are introduced (nucleation of microphase separated regions) or removed by growth kinetics. Through the annealing cycle phase separated regions will nucleate in various places, grow and increase in order (Segalman et al., 2003). This growth will produce the classical structural motif of a 'polycrystalline' type grain structure where local ideal arrangements are separated from one another by extended defects akin to grain boundaries. Grains will grow by consumption of smaller grains and this process will be kinetically limited as described below. Other non-equilibrium or extrinsic defects can be present and these include defects induced by surface flaws, poor polymer size dispersion and impurity inclusion.

Theoretically, non-equilibrium defects resulting from pattern errors (as distinct from those precipitated by surface defects, impurities etc.) can be removed by annealing; providing enough thermal energy to allow ideal configurations to be sampled thereby increasing the size of the regions of the ideal arrangement. However, due to the nature of the chemical interactions between blocks (which can be rather small) and the relatively high glass transition temperatures (which limit polymer chain mobility needed to sample the ideal ordered arrangement) coupled to low meting points and low order-disorder temperatures, the temperature window for annealing out these non-equilibrium defects may be rather small and the defects may be essentially kinetically metastable. In many examples in the literature, BCP films are ordered at just above the glass transition temperature conferring enough chain mobility but as far as practically possible from the order-disorder temperature. This methodology may necessitate very extended heating times to remove defects and practically (because of local and large area mass transport limitations) equilibrium may not be achieved even after inordinately long annealing periods and non-equilibrium defects will still be present. This is largely due to the requirement for defect annihilation associated with coarsening of the randomly orientated grain structure described above that results from the kinetics of nucleation and growth of phase separated regions (Harrison et al., 2000). Thus, although ΔG_{DF} may be positive for many BCP systems, implying no defects should be formed if the system attains complete equilibrium, in practice this is unlikely. In many cases a clear distinction of equilibrium defects and non-equilibrium defects cannot be practically achieved. The advent of advanced force microscopy methods facilitates defect studies without causing damage to the sample. Of particular importance are *in situ* AFM methods that allow real time data collection during pattern evolution.

3.3 Experimental studies of defect reduction in block copolymer systems

The defects that can occur in BCP nanopatterns can take several forms and it is beyond the scope of this chapter to detail these in full, however, it is worth providing a general overview. They take the form of many structural defects in other systems and can be broadly described as dislocations and disclinations and a good review is provided elsewhere (Krohner and Antony, 1975). In the simplest explanation, a dislocation is a defect that affects the positional order of atoms in a lattice and the displacement of atoms from their ideal positions is a symmetry of the medium. Screw and edge dislocations representing insertion of planes or lines of atoms are typical of dislocations. For a disclination the defects (lines, planes or 3D shapes) the rotational symmetry is altered through displacements that do not comply with the symmetry of the environment. Kleman and Friedel give an excellent review of the application of these topics to modern materials science (Kleman and Friedel, 2008).

A number of di-block BCP patterns (more complex BCPS are beyond the scope of this article) exist as a function of composition and temperature and these have been fully described elsewhere (Morris et al., 2009).The two most important phases from an application point are a lamellar phase (at a composition of around 50:50) and a hexagonal phase (composition around 33:66). The lamellar phase exists as sheets of each block arranged in a stripe pattern. A typical example is shown in figure 3. Lamellar structures can be seen in figure 3 (A and B). Normally, they adopt a 'fingerprint pattern' with a complex series of swirls and regions of parallel lines as shown in figure 3 (A) for the diblock BCP polystyrene-b-polymethylmethacrylate (PS-b-PMMA with a molecular weight of around 18,000 g mol^{-1} for each block) . The lamellae (sheets that form stripes) can be vertical to the surface plane as shown or horizontal depending on the surface chemistry. They adopt this complex fingerprint structure because this structure, since the sheets are largely parallel even though they do curve, allows almost complete minimization of the intermolecular force derived enthalpy factors driving self-assembly. However, entropy is increased and the pattern therefore allows minimisation of free energy. In certain cases where the structure can be directed (i.e. self-assembly) by interaction with pre-patterned chemistries known as chemical patterning (Nealey, 2000). The pre-patterns have a preferred chemistry with one block (e.g. hydrophobic - hydrophobic) that constrains the BCP pattern to the underlying chemical pattern. Another form of directed self-assembly is using surface topography to confer preferential pattern alignment to, e.g., the sidewall within a trench or similar. This is known as graphoepitaxy and is detailed further below. An example of the same BCP used in figure 3 (A) that has been directed into a more regular structure is shown in figure 3 (B). In this case the pattern was directed by surface strain. Figure 3 (C and D) shows patterns for a PS-b-PMMA BCP of block molecular weights around 42,000 and 21,000 g mol^{-1} respectively. This is a cylinder forming system where PMMA cylinders are distributed in an hexagonal arrangement through a PS matrix. The orientation of the cylinders, i.e. parallel or vertical to the surface plane, is determined by the surface chemistry (Hawker et al., 1997). Surfaces that are neutral, i.e. they interact with both blocks to a similar extent will cause vertical orientation of the pattern so that the PMMA cylinders are vertical to the surface plane as shown in figure 3 (C). If the surface chemically favours the matrix block, PS in this example, the cylinders will be parallel to the surface plane as shown in figure 3 (D). This arrangement forms fingerprint patterns similar to those exhibited by the lamellar structure and it can be difficult to distinguish these phases by top-down imaging alone. Surface chemistry manipulation is of great concern in controlling polymer patterns and in other forms of self-assembly.

Also shown in figure 3 are some typical pattern defects. In figure 3 (A) the circle marks an area that encloses an edge dislocation and an extra 'line' has been inserted into the pattern. The boxes in the same image mark areas enclosing disclinations which are very common in this 'fingerprint' structure. In figure 3 (C) the box marks what can be described as a ground boundary separating two distinct alignments of the hexagonal pattern. This sort of grain boundary is made up of a number of dislocations and disclinations. As discussed above, these defects may originate from equilibrium or non-equilibrium effects although absolute assignment can be difficult. The majority of dislocations and disclinations are probably equilibrium in nature but may also arise from imperfections imposed by surface flaws and impurities. Grain boundaries may similarly be of both types. Other types of defects can exist. Mis-orientation is common particularly if the substrate surface chemistry is not isotropic and variation in height etc can occur unless coating procedures and surface chemistry are very carefully controlled (Fitzgerald et al., 2009)

The thermodynamic and kinetic limitations of forming ideal self-assembled patterns are clear. In many self-assembled systems such as mesoporous silicates any defects are frozen in during synthesis because the self-assembled structure acts as a template for the formation of the final ordered structure. In this case an ordered micellar arrangement is a framework around which an inorganic framework condenses. As mentioned above, one of the key advantages of the BCP microphase separation self-assembly is the ability to anneal and reduce defect densities towards their equilibrium value. Optimum annealing temperatures have generally not been determined and are likely to vary as a function of composition and molecular weight of the BCP (since these determine the magnitude of the glass transition and melting temperature). Choice of annealing temperatures is not simple. The optimum temperature is one where ordering is achieved in a practical time but is low enough to reduce the equilibrium defect concentration to a level demanded by the application for which the materials will be used. Cooling rates are important because films may reach an equilibrium at elevated containing more equilibrium defects than desirable (but thereby allow high concentrations of extrinsic to be annealed out) but cooling at an appropriate rate would allow the equilibrium concentration to be reduced. Cooling too quickly will effectively 'freeze-in' a non-equilibrium defect concentration. The sensitivity of BCP microphase separation to temperature in thin films is illustrated below. This also shows some of the essential elements of this self-assembly mechanism.

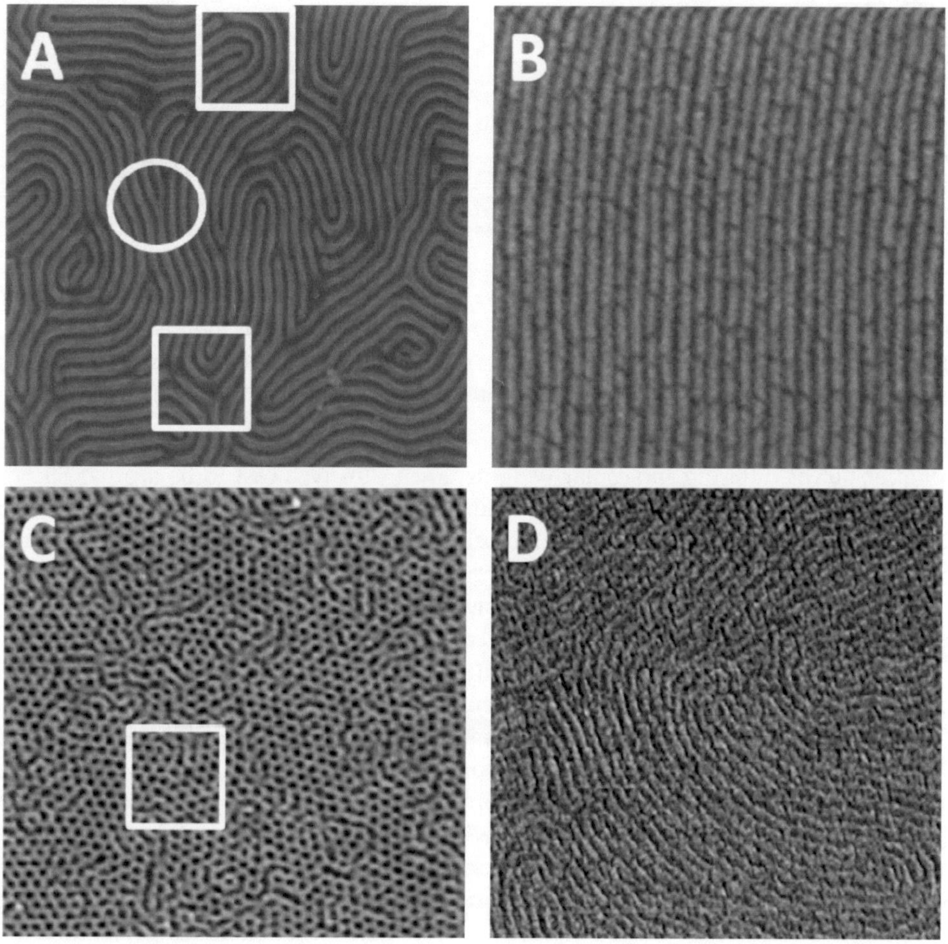

Fig. 3. Typical PS-b-PMMA BCP patterns formed via microphase separation. (A) and (B) are representations of a lamellar self-assembly (18,000-18,000 g mol^{-1} BCP composition). (C) and (D) are representations of a hexagonal arrangement from a PS-b-PMMA BCP of composition 42,000 and 21,000 g mol^{-1} respectively. Areas marked are described in the text. Images shown are representations of 1 µm x 1 µm (upper images) and 4 µm x 4 µm (lower images). Images were taken by secondary electron microscopy after selective removal of the PMMA block.

Fig. 4. PS-b-PEO thin films on silicon substrates. (A) PS-b-PEO of molecular weight 42,000–11,500 g mol^{-1} and (B) PS-b-PEO of molecular weight 32,000–11,000 g mol^{-1}. Images were taken by secondary electron microscopy after selective removal of the PEO block. It is relatively easy (in the left hand image) to see grain boundary type structures as the areas where one alignment of the structure exist are quite large.

The polymers used in figure 4 were polystyrene–polyethylene oxide (PS-b-PEO) block copolymer but of differing molecular weights. Each film was cast to be around 50 nm thick and the composition is such is to form a regular hexagonal arrangement of PEO cylinders in a PS matrix. Good microphase separation in each was achieved by identical solvent annealing in a mixture of toluene/water. Solvent annealing is an alternative to simple annealing where a solvent atmosphere allows the polymer to swell (toluene swells PS and water PEO). During swelling, the glass transition temperature decreases because the polymer chains are separated by the solvent molecules allowing low temperature treatments to bring about annealing [Fitzgerald et al. 2009). Although the molecular weights are quite similar, the difference in long-range order between the two polymers is remarkable. For the higher molecular weight the polymer has a classic grain-type structure where regions of a well-ordered hexagonal phase are separated by boundaries between different rotations. For the lower molecular weight, the entire image is a single grain with no grain-boundaries present (note that this terminology is used loosely as this is not a true grain boundary in the strictest sense since this is not a crystalline structure). The difference in the degree of long-range order probably results from the lower molecular weight having a lower glass transition temperature and hence having greater chain mobility. In this polymer, the structural changes are probably occurring during cooling because it has been shown that fast cooling rates can produce re-orientation of the cylinders in the film form this vertical alignment (i.e. cylinders normal to the surface plane) to cylinders parallel to the surface plane. This represents an important point in many cases of self-assembly. Even though the self-assembled organisations are formed at or close to equilibrium, they are often removed from the equilibrium conditions for further processing and characterisation. Examples of this include: solvent evaporation, cooling, pressure reduction, dehydration and chemical reactions and their effects.

As mentioned above, these BCP films have the advantage that they can be progressively improved by annealing to reach an equilibrium condition where the number of defects can be minimised. Many other forms of self-assembly are processes where the structure is a representation of the minimum free energy configuration in the presence of a solvent. E.g. in the assembly of nanoparticles, the particles are brought together to the arrangement of lowest free energy via diffusion within a solvent. When the solvent is removed to produce a film for example, the structure cannot readily be refined because of the rigidity of the product. BCP thin films are normally cast from solution (either spin-coated or dip-coated usually) and solvent is removed to produce a non-equilibrium structure. Frequently, this structure may be partly microphase separated but it is unusual for regular arrangements to be achieved during processing of the film because solvent evaporation rates are fast. Microphase separation is then promoted through an ageing (if chain movement is rapid enough around room temperature) or annealing step. During annealing (or ageing) the film will move towards the equilibrium structure at that temperature before being cooled for use or characterisation. The final structure may represent the equilibrium structure at the annealing temperature, the temperature it is reduced to or an intermediate temperature depending on the cooling rate.

Of course, during annealing, the copolymer system will move towards equilibrium with the concentration of defects given by a Boltzmann type distribution function, i.e. as shown in equation 15. However, it may be practically impossible to achieve the equilibrium as there are severe mass transport limitations to the movement of the polymer chains. Dis-entanglement requires many coherent chain movements and there will be considerable kinetic barriers to achieving equilibrium which is a structure of a single domain extending across the substrate.

Since, the microphase separated block copolymer arrangement can be nucleated at many sites across the substrate (as discussed above) the progress towards the equilibrium structure can be viewed as a type of grain coarsening akin to that seen in metallurgy. In this way, non-equilibrium defect structures formed after coating consist of poly grain structures whose size can be extended by lengthening of the anneal time and annihilation of the defects. As outlined below, the growth of domains is kinetically limited and the process slow. Grain growth in these systems has been shown to follow a $t^{0.25}$ power law (Harrison et al., 2000). If this law is generally obeyed then a plot of the number of defects against $1/t^4$ should be straight line with an intercept on the y-axis of the equilibrium number of defects. Note, however, characterising the number of defects at elevated temperature is experimentally difficult to quantify and caution must be applied in studies of this type. Atomic force microscopy (AFM) and secondary electron microscopy (SEM) are usually used to observe these patterns but are difficult techniques at high temperatures particularly when significant pressures of solvent. Thus, the number of defects resulting from an annealing step is observed at room temperature in conditions well removed from the annealing process. Thus, as explained above, the actual number of defects may not represent an equilibrium value at the annealing temperature.

Typical kinetic studies are shown in figures 3 to 6 three BCP systems showing hexagonal arrangements of the minority phase in a matrix of the major phases. These polymers were polystyrene-polyethylene oxide (PS-*b*-PEO, 32,000–11,000 g mol^{-1}) polystyrene-polymethylmethacrylate (PS-*b*-PMMA, 42,000-21,000 g mol^{-1}) and polystyrene-polyferrocenyl dimethylsilane (PS-*b*-PFS, 60,000-30,000 g mol^{-1}). The samples were cast onto standard cleaned silicon (100) substrates and then annealed. As cast films are generally very poorly ordered as shown by typical AFMs in figure 4. Annealing brings about considerable improvement in the film patterns as shown in figures 5-7. It should be noted that for the PS-b-PMMA film that both pattern defects and film defects are present. The film defects originate from poor wetting of the silicon surface by the BCP. However, in all cases the pattern defects do show a linear decrease with 1/anneal time[4] in agreement with the general model. In all cases, the number of defects is visually counted with a 2 µm x 2 µm image. In all cases, the plot of number of defects versus time show that the defect annihilation rate is very slow at extended times. In no case shown is the number of defects at equilibrium equal to zero as indicated by the intercept in the linearised form of the data. This is an important point in the study of self-assembly; the self-assembled pattern is likely to have a considerable number of defects present regardless of the care taken in synthesis or preparation. The potential use of self-assembly as a nanofabrication tool has been limited by the fact that they cannot rival techniques such as photolithography where defect concentrations close to zero can be engineered.

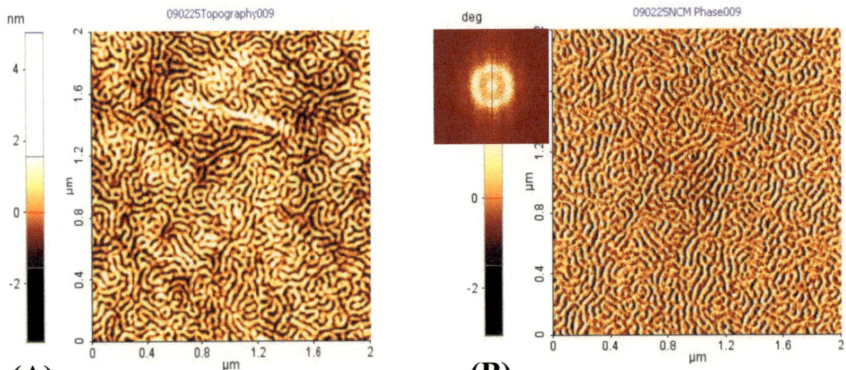

(A) (B)

Fig. 5. AFM topography (A) and phase (B) images of PS-*b*-PFS thin films prepared from 1.0 wt% solution of polymer in toluene prior to any annealing. Some limited short-range order is present as indicated by the Fourier transform data shown as an insert in (A).

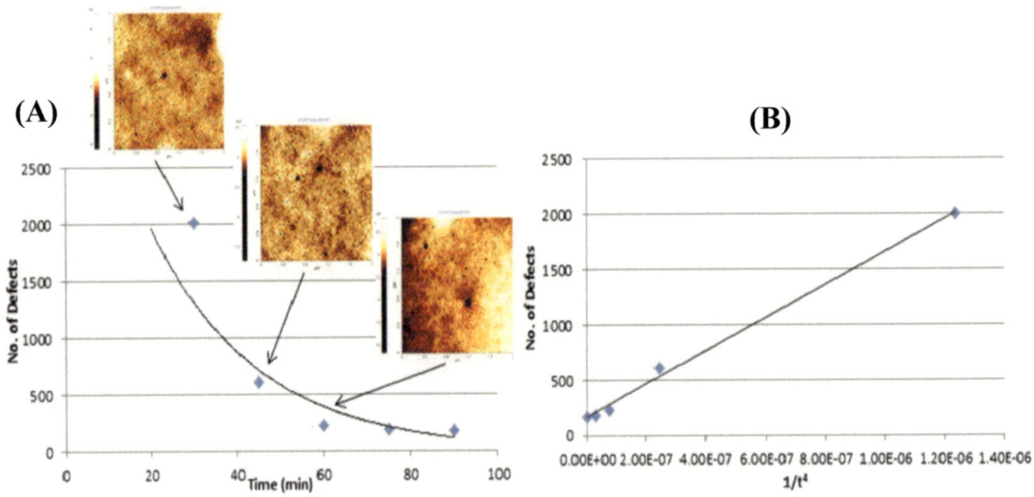

Fig. 5. Graphs plotting (A) the no. of defects vs. anneal time and (B) no. of defects vs. $1/t^4$ (where t is the anneal time) of PS-b-PEO thin films (solvent annealed in a toluene/water atmosphere at 50^0C for various times)

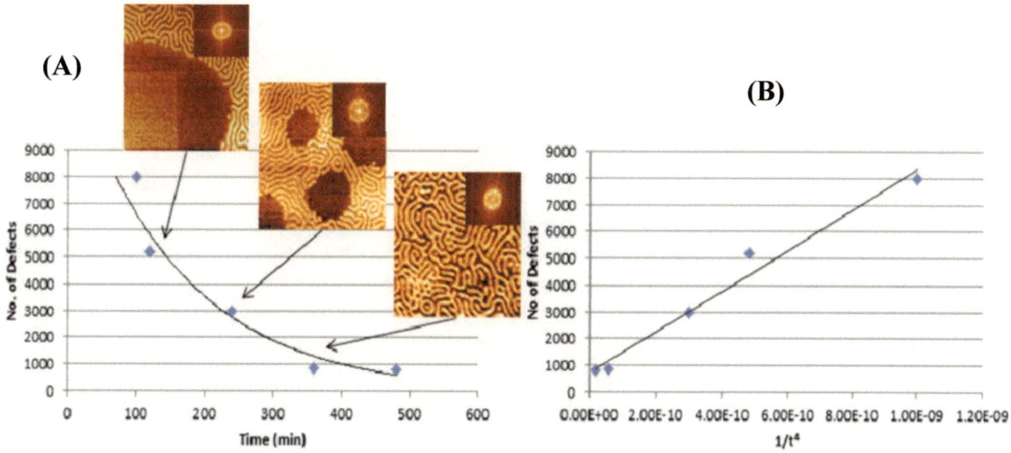

Fig. 6. Graphs plotting (A) the no. of defects vs. anneal time and (B) no. of defects vs. $1/t^4$ (where t is the anneal time) of PS-b-PMMA thin films (thermal annealed at 170ºC for various times)

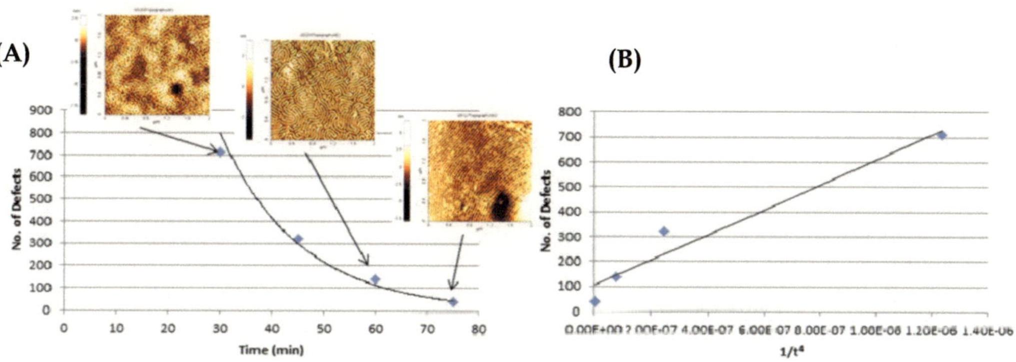

Fig. 7. Graphs plotting (A) the no. of defects vs. anneal time and (B) no. of defects vs. $1/t^4$ (where t is the anneal time) of PS-b-PFS thin films (solvent annealed in a THF atmosphere at room temperature for various times)

4. Graphoepitaxy

As was briefly mentioned above, one method used to reduce defect concentration and control alignment is graphoepitaxy. Graphoepitaxy is where surface topography is used to direct the BCP structure. The term graphoepitaxy was originally coined to describe how a substrate topographic periodicity can be used to control the crystallographic alignment of thin films and the technique evolved to become a popular method for defining highly crystalline polymer films. It is generally accepted that strain imposed by the topography is the origin of the alignment effects; however, chemical interactions of the BCP with the topography (as outlined below) probably play a more significant role in influencing the alignment process. As discussed further in the following sections, it will be seen how by engineering the preferential interaction of one block with the topography can ordain pattern alignment eliminating the fingerprint patterns described above. Of equal importance is the reduction in defect density within these aligned patterns that is also provided by the topography. This seems to derive from a number of factors. Firstly, the strong polymer-sidewall interactions that increase the enthalpic driving force for regular assembly. Secondly, and associated with this enthalpic effect is the increased energy cost of including defects which are also associated with localisation of higher strain energies around the defect. Finally, there is the spatial constraint of the pattern which makes inclusion of defects statistically less likely.

The advantage of graphoepitaxial techniques for BCP nanopattern development is that a single relatively large substrate feature such as a channel can be used to direct the BCP nanopattern with precise alignment into almost single crystal-like periodicity within the topographically defined feature. Fasolka et al. (Fasolka et al., 1997) were the first researchers to show that corrugated substrate surfaces could be used to direct the development of microphase separated block copolymers. These authors used a simple off-cut silicon substrate to generate a sawtooth topography and this was sufficient to generate regular BCP periodicity. Segalman was the first author to demonstrate that designer topography (in this work channels or rectangular cross-section separated by flat terraces or mesas and examples provided here will refer only to this shape) could be used to generate aligned, to the edge of the channel, nanopatterns of extremely high periodicity (Segalman et al., 2001). Segalaman's ground-breaking work not only demonstrated the possibility of this methodology for control of BCP structures but also reported the possibility of unusual edge effects due to varying film thickness as well as proposing a mechanism for alignment. It was suggested that alignment occurs *via* nucleation at the channel walls and that, below a critical channel width, a single domain structure could be formed.

Graphoepitaxy represents a means to combine established methods of surface engineering such as uv-lithography (to generate topography) and chemical functionalisation (to define the interaction of the BCP with the topography formed) to impose alignment on the walls. Chemical functionalisation, as mentioned above, is critical and a recent paper by Nealey et al. (Nealey et al., 2010) described how polymer brushes can be used to fine-tune the polymer-topography interactions. A polymer brush is a random co-polymer of the two blocks used in the self-assembling BCP. Changing the composition of the random brush allows the interactions to be modified. Efforts to control the polymer-topography interactions have led to the development of a technique known as soft-graphoepitaxy where the topography is generated from polymer materials (usually lithographic resist materials) that allow specific interactions with one block. The surface engineering must also be carefully controlled. If simple rectangular, cross-section channels are used the width of the channels should be a simple geometric ratio to the spacing of the pattern allowing for preferential wetting of one block at the sidewall or else defects will be precipitated as discussed below. Practically, this can be difficult to achieve and Nealey et al. have shown how the use of mixtures of the BCP with homopolymers can be used to modify the feature size of the BCP pattern to match the surface topography.

Segalman's original work on BCP graphoepitaxy was based around aligned sphere forming polystyrene-*b*-poly(2-vinylpyridine) (PS-*b*-PVP) di-block copolymers. The work has progressed very quickly and reached a high level of sophistication (see for example the review by Segalman et al.). Work reported to date has demonstrated alignment of both horizontal and vertical orientations of cylinder forming systems and sphere forming systems. One area of considerable importance has been the development 'sparse' surface topographies which minimise the size of the topographical features and considerably reduce the mesa contribution. Ross and co-workers have developed this technique to align vertical cylinders or spheres so that a low density of 'posts' guide the structure whilst being almost indistinguishable in terms of position, size and chemistry from a feature in the BCP nanopatterns (Bita et al., 2008). These sphere and vertical cylinder structures can be used to create column or nanodot structures by pattern transfer or templating methods.

In terms of emerging electronic structures or interconnect arrangements, the formation of parallel nanowires at a substrate has become an important challenge. So-called FIN-FET structures consisting of several nanowires controlled through a single gate has become an important topic of research because it may provide an alternative form of transistor to the well-established metal oxide semiconductor structure at very small transistor sizes. There has, therefore, been considerable work on the controlled alignment of lamellar (stripes orientated vertically to surface plane, Figure 3B) and cylinder (parallel to surface plane, Figure 3D) forming BCPs where coupling these techniques with templating (selective deposition into the structure) and/or pattern transfer (where the pattern is selectively etched chemically or physically can transform these structures into nanowire arrays. PS-*b*-PMMA is of particular interest because of the well-established etch characteristics of this system and has been shown to successfully produce transistor-type device structures (Black et al., 2007). The development of ultra-small circuitry (not only devices but also interconnects, vias, magnetics and capacitors has provided the motivation of much of the work into both graphoepitaxy and chemical patterning briefly described here. This is because, in practical and useful structures, it is not only necessary to define an arrangement, but also accurately position the structures so that wiring can be overlaid to define their function.

It is also emphasised that forming well-aligned and positioned parallel (to the surface plane) arrangements of BCPs offers some further challenges compared to vertical cylinder arrangements e.g. Film thickness and transfer of polymer to the channels can present a number of experimental difficulties. Suh et al. studied orientation effects in cylinder forming block copolymer films in detail and it is now generally accepted that vertical orientation of the cylinders is favoured for very thin films because the parallel arrangement can only be sustained with inclusion of elastic strain in the structure at low dimension (this strain effectively reduces with thickness). The reason for this can be viewed thermodynamically because there will always be a finite difference in the free energy of a parallel and vertical structure because of interactions with the substrate modifying the enthalpy of the pattern. If e.g. the film is below a thickness where a whole number of layers can be formed in a parallel arrangement of cylinders, the film must either deform to allow an integer number or layers or allow the formation of incomplete layers which increases surface roughness. Both of these effects are associated with increased surface strain and reduction of the total free energy. Although this strain is effectively reduced with layer thickness (spread over a larger volume of the BCP), it is extremely important to control the orientation in the cylinder forming patterns because only single layers of cylinders are required if the pattern is to be used for creation of devices since multilayer structures cannot be readily filled or transferred to the substrate by etch or template methods. The importance of controlling film thickness within graphoepitaxial channels has been recently discussed in detail (Fitzgerald, 2009)

4.1 Defects and graphoepitaxy

It should be immediately apparent that graphoepitaxy will have a profound effect on defect formation. Ideally, topography-polymer interactions and topographical dimensions will be such to allow the BCP to form a 'perfect' structure (i.e. with an equilibrium number of defects only). Confining the polymer to distinct and small regions of a substrate should also decrease kinetic limitations defined by mass transport over large substrate surfaces. This may also have significant practical implications because, as can be seen above in figures 5-7, many hours are needed to produce well-ordered arrangements and such long periods may not be compatible with commercial manufacturing processes.

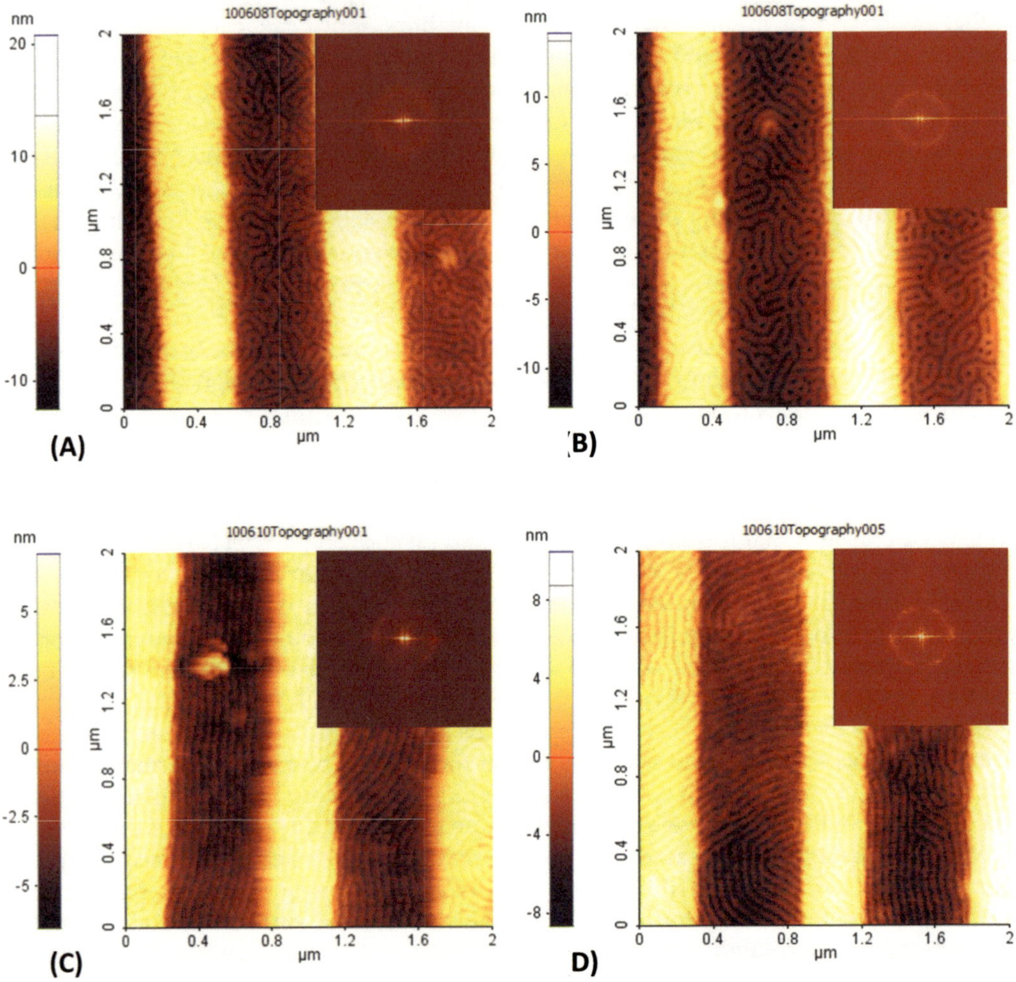

Fig. 8. AFM topographic images of PS-PFS thin films prepared from 0.5 wt% solutions of polymer in toluene on channel cut topographically defined substrates (600 nm channels) after solvent annealing in THF at room temperature for (A) 0 min, (B) 15 min, (C) 30 min, (D) 60 min

Some recent data from our laboratories (figure 8) show the importance of the relationship between the anneal time and alignment/defect numbers in topographical substrates and the necessity to control the placement of the polymer very carefully. The cylinder forming PS-b-PFS (60000-30000 g mol^{-1}) system on Si(100) substrates has a strong tendency to form parallel arrangements of cylinders. We assert that this is because chemical interactions with both the substrate and the surface interface strongly favour PS wetting layers and vertical arrangements where the PFS component contacts the substrate are not thermodynamically favourable. If this pattern forming BCP is spun-cast onto channel cut topographically defined substrates (600 nm channels), this results in a non-uniform film adopting the corrugation of the substrate where polymer is located on the top of the crests (or mesas) as well as within the channels as can be clearly seen from patterns over the entire surface. At this point the film is highly disordered and exhibits a random dispersion of PFS domains in the PS matrix (figure 8A). Solvent annealing in a THF-dominant atmosphere allows chain mobility and after 15 min there is an increase in the level of ordering though there is no directional effects from the sidewalls identified (figure 8B). It should be noted this level of ordering is significantly greater than seen in figure 7C because of the constraint of the polymer and reduction in kinetic limitations described above. Further annealing periods can lead to the eventual formation of a well-ordered parallel arrangement of cylinders (figure 8C and D). The alignment of the pattern to the sidewalls is not ideal over the entire surface because it is not possible to locate polymer only in the trenches and where the BCP covers the mesas, a more random orientation can occur (figure 8D) as the cylinders are able to escape the constraining effects of the channels. However, it is important to note that alignment and orientation is nucleated at the channel edge (figure 8C) and can be passed into the mesa structures. Presumably this is because of the favourable interactions of the PS component with the side wall.

Whilst it is highly unlikely that a true minimum energy configuration of a BCP film on a flat substrate (i.e. only an equilibrium number of defects present) can be achieved over a large area flat substrate because of all the effects outlined above, in confined systems it may be possible to achieve close to this minimum energy arrangement in highly localised regions of topography if film coating can be suitably controlled. In these cases (and if the pattern can be aligned to a substrate feature through favourable chemical interactions) then random grain orientations can be avoided and the requirement for defect annihilation during grain coarsening via very long period anneals can be decreased or essentially eliminated. Thus, close to ideally ordered arrangements may be achieved. Graphoepitaxy as described above, therefore, offer a means of aligning patterns and thus potentially provide a solution to problems associated with forming defect free films for manufacturing purposes. It, thus, seems necessary that these bottom-up techniques for patterned surface formation are combined with a top-down lithographic method in order to achieve ideal arrangements. One innovative approach to the problem of developing and use of advanced lithography is to use the assembly of another block copolymer film which can be readily aligned through favourable interactions with substrate features (see below). This polymer film is then subsequently used to chemically pattern the BCP of interest. This approach has been used with a cylinder forming PS-*b*-PMMA system to form a chemical pattern for development of well-ordered lamellar forming PS-*b*-PMMA (Black et al, 2007).

Low defect concentrations in BCP phase separated structures have been reported using graphoepitaxial methods. As discussed above, in favourable circumstances the topographically patterned surfaces align and orientate the phase separated BCP structure through interactions between the surfaces and one or both blocks. These interactions force the BCP structure into registry and single grain structures. There are many examples in the literature of graphoepitaxial defined single grain structures (Fitzgerald et al., 2007). Various authors report that the defect nature of these directed structures are largely insensitive to the match between polymer feature spacing and channel width (commensurability) except that as width increases there is a corresponding increase in the number of polymer features within the topography. It is of course noted, as outlined above, that the polymer structure will exhibit strain (*i.e.*, the spacing between features will stretch or compress) so as to fit an integer number of features across the topography such that strain is minimised. As shown by Ross and co-workers (Ross et al., 2004) the energy of the system will be minimum at the trench width at which commensurability occurs (*i.e.*, width = n polymer feature spacings, n = an integer) but will increase at higher or lower values. The strain energy will be at a maximum when the trench width is equivalent to n + ½ polymer feature spacings).

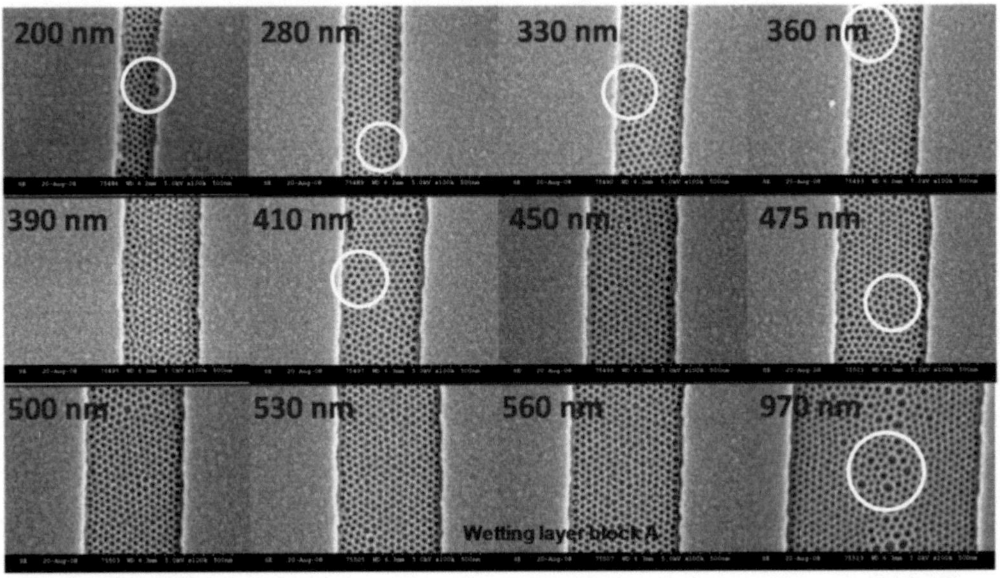

Fig. 9. SEM images of etched cylinder forming PS-b-PEO (PEO removed) in rectangular trenches (60 nm depth and width as shown) of various widths. The white circles show various defects present including grain boundaries, dislocations and point defects. See text for further details.

Some of the data from these laboratories illustrates some of the important defect chemistry of BCP systems in topographical patterns when studying the cylinder forming PS-*b*-PEO system. These data are summarised as in figure 9 and shows the patterns formed as a function of width of a channel generated in Si(100) wafers by uv-lithography. In this data, great care was taken to use polymer amounts that just provided enough material to fill the channels. In all cases, defect free, single grain structures were a rarity and clear grain boundary type and other types of defect structures can be seen with some marked for convenience (white circles). There is an obvious reduction in defect density at channel widths of 390 and 560 nm and we suggest that these values are commensurate with the pattern feature size as discussed above. Analysis of the data in figure 9 suggests a number of other points. Firstly, the sidewall structure is not ideal and can be readily seen around narrower trench widths. These cause a local variation in channel width that is of the same order of magnitude as the polymer pattern feature size. This variation must by necessity introduce high local strain energies and consequently causes the precipitation of defects and, thus, defect concentrations are relatively high. As the width increases the defect concentration observed apparently decreases until at very high widths it increases again. We explain these observations in the following manner. The BCP energy is dominated by block-block and block-interface interactions so that filling of the topography and maximising the number of features within the topography are the most important factors. When the channel width and phase separated feature spacing is incommensurate, strain energy (highly dependent on the polymer properties) results and an ideal pattern can only be achieved if this strain energy is less than the total energy recoverable from changes in structure and defect formation. Thus, as width increases it becomes easier to maintain ideal single grain, defect free structures because the strain is distributed over a larger polymer volume and so is proportionally less. At very large channel widths nucleation can occur at both side walls leaving an area at centre where defects must form to allow volume fill.

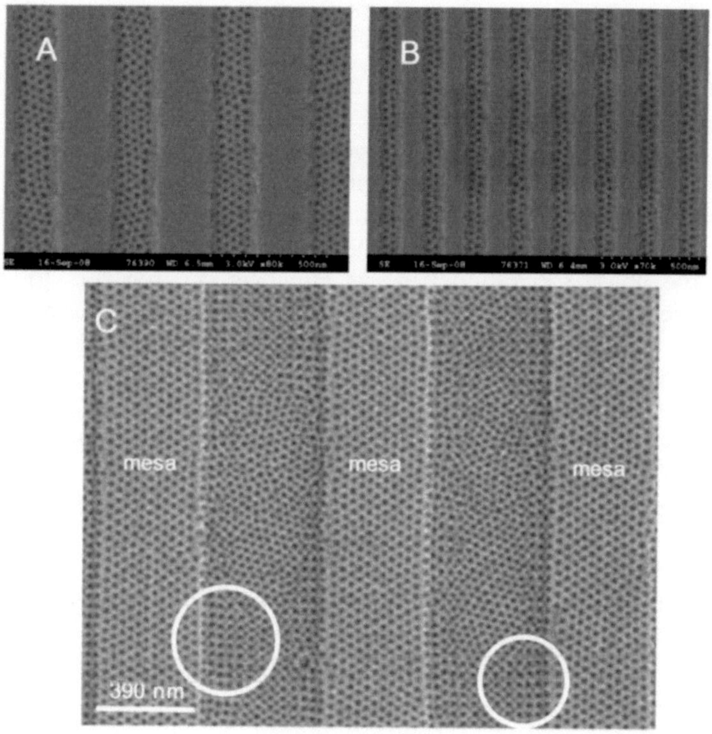

Fig. 10. SEM images of etched cylinder forming PS-b-PEO (PEO removed) in rectangular trenches (60 nm depth and widths of 175 (A), 120 (B) and 390 (C) nm). Polymer was deposited to just fill channels in (A) and (B) but was overfileld in (C) to allow phase separation at the mesas.

If this description is correct there should be a minimum channel width where no equilibrium defects should be observed since the introduction of strain energy would raise the total energy of the system excessively. This can be modelled in the same way as described above in figure 2 except that the activation energy for defect formation (a value of 16 kJ mol^{-1} was used) is effectively increased by an incommensurate strain energy of 4 kJ mol^{-1}. It was further assumed that this incommensurate strain was dispersed over the volume of polymer in the channel. The results are shown in Figure 2(D) which illustrates that even when defect formation is favourable, that there is a critical dimension when there should be no equilibrium defects formed. Figure 10 provides experimental support for this model. In the narrowest channel widths of 20 nm, there is an almost ideal arrangement of BCP structure. Figure 10(C) provides clear evidence that the majority of defects observed in graphoepitaxy result from the strain introduced by the channels. Here, a cylinder forming PS-*b*-PEO polymer was spin-coated to create material at the channel mesas and within the channel. The BCP structure within the channel is highly defective with the defect motifs seen above clearly visible. Also the expected hexagonal pattern is not exhibited uniformly and an unexpected cubic arrangement of the cylinders is observed in places as a result of the imposed strain. This cubic pattern clearly has a feature spacing that is commensurate with the channel width. However, at the mesas, an almost ideal structure is seen. This can be explained by the fact that any strain caused by incommensurity can be relieved by a slight expansion of the film.

5. Conclusions

The microphase separation of block copolymers shows a great deal of promise as a means of generating regular nanopatterns at surfaces. They may, therefore, find application as a means to novel nanomaterials and nanoelectronics device structures. The possible formation of these patterns is thermodynamically determined by the strength of the chemical interactions which is balanced by entropy considerations. Polymer composition determines the structural arrangement whilst molecular weight and the physical properties determine the kinetics of the phase separation process. However, in thin film form the chemical interactions between the blocks and their environment, *i.e.*, the interfaces that surround them, must also be carefully considered so that the microphase structure exhibits controlled alignment and orientation. The use of surface engineering to control the chemical interactions with the surface and chemical pre-patterning are strict needs if the requirements for long-range order and periodicity are to be met. Like all self-assembly processes, defect concentration is of vital important and these systems can exhibit a number of defects originating from thermodynamic and kinetic limitations. Recent work, as featured above, suggests that defect-free patterns over macroscopic dimensions may be achievable. However, a much more detailed understanding of the origin of such defects will be required before processes that reduce defect concentration to the number required by the microelectronics industry can be put in place.

6. Acknowledgements

We would thank SFI though funding via the CRANN CSET scheme as well as for funding Prof. Morris and Dr. Holmes under the Principal Investigator Scheme. The authors would like to acknowledge the EU FP7 programme for funding the LAMAND consortium under the NMP scheme and in particular for support of Dr. O'Mahony. The authors would like to thank Intel Ireland and Intel Components Research for provision and development of patterned silicon wafers under the Adaptive Grid Substrate CRANN programme. Staff at the CRANN Advanced Microscopy Laboratories and Central Equipment Facility as well as staff at The Tyndall National Institute are thanked for their expertise and access to facilities.

7. References

Whitesides, G.M.; Grzybowski, B. Self-assembly at all scales. *Science* 2002, *295*, 2418-2421.

Whitesides, G.M.; Mathias, J.P.; Seto, C.T. Molecular self-assembly and nanochemistry: A chemical strategy for the synthesis of nanostructures. *Science* 1991, *254*, 1312-1319

Misteli, T. The concept of self-organization in cellular architecture. *J. Cell Biol.* 2001, *155*, 181–185.
Jones, R.L., *Soft Machines*, Oxford University Press, Oxford, New-York, 2004.
Bensaude-Vincent, B., Self-Assembly, Self-Organization: Nanotechnology and Vitalism. *Nanoethics* 2009, *3(1)*, 31-42.
Grzybowski B. A., Wilmer, C. E., Kim, J., Borowne, K. P. and Bishop, K. J. M., *Soft Matter*, 2009, *5*, 1110-1128.
Halley, J. D. and Winkler, D. A., *Complexity*, 2008, *14*, 10-17.
Gerhart, J. and Kirschner, M., Embryos and Evolution, *Blackwell Science*, MA, USA, 1997.
Nicolas, G. and Prigogine, I., Self-Organization in Non-Equilibrium Systems, *Weinheim*, New York, USA, 1995.
Capone, B., Pierleoni, C., Hansen, J-P. and Krakoviack, V., *J. Phys. Chem. B*, 2009, *113 (12)*, 3629–3638.
Adams, M., Dogic, Z., Keller, S. L. and Fraden, S., *Nature*, 1998, *393*, 349-352.
Fraden, S., Maret, G., Caspar, D.L.D and Meyer, R. B., *Phys. Rev. Lett.*, 1989, *63*, 2068-2071.
Bai, G., Wang, Y., Yan, H., Li, Z. and Thomas, R.K., *J Colloid Interface Science*, 2001, *240*, 375-377.
Lindoy, L.F. and Atkinson, I.M., Self-Assembly in supramolecular systems, *The Royal Society of Chemistry*, Cambridge, UK, 2001.
Rice, R.; Arnold, D.C.; Shaw, M.T.; Lacopina, I.; Quinn, A.J.; Amenitsch, H.; Holmes, J.D.; Morris, M.A. Ordered mesoporous silicate structures as potential templates for nanowire growth. *Adv. Funct. Mater*. 2007, *17*, 133-141.
Fang, Y. and Leddy, J., *J. Phys. Chem.*, 1995, *99*, 6064.
Petkov, N.; Platschek, B.; Morris, M.A.; Holmes, J.D.; Bein, T. Oriented growth of metal and semiconductor nanostructures within aligned mesoporous channels. *Chem. Mater.* 2007, *19*, 1376-1381.
Pileni, M.P. Nanocrystal Self-Assemblies: Fabrication and Collective Properties. *J. Phys. Chem.* B 2001, *105*, 3358–3371.
Borah, D., Shaw, M.T., Rasappa, S., Farrell, R.A., O'Mahony, C., Faulkner, C.M., Bosea, M., Gleeson, P., Holmes, J.D. and Morris, M.A., J. Phys. D: Appl. Phys., 2011, *44*, 174012.
Farrell, R.A., Petkov, N., Shaw, M.T., Djara, V., Holmes, J.D. and Morris. M.A., Macromolecules, 2010, *43 (20)*, 8651–8655.
Lodge, T.P. Block Copolymers: Past Successes and Future Challenges. *Macromol. Chem. Phys.* 2003, *204*, 265-273.
Soo, P.P.; Huang, B.Y.; Jang, Y.I.; Chiang, Y.M.; Sadoway, D.R.; Mayes, A.M. Rubbery block copolymer electrolytes. *J. Electrochem. Soc.* 1999, *146*, 32-37.
Ulbricht, M. Advanced functional polymer membranes. *Polymer* 2006, *47*, 2217-2262.
Pease, R.F.; Chou, S.Y. Lithography and other patterning techniques for future electronics. *Pro. IEEE.* 2008, *96*, 248-270.
Bloomstein, T.M.; Marchant, M.F.; Deneault, S.; Hardy, D.E.; Rothschild, M. 22-nm immersion interference lithography. *Opt. Express* 2006, *14*, 6434-6443.
ITRS roadmap 2005 [Online.] Available: http://www.itrs.net/.
Jeong, S.J.; Xia, G.; Kim, B.H.; Shin, D.O.; Kwon, S.H.; Kang, S.W.; Kim, O.S. Universal block copolymer lithography for metals, semiconductors, ceramics, and polymers. *Adv. Mater.* 2008, *20*, 1898-1904.
D'Errico, G. In *Encyclopaedia of Surface and Colloid Science*, Somasundaran P., Ed.; CRC Press: Boca Raton, FL, USA, 2006; p. 3840-3848.

Hamley, I.W. *Developments in Block Copolymer Science and Technology*; Wiley: Hoboken, New Jersey, 2004.

Kim, G.; Libera, M. Microstructural development in solvent-cast polystyrene-polybutadiene-polystyrene (SBS) triblock copolymer thin films. *Macromolecules* 1998, *31*, 2569-2577.

Bates, F.S. Polymer-polymer phase behaviour. *Science* 1991, *251*, 898-905.

Matsen, M.W.; Bates, F.S. Unifying weak- and strong-segregation block copolymer theories. *Macromolecules*. 1996, *28*, 1091-1098.

Leibler, L. Theory of microphase separation in block copolymers. *Macromolecules* 1980, *13*, 1602-1617.

Grason, G.M. The Packing of Soft Materials: Molecular Asymmetry, Geometric Frustration and Optimal Lattices in Block Copolymer Melts. *Phys. Rep.* 2006, *433*, 1-64.

Hammond, M.R.; Cochran, E.; Fredrickson, G.H.; Kramer, E.J. Temperature dependence of order, disorder, and defects in laterally confined di-block copolymer cylinder monolayers. *Macromolecules* 2005, *38*, 6575-6585.

Segalman, R.A.; Hexemer, A.; Kramer, E.J. Edge effects on the order and freezing of a 2D Array of block copolymer spheres. *Phys. Rev. Letts.* 2003, 91, 196101/1-4.

Harrison, C.; Angelescu, D.E.; Trawick, M.; Zhengdong, C.; Huse, D.A.; Chaikin, P.M.; Vega, D.A.; Sebastian, J.M.; Register, R.A.; Adamson, D.H. Pattern coarsening in a 2D hexagonal system. *Europhys. Lett.* 2004, *67*, 800-806.

Kröner, E. and Anthony, K-H., *Annual Review of Material Science*, 1975, *5*, 43-72.

Kleman, M. and Fridel, J., *Reviews of Modern Physics*, 2008, *80* 61-115.

Farrell, R.A., Fitzgerald, T.G., Borah, D., Holmes, J.D. and Morris M.A., *J. Mol. Sci.*, 2009, *10*, 3671-3712.

Peters, R.D., Yang, X.M., Wang, Q., de Pablo, J.J. and Nealey, P.F. *J. Vac. Sci. Tech. B*, 2000, *18*, 3530-3534.

Mansky, P., Liu, Y., Huang, E., Russell, T.P. and Hawker, C.J., Controlling polymer-surface interactions with random copolymer brushes. *Science*, 1997, *275*, 1458-1460.

Fitzgerald, T.G., Farrell R.A., Petkov, N., Bolger, C.T., Shaw, M.T., Charpin, J.P.F, Gleeson, J.P., Holmes, J.D. and Morris M.A., *Langmuir*, 2009, *25*, 13551.

Fasolka, M.J.; Harris, D.J.; Mayes, A.M.; Yoon, M.; Mochrie, S.G.J. Observed substrate topography-mediated lateral patterning of di-block copolymer films. *Phys. Rev. Lett.* 1997, *79*, 3018-3021.

Bita, I.; Yang, J.K.W.; Jung, Y.S.; Ross, C.A.; Thomas, E.L .; Berggren, K.K Graphoepitaxy of self-assembled block copolymers on two-dimensional periodic patterned templates. *Science* 2008, *321*, 939-943.

Black, C.T.; Ruiz, Z.; Bretya, G.; Cheng, J.Y.; Colburn, M.E.; Guarini, K.W.; Kim, H.C .; Zhang, Y. Polymer self assembly in semiconductor microelectronics. *IBM J. Res. Dev.* 2007, *51*, 605-633.

Fitzgerald, T.G.; Borsetto, F.; O'Callaghan, J.M.; Kosmala, B.; Shaw, M.T.; Holmes, J.D.; Morris, M.A. Polymer nanostructures in sub-micron lithographically defined channels: film-thickness effects on structural alignment of a small feature size polystyrene-polyisoprene-polystyrene block copolymer. *Soft Matter.* 2007, *2*, 916-921.

Cheng, J.Y.; Mayes, A.M.; Ross, C.A. Nanostructure engineering by templated self-assembly of block copolymers. *Nat. Mater.* 2004, *3*, 823-828.

Han, E., Kang, H., Liu, C-C., Nealey, P.F. and Gopalan, P., *Advanced Materials*, 2010, *22*, 4325-4329.